Metabolic Engineering

Metabolic Engineering

Edited by **Ralph Becker**

SYRAWOOD
PUBLISHING HOUSE

New York

Published by Syrawood Publishing House,
750 Third Avenue, 9ᵗʰ Floor,
New York, NY 10017, USA
www.syrawoodpublishinghouse.com

Metabolic Engineering
Edited by Ralph Becker

International Standard Book Number: 978-1-68286-153-0 (Hardback)

Printed in the United States of America.

Contents

Preface

The purpose of the book is to provide a glimpse into the dynamics and to present opinions and studies of some of the scientists engaged in the development of new ideas in the field from very different standpoints. This book will prove useful to students and researchers owing to its high content quality.

Metabolic engineering is a multidisciplinary field which uses techniques of genetic engineering and mathematical modeling to regulate the metabolic processes of a cell. The regulation of these genetic processes and modification of metabolic networks are used for production of specific substances. This discipline incorporates principles from computational sciences, molecular biology and chemical engineering to alter specific biochemical reactions. This book focuses on recombinant DNA technology, analytical methods and other tools to understand the applications of metabolic engineering. It is a wonderful source of reference for students, researchers and academicians working in this field.

At the end, I would like to appreciate all the efforts made by the authors in completing their chapters professionally. I express my deepest gratitude to all of them for contributing to this book by sharing their valuable works. A special thanks to my family and friends for their constant support in this journey.

Editor

Proof of Evidence: PPAR-induced ANGPTL4 in Lipid and Glucose Metabolism

Kenichi Yoshida

Department of Life Sciences, Meiji University School of Agriculture, 1-1-1 Higashimita, Tama-ku, Kawasaki, Kanagawa 214-8571, Japan. E-mail: yoshida@isc.meiji.ac.jp.

Angiopoietin-like protein 4 (ANGPTL4) was identified as a peroxisome proliferator-activated receptor (PPAR)-induced gene. The genetic finding that mutation in ANGPTL3 causes hypolipidemia in mice moved us to test whether ANGPTL4 could also regulate lipid metabolism *in vivo*. We successfully proved that the introduction of ANGPTL4 as well as ANGPTL3 protein into mice rapidly induced hyperlipidemia. This suggests that the identification of novel PPAR-induced secreted proteins would contribute greatly to the elucidation of the molecular mechanisms of metabolic syndrome, including cardiovascular disease. In addition to lipid metabolism, ANGPTL4 is now regarded as a regulator of glucose metabolism. Emerging biochemical and genetic studies are expected to establish proof-of-evidence of ANGPTL4 as a promising drug development target.

Key words: Angiopoietin-like protein 4, peroxisome proliferator-activated receptor, lipid metabolism, drug target

Table of contents

INTRODUCTION

ANGPTL4 (angiopoietin-like protein 4) was first identified a protein whose expression is induced by peroxisome proliferator-activated receptor (PPAR) gamma ligands (Yoon et al., 2000). The expression of ANGPTL4 is also elevated in genetic models of obesity. Almost simultaneously, ANGPTL4 was also reported as a FIAF (fasting-induced adipose factor) using subtractive hybridization comparing liver mRNA from wild-type and PPAR alpha null mice (Kersten et al., 2000). ANGPTL4/FIFA mRNA is predominantly detected in adipose tissue and is strongly up-regulated in white adipose tissue and the liver during fasting. These evidences suggest that ANGPTL4 is a circulating protein predominantly secreted from adipose tissue and liver. Overexpression of ANGPTL4 reduced hyperglycemia to a normal level and markedly alleviated

glucose intolerance and hyperinsulinemia in db/db diabetic mice (Xu et al., 2005). It has been reported that ANGPTL3 is a target gene of liver X receptor (LXR), while ANGPTL4 expression is not (Ge et al., 2005; Kaplan et al., 2003). ANGPTL4 could exert distinct effects on lipid and glucose metabolism mainly through PPAR signaling but not through LXR, because ANGPTL4 mRNA was up-regulated by PPARalpha, PPARgamma, and PPARbeta/delta agonists (Mandard et al., 2004). PPARalpha plays an important role during fasting via the ligand-dependent transcriptional activation of target genes, while PPARgamma regulates systemic insulin signaling. It is also known that ANGPTL4 decreases hepatic glucose production and enhances insulin-media-ted inhibition of gluconeogenesis in primary rat hepatocytes

(Xu et al., 2005). Therefore, these findings suggest a possi-ble role for ANGPTL4 in the regulation of glucose homeo-stasis as well as lipid metabolism. In this opinion, we intro-duce the nature of ANGPTL4 and established proof-of-evi-dence about ANGPTL4.

Biochemistry of ANGPTL4 protein

ANGPTL4 is a member of the angiopoietin family of secreted proteins. Among this family, ANGPTL4 is most closely related to ANGPTL3 protein. Surprisingly, the positional cloning of KK/Snk, an obese and diabetic mouse model with a unique hypotriglyceridemia phenol-type, revealed that ANGPTL3 regulates lipid metabolism in mice (Koishi et al., 2002). Apolipoprotein E knockout-KK/Snk mice developed three-fold smaller atherogenic le-sions in the aortic sinus compared with apolipoprotein E knockout mice, indicating that ANGPTL3 could affect arteriosclerosis (Ando et al., 2003). We, for the first time, demonstrated that the intravenous injection of ANGPTL4 protein in KK/Snk mice rapidly increased the circulating plasma lipid levels (Yoshida et al., 2002). Moreover, ANGPTL4 as well as ANGPTL3 increased the plasma lipid levels by inhibiting the lipoprotein lipase (LPL) acti-vity (Shimizugawa et al., 2002; Yoshida et al., 2002). LPL has a central role in lipoprotein metabolism to maintain normal lipoprotein levels in blood and also in certain tis-sue (Otarod and Goldberg, 2004).

Studies then shifted to the biochemical characterization of ANGPTL4 proteins. Similar to other angiopoietin-like proteins, ANGPTL4 consists of an N-terminal coiled-coil domain and a C-terminal fibrinogen-like domain. The N-terminal coiled-coil domain of ANGPTL4 is sufficient to mediate its oligomerization. Oligomerized ANGPTL4 un-dergoes proteolytic processing to release its carboxyl fibrinogen-like domain (Ge et al., 2004a). Treatment with fenofibrate, a potent PPARalpha agonist, markedly inc-reased the plasma levels of truncated ANGPTL4, but not of full-length ANGPTL4, in humans. Together, these data suggested that the oligomerization and proteolytic pro-cessing of ANGPTL4 may regulate its biological activi-ties *in vivo* (Mandard et al., 2004). Indeed, a loss of oligo-merization decreased the stability of the N-terminal coiled-coil domain of ANGPTL4 and also reduced its abili-ty to increase plasma triglyceride levels (Ge et al., 2004b). The N-terminal coiled-coil domain of ANGPTL4 binds transiently to LPL, and this interaction results in the conversion of the enzyme from catalytically active dimers to inactive monomers (Sukonina et al., 2006).

ANGPTL4 mice model

Using genetically engineered mice, we were able to elucidate the importance of ANGPTL4 in lipid and glucose metabolism *in vivo*. Transgenic mice that overexpress

ANGPTL4 exclusively in the heart exhibited a restricted inhibition of cardiac LPL activity and developed left ventri-cular dysfunction (Yu et al., 2005). This outcome was ex-plained by the inhibition of lipoprotein-derived fatty acid delivery as a result of the induction of ANGPTL4 in the heart. In addition, transgenic mice overexpressing ANGPTL4 in the liver displayed elevated plasma trigly-ceride levels. In contrast to the transgenic mice, ANGPTL4- as well as ANGPTL3-deficient mice displayed hypotriglyceridemia; however, we now know that ANGPTL4 and ANGPTL3 function to regulate circulating triglyceride levels during different nutritional states caused by feeding/fasting through the differential inhibit-tion of LPL (Koster et al., 2005; Li, 2006).

Concluding remarks

We have learned much about the nature of ANGPTL4 and have established proof-of-evidence that ANGPTL4 is involved in the regulation of fat, lipid and glucose meta-bolic homeostasis. However, the exact roles of ANGPTL4 in regard to physiology and pathology in humans remain uncertain. Recently, ANGPTL3 has been shown to be associated closely with arterial wall thickness in human subjects (Hatsuda et al., 2007). Genetic association studies, such as assigning single nucleotide polymer-phisms (SNP) in the ANGPTL4 gene and ELISA to moni-tor plasma concentrations of ANGPTL4 protein during changes in nutritional status or the pathogenesis of meta-bolic syndrome, are needed to predict cardiovascular and other disease risk. In patients with type 2 diabetes, serum levels of ANGPTL4 protein were significantly lower than those in healthy subjects (Xu et al. 2005). Recently, sequencing of a large population (n = 3,551) to examine the role of the ANGPTL4 in lipid metabolism revealed that nonsynonymous variants in ANGPTL4 are prevalent in individuals with triglyceride levels (Romea et al., 2007). We suggest that ANGPTL4 is a fasting-induced regulator of LPL especially in adipose tissue. Moreover, ANGPTL4 sits on a unique situation where ANGPTL4 regulates glu-cose metabolism as well as lipid metabolism. From these evidences, ANGPTL4 is now regarded as a promising drug development target.

ACKNOWLEDGEMENTS

We acknowledge financial support from the "High-Tech Research Center" Project for Private Universities: a matching fund subsidy from MEXT (Ministry of Education, Culture, Sports, Science and Technology) of Japan, 2006-2008.

REFERENCES

Ando Y, Shimizugawa T, Takeshita S, Ono M, Shimamura M, Koishi R,

Furukawa H (2003). A decreased expression of angiopoietin-like 3 is protective against atherosclerosis in apoE-deficient mice. J. Lipid Res. 44: 1216-1223.

Ge H, Yang G, Huang L, Motola DL, Pourbahrami T, Li C (2004a). Oligomerization and regulated proteolytic processing of angiopoietin-like protein 4. J. Biol. Chem. 279: 2038-2045.

Ge H, Yang G, Yu X, Pourbahrami T, Li C (2004b). Oligomerization state-dependent hyperlipidemic effect of angiopoietin-like protein 4. J. Lipid Res. 45: 2071-2079.

Ge H, Cha JY, Gopal H, Harp C, Yu X, Repa JJ, Li C (2005). Differential regulation and properties of angiopoietin-like proteins 3 and 4. J. Lipid Res. 46: 1484-1490.

Hatsuda S, Shoji T, Shinohara K, Kimoto E, Mori K, Fukumoto S, Koyama H, Emoto M, Nishizawa Y (2007). Association between plasma angiopoietin-like protein 3 and arterial wall thickness in healthy subjects. J. Vasc. Res. 44: 61-66.

Kaplan R, Zhang T, Hernandez M, Gan FX, Wright SD, Waters MG, Cai TQ (2003). Regulation of the angiopoietin-like protein 3 gene by LXR. J. Lipid Res. 44: 136-143.

Kersten S, Mandard S, Tan NS, Escher P, Metzger D, Chambon P, Gonzalez FJ, Desvergne B, Wahli W (2000). Characterization of the fasting-induced adipose factor FIAF, a novel peroxisome proliferator-activated receptor target gene. J. Biol. Chem. 275: 28488-28493.

Koishi R, Ando Y, Ono M, Shimamura M, Yasumo H, Fujiwara T, Horikoshi H, Furukawa H (2002). Angptl3 regulates lipid metabolism in mice. Nat. Genet. 30: 151-157.

Koster A, Chao YB, Mosior M, Ford A, Gonzalez-DeWhitt PA, Hale JE, Li D, Qiu Y, Fraser CC, Yang DD, Heuer JG, Jaskunas SR, Eacho P (2005). Transgenic angiopoietin-like (angptl)4 overexpression and targeted disruption of angptl4 and angptl3: regulation of triglyceride metabolism. Endocrinology 146: 4943-4950.

Li C (2006). Genetics and regulation of angiopoietin-like proteins 3 and 4. Curr. Opin. Lipidol. 17: 152-156.

Mandard S, Zandbergen F, Tan NS, Escher P, Patsouris D, Koenig W, Kleemann R, Bakker A, Veenman F, Wahli W, Muller M, Kersten S (2004). The direct peroxisome proliferator-activated receptor target fasting-induced adipose factor (FIAF/PGAR/ANGPTL4) is present in blood plasma as a truncated protein that is increased by fenofibrate treatment. J. Biol. Chem. 279: 34411-34420.

Otarod JK, Goldberg IJ (2004). Lipoprotein lipase and its role in regulation of plasma lipoproteins and cardiac risk. Curr. Atheroscler. Rep. 6: 335-342.

Romeo S, Pennacchio LA, Fu Y, Boerwinkle E, Tybjaerg-Hansen A, Hobbs HH, Cohen JC (2007). Population-based resequencing of ANGPTL4 uncovers variations that reduce triglycerides and increase HDL. Nat Genet. 39: 513-516.

Shimizugawa T, Ono M, Shimamura M, Yoshida K, Ando Y, Koishi R, Ueda K, Inaba T, Minekura H, Kohama T, Furukawa H (2002). ANGPTL3 decreases very low density lipoprotein triglyceride clearance by inhibition of lipoprotein lipase. J. Biol. Chem. 277: 33742-33748.

Sukonina V, Lookene A, Olivecrona T, Olivecrona G (2006). Angiopoietin-like protein 4 converts lipoprotein lipase to inactive monomers and modulates lipase activity in adipose tissue. Proc. Natl. Acad. Sci. USA 103:17450-17455.

Xu A, Lam MC, Chan KW, Wang Y, Zhang J, Hoo RL, Xu JY, Chen B, Chow WS, Tso AW, Lam KS (2005). Angiopoietin-like protein 4 decreases blood glucose and improves glucose tolerance but induces hyperlipidemia and hepatic steatosis in mice. Proc. Natl. Acad. Sci. USA 102: 6086-6091.

Yoon JC, Chickering TW, Rosen ED, Dussault B, Qin Y, Soukas A, Friedman JM, Holmes WE, Spiegelman BM (2000). Peroxisome proliferator-activated receptor gamma target gene encoding a novel angiopoietin-related protein associated with adipose differentiation. Mol. Cell. Biol. 20: 5343-5349.

Yoshida K, Shimizugawa T, Ono M, Furukawa H (2002). Angiopoietin-like protein 4 is a potent hyperlipidemia-inducing factor in mice and inhibitor of lipoprotein lipase. J. Lipid Res. 43: 1770-1772.

Yu X, Burgess SC, Ge H, Wong KK, Nassem RH, Garry DJ, Sherry AD, Malloy CR, Berger JP, Li C (2005). Inhibition of cardiac lipoprotein utilization by transgenic overexpression of Angptl4 in the heart. Proc. Natl. Acad. Sci. USA 12: 1767-1772.

The development of the pitcher plant *Sarracenia purpurea* into a potentially valuable recombinant protein production system

Bruce A. Rosa[1,2], Lada Malek[2] and Wensheng Qin[1,2]

[1]Biorefining Research Initiative, Lakehead University, 955 Oliver Road, Thunder Bay ON, Canada, P7B 5E1.
[2]Department of Biology, Lakehead University, 955 Oliver Road, Thunder Bay ON, Canada, P7B 5E1.

The unique inducible system of protein secretion by the carnivorous pitcher plant *Sarracenia purpurea* may be an ideal system for recombinant protein farming. *S. purpurea* is relatively uncommon and difficult to grow *in vitro*, so it has not been explored as a potential source of recombinant proteins. However, it naturally secretes large amounts of proteins into a liquid found in the leaf pitchers, so it may be an ideal way to collect recombinant proteins in leaf pitchers. Here, the advantages of transgenic *S. purpurea* systems over traditional transgenic plant systems for the production of recombinant pharmaceutical proteins are explored, and the steps necessary to produce such a system are discussed.

Key words: Transgenic plants, recombinant protein farming, carnivorous plants, gene technology.

INTRODUCTION

The purple pitcher plant *Sarracenia purpurea* is a carnivorous rosette-forming plant that grows in sphagnum bogs, nutrient poor fens, seepage swamps, and pine savannas of the eastern United States and Canada (Ellison et al., 2004; Schnell, 2002). Many pitcher plants can be found in the region surrounding Lakehead University in Thunder Bay, Ontario, Canada, where some research on this plant is taking place. *S. purpurea* has modified pitcher-shaped leaves which serve as reservoirs to collect rainwater, into which the plant secretes its own liquid, as well as hydrolases and other proteins for prey digestion (Figure 1).

Secretions are produced at the lip (or "hood") of the leaves, which attract prey to the plant. Prey attracted to the plant are directed downward by hair-like spines on the inside of the hood, and eventually contact and drown in the liquid contained in the pitchers (the "trap") (Ellison and Gotelli, 2002). There, prey is digested by a number of hydrolases including RNases, nucleases, phosphatases, and proteases (Gallie and Chang, 1997). The trapping mechanism in *S. purpurea* is so efficient that higher animals such as frogs are often found partially digested inside the pitchers, which is why the plant is classified as carnivorous rather than insectivorous (Lindquist, 1975). There has been considerable debate over whether digestive enzymes in the pitcher are produced by microorganisms living in the plants or by the plants themselves, but many studies indicate that microorganisms may just play an incidental role in prey digestion, and instead play a critical role only in nitrogen acquisition and fixation (Bradshaw and Creelman, 1984; Gallie and Chang, 1997). Some studies state that *S. purpurea* relies on bacteria for essential digestive processes, but these studies present no data from enzymatic tests on sterile pitcher fluid, and contain no references that do (Adams and Smith, 1977; Harvey and Miller, 1996; Heard, 1994). In addition, nitrogen demand by the plant strongly affects the expression of carnivory, and the pitchers continue to accumulate digestive enzymes even in the absence of any microrganisms (Ellison and Gotelli, 2002; Gallie and Chang, 1997; Givnish et al., 1984).

CARNIVORY IN *SARRACENIA PURPUREA*

Carnivory in pitcher plants is thought to have evolved primarily for nitrogen acquisition, as they typically grow on nitrogen-poor peat bogs and cease expression of carnivory in nitrogen rich environments (Ellison and Gotelli,

*Corresponding author. E-mail: wqin@Lakeheadu.ca.

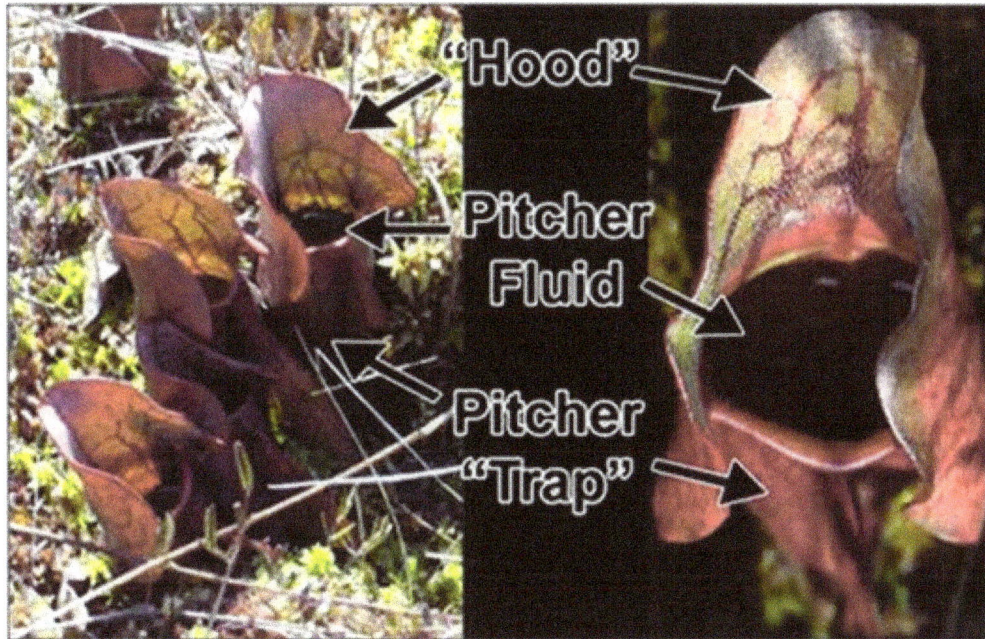

Figure 1. The basic anatomy of the North American pitcher plant *Sarracenia purpurea*. Pictures were taken by Bruce Rosa at William's Bog, near Kingfisher Lake in Thunder Bay, Ontario, Canada.

2002; Givnish et al., 1984). Phosphate and potassium acquisition is also primarily achieved through carnivory (Chapin and Pastor, 1995; Christensen, 1996; Jaffe et al., 1992).

Upon first opening, the traps secrete and accumulate hydrolases for approximately one week, after which time hydrolase secretion reduces to low levels until the trap is induced by prey (Gallie and Chang 1997). RNase, nuclease and protease activity all appear to be induced by the presence of either protein or nucleic acids, suggesting a coordinated inducible response to prey for these hydrolases (Gallie and Chang, 1997). However, phosphatase secretion is only induced by proteins and not nucleic acids, suggesting additional complexity to the induced response (Gallie and Chang 1997). Mechanical stimulation of the plants has no effect on prey recognition and the subsequent release of hydrolases, as it does in other carnivorous plants such as the Venus' flytrap *Dionaea muscipula* (Robins, 1976). Further study is necessary in order to understand the complex pathways in the carnivory response of *S. purpurea*.

The ability to produce hydrolases only in the presence of prey significantly reduces the metabolic cost of carnivory to the plant. This is necessary because prey capture efficiency is commonly very low, with some studies reporting an average of 0.070 prey leaf^{-1} day^{-1} (Newell and Nastase, 1998). The detection of prey in one leaf does not induce hydrolase secretion in another leaf on the same plant (Gallie and Chang, 1997). This trap-specific response is another apparent adaptation to reduce the metabolic cost of carnivory.

Other compromises are made by *S. purpurea* in order to balance the cost and the benefit of carnivory. It has been observed that the presence of a strong red coloration in the leaves, indicating the presence of nectar, serves as a strong attractant for prey at the cost of the energy required to produce the nectar (Cresswell, 1993). Other studies indicate that having larger traps increases the ability of the plants to successfully trap prey, at the cost of additional nitrogen spent to increase the size of the traps (Cresswell, 1993; Gibson, 1991). Perhaps because of these trade-offs, there is pronounced morphological variability in terms of the size, shape, and color of leaves between *S. purpurea* across its geographical range (Buckley et al., 2003; Schnell, 1979; Schnell, 2002). Even in one location, these characteristics can change dramatically in response to local environmental conditions, particularly temperature and moisture (Ellison and Gotelli, 2002; Mandossian, 1996). The pronounced variability in these visual characteristics in response to the pitcher plant's surroundings and prey suggests that the plants may also develop distinct hydrolase induction mechanisms and secretion responses. Little, however, is known about the exact amounts and specific types of all the hydrolases secreted by *S. purpurea*.

THE ROLE OF MICROORGANISMS IN PITCHER PLANT DIGESTION

The liquid in the trap of *S. purpurea* contains a suite of decomposers including bacteria, protists, rotifers, and fly larvae, which may aid in the digestive process, but likely only play an incidental role (Bradshaw and Creelman, 1984; Ellison and Gotelli, 2002; Gallie and Chang, 1997;

Harvey and Miller, 1996; Whitman et al., 2005). Immature pitchers are closed and sterile, but once opened and matured, the species complement of microorganisms present in them varies considerably even in the same geographical region, due to random introductions by insects and wind, competition between microorganisms and due to the differential availability of nutrients (Hepburn and St. John, 1927; Whitman et al., 2005). Microorganisms appear to be introduced to the pitcher most often by prey, which may bring bacteria from any outside source, or may cross contaminate pitchers when they successfully escape one trap and then enter another (Gibson, 1991; Whitman et al., 2005). The types of bacteria isolated from pitchers have been also been found to be very similar to the bacteria harbored in the complex microbial communities of the exoskeletons and guts of common prey insects (Siragusa et al., 2007).

It is unlikely that microorganisms play a critical role in prey digestion, as S. purpurea continues to produce hydrolytic enzymes after treatment with the antibiotics ampicillin, carbenicillin and cefotaxime at quantities sufficient to effectively inhibit all microbial activity (Gallie and Chang, 1997). Pitchers also secrete liquid containing high concentrations of hydrolases even in the absence of water and microorganisms in the pitcher (Gallie and Chang, 1997). It has also been suggested that the digestion of proteins by bacteria is too slow to be beneficial to the plant (Hepburn and St. John, 1927). Although microorganisms do not appear to play an important role in prey digestion, S. purpurea does appear to rely on bacteria for aiding in the acquisition and conversion of nitrogen sources, which is essential for the plant's growth (Butler et al., 2008; Harvey and Miller, 2006).

The microorganisms present in the traps of S. purpurea may be an interesting subject for further study, as any organism living inside the pitcher traps are somehow protected against the harsh hydrolytic environment created by the plants. This protection may be due to the development of cell membranes immune to degradation by hydrolases, or due to the presence of inhibitors of excreted proteases and hydrolases, which would neutralize threats from the pitcher's digestive processes.

The active protein excretory system in S. purpurea is a unique trait that may be exploited in order to design of a recombinant protein farming system that is more profitable than traditional transgenic plant systems.

Traditional transgenic plant systems

Production of recombinant proteins in transgenic plants and plant cell cultures results in high quality proteins that are generally safer for use as drugs than proteins harvested from bacterial or mammalian sources due to the low risk of contamination by mammalian viruses, pathogens and toxins (Table 1) (Ferrante and Simpson, 2001; Ma et al., 2003).

In whole-plant systems, proteolytic activity within transgenic plant cells adversely affects recombinant protein yields (Ma et al., 2003). Subcellular targeting of recombinant proteins in plant systems is a common method of isolating the proteins from proteases without inhibiting the critical functions of these proteases in the plants (Ma et al., 2003). Even in these systems, however, it is difficult to obtain high yields of recombinant proteins.

Targeting recombinant proteins to the cytoplasm is inefficient because it exposes proteins to 20S and 26S proteasomes which rapidly degrade the protein with the help of ATP hydrolysis via the ubiquitination system (Callis, 1995). The plant cell vacuole contains serine, cysteine and aspartic acid proteases, metalloproteinases and other unidentified proteases which degrade proteins rapidly despite the vacuole's capacity for storage space (Callis, 1995; Hara-Nishimura et al., 1991). The chloroplast lacks the ubiquitin system and produces high levels of protein, but contains other proteases, some of which are similar to prokaryotic proteases and some of which are transported into the chloroplast from the cytoplasm (Adam, 1996; Malek et al., 1984; Watson et al., 2004).

One of the more efficient methods for recombinant protein production is targeting proteins for retention in the endoplasmic reticulum. This is an effective strategy due to its oxidizing environment, the presence of molecular chaperones and a lack of proteases (Ma et al., 2003). Novel cellular targets have emerged recently, including targeting recombinant proteins to the oil bodies of seeds, which are relatively easy to purify from the other proteins in the plant cell (Kiihnel et al., 1996; Nykiforuk et al., 2006). However, despite the fact that these systems manage to produce relatively high yields of properly folded proteins, all present transgenic whole-plant protein harvesting systems require destruction of the plants or seeds, which results in the need to wait for plant growth, and to pay for growing materials.

Plant cell culture secretion systems circumvent the typical expensive extraction process because purification is performed on the liquid medium instead of cell lysate. Excreted recombinant proteins have been shown to fold more efficiently than proteins targeted intracellularly, resulting in more active forms of recombinant proteins that degrade less due to a lack of proteases (Schillberg et al., 1999). Purification of the liquid medium is easier because there is much lower protein contamination from cellular components, and intracellular proteases are not exposed to the foreign protein. However, purification is complicated by the severe dilution of the target protein. The total soluble protein content of the recombinant protein is highly variable in plant cell culture systems, but is often as low as 0.05% (Doran, 2000; Hellwig et al., 2004). In addition to producing low concentrations of recombinant proteins, production and maintenance of plant cell culture systems is expensive, so commercial protein farming companies typi-cally utilize whole-plant systems instead.

Table 1. Comparison of production systems for recombinant pharmaceutical proteins, adapted from Ma et al., 2003.

System	Overall cost	Scale-up capacity	Production timescale	Product quality	Glycosylation	Contamination risks
Transgenic Plants	Very Low	Very High	Long	High	Minor Differences	Low Risk
Plant Cell Cultures	Medium	Medium	Medium	High	Minor Differences	Low Risk
Transgenic *S. purpurea* [a]	Low	High	Short [b]	Unknown	Unknown	Low Risk
Bacteria	Low	High	Short	Low	None	Endotoxins
Yeast	Medium	High	Medium	Medium	Incorrect	Low Risk
Mammalian Cell Culture	High	Very Low	Long	Very High	Correct	Viruses, Prions and Oncogenic DNA
Transgenic Animals	High	Low	Very Long	Very High	Correct	Viruses, Prions and Oncogenic DNA

[a] - Projected.
[b] -The initial production of the system would take a long time, but once developed, protein production would be relatively fast because the plants would not need to be destroyed to harvest the proteins.

Newer and more efficient transgenic plant technology has allowed companies to patent and produce more sophisticated and profitable recombinant protein systems. Recombinant pharmaceutical antibody production plants have been developed for commercial use from transgenic tobacco, alfalfa, corn, rice and other crops (Stoger et al., 2002). Sigma Inc, a chemical and molecular bio-logy supply company, has been selling recombinant β-glucuronidase derived from the cytoplasm of transgenic corn seeds for several years (Ma et al., 2003). Using newer technology, Sembiosys Inc has produced transgenic safflower seeds which produce insulin bound to oil bodies (Stephan, 2008).

These recombinant proteins are more easily purified from the rest of the seed than recombinant proteins expressed in seed cytoplasm (Nykiforuk et al., 2006; Stephan, 2008). The purified insulin product is expected to be commercially available by 2012 (Stephan, 2008).

In most intracellular transgenic plant protein systems, protein expression is limited by post-transcriptional gene silencing (PTGS). Viral-based transient expression vector systems such as

tomato bushy stunt virus can prevent the onset of PTGS, resulting in higher levels of active intra-cellular protein expression (Voinnet et al., 2003).

Transgenic rhizosecretion systems, in which transgenic plant roots secrete recombinant antibodies which can later be harvested, are under development and have the potential to become profitable biotechnology systems (Drake et al., 2003). These secretory systems have the advantage of producing relatively pure recombinant proteins, without the difficulty of maintaining cell cultures (Drake et al., 2003).

TRANSGENIC SARRACENIA PURPUREA SYSTEMS

Aseptically grown *S. purpurea* has the potential to be developed into a novel transgenic whole-plant system in which recombinant proteins are se-creted into sterile liquid in the pitcher. This system would circumvent the need to destroy the plant or the seeds, as is currently required for whole-plant recombinant protein systems. Because *S. purpurea* is naturally adapted to secrete high

levels of com-plex proteins, it has the potential to produce much higher total yields of recombinant proteins than rhizosecretion systems. As well, *S. pur-purea's* specialized secretory cells may be a uni-que target for developing cell suspension systems that produce much higher total soluble protein content than traditional plant cell suspension cultures.

The development of a transgenic *S. purpurea* system

The hydrolytic enzymes secreted by pitcher plants must be identified and quantified in order to identify genes with high expression levels in secretory path-ways. Using a gene vector method such as *Agro-bacterium tumefaciens*, these target genes could theoretically be replaced with genes for commer-cially important recombinant proteins such as insulin, interferon or kinase C. However, genetic transformation systems have not yet been tested in *S. purpurea*.

Proteolytic enzymes secreted in the pitchers will be responsible for some degradation of the target recombinant proteins in this proposed system unless they are inactivated. It is not currently known how many specific proteases are produced by the pitchers of *S. purpurea*. However, once identified, these genes could be knocked out using gene technology such as transposon muta-genesis, TILLING (targeted, induced local lesions in genomes), which uses chemicals targeted for specific gene sequences, or Deleteagene, which induces deletion mutations using fast neutrons (Feng, 2006). These genes could also be inhibited by the use of specific enzyme inhibitors, but these would reduce the economic feasibility of the experiments, as they are relatively expensive and would need to be continually added to the pitcher fluid to ensure complete inhibition. It is not currently known how many specific proteases are produced by the pitchers of *S. purpurea*.

It is unclear how stable recombinant proteins in pitcher fluid will be, even in the absence of proteases. However, pH-adjusted sterile buffers optimized for each recombinant protein could theoretically be utilized to ensure protein stability without inhibiting secretion rates.

Since carnivory is only expressed in pitcher plants that are low on nitrogen, completely eliminating protease activity in the medium could be harmless to the plant if nitrogen and phosphate (in the form of NH_4NO_3) is added after harvesting the recombinant protein (Ellison and Gotelli, 2002). This is far different from traditional whole-plant systems, in which protein expression is intracellular and proteolytic activity cannot be eliminated because proteases are required in many cellular pathways (Bond and Butler, 1987).

High rates of recombinant protein secretion by the plants could be induced by nitrogen starvation and by signaling from target proteins and nucleotides in the pitcher (Ellison and Gotelli, 2002; Gallie and Chang, 1997). In order to optimize the protein secretion rates it will be necessary to identify the specific amino acid or nucleotide sequences that result in the highest secretion rates in the plants. Optimal levels of nitrogen supplementation would also need to be identified, as nitrogen would be depleted to create the recombinant proteins, but supplementing too much nitrogen would result in a loss of carnivory.

The overall design for a transgenic whole-plant *S. purpurea* system for recombinant protein farming would therefore involve supplementing the transgenic pitchers (containing the recombinant gene and knocked out secretory enzyme genes) with sterile water containing the inducer molecule and small amounts of nitrogen, then collecting the liquid after a period of time and purifying the relatively pure secreted recombinant protein from it.

Once optimum induction and secretion rates are identified, a continuous harvesting system should be developed in order to quickly purify recombinant proteins, which would ensure minimal degradation.

S. purpurea cell suspension cultures should also be considered for use in a recombinant protein harvesting system. Secretory cells from the inside of the pitcher could be isolated and cultured in order to develop a novel cell suspension system, in which the cells are inherently programmed for high rates of protein secretion. These cultures may produce much higher overall recombinant protein yields than traditional plant cell suspension cultures. Once developed, this system would also circumvent the difficulty of growing whole *S. purpurea* plants aseptically.

CONCLUSION AND DISCUSSION

Current literature on the topic of enzyme secretion in *S. purpurea* is complicated by the presence of a variety of microorganisms found in the natural environment. Growing *S. purpurea in vitro* and in sterile conditions is difficult, and has contributed to the lack of knowledge about the molecular pathways inside the pitchers. However, *S. purpurea* has the potential to become a profitable transgenic plant system for recombinant protein production. It has a unique natural system of protein secretion that may be utilized to develop cell suspension systems with high recombinant protein yields, or whole-plant systems in which no tissues or seeds need to be destroyed in order to harvest the recombinant protein. This whole-plant system would reduce costs associated with growing the plants as well as with purifying the target protein, so it may present a more profitable and commercially successful alternative to current transgenic systems.

A significant amount of research will have to be performed in order to identify the enzymes that are secreted at high rates in response to the presence of prey in the pitchers, which would serve as targets for recombinant protein genes. The actual amount of any specific enzyme secreted by the plants is not known at this time, but this measurement would be relatively simple and would help to determine the economic feasibility of a transgenic whole-plant *S. purpurea* system. The remaining secreted enzymes and the gene regulatory pathways leading to specific hydrolase induction and excretion would also need to be characterized and then eliminated by gene knockout technology. This research could also potentially lead to the discovery of novel enzymes secreted by these relatively poorly known plants.

REFERENCES

Adam Z (1996). Protein stability and degradation in chloroplasts. Plant Mol. Biol. 32: 773-783.

Adams RM II, Smith GW (1977). An S.E.M. survey of the five carnivorous pitcher plant genera. Am. J. Bot. 64: 265-272.

Bond JS, Butler PE (1987). Intracellular proteases. Ann. Rev. Biochem. 56: 333-364.

Bradshaw WE, Creelman RA (1984). Mutualism between the carnivorous purple pitcher plant and its inhabitants. Am. Midl. Nat. 112: 294-303.

Buckley HL, Miller TE, Ellison AM, Gotelli NJ (2003). Reverse latitudinal trends in species richness of pitcher-plant food webs. Ecol. Lett. 6: 825-829.

Butler JL, Gotelli NJ, Ellison AM (2008). Linking the brown and green:

nutrient transformation and fate in the *Sarracenia* Microecosyst.. Ecol. 89: 898-904.

Callis J (1995). Regulation of protein degradation. The Plant Cell. 7: 845-857.

Chapin CT, Pastor J (1995). Nutrient limitations in the northern pitcher plant *Sarracenia purpurea*. Can. J. Bot. 73: 728-734.

Christensen NL (1996). The role of carnivory in *Sarracenia flava L.* with regard to specific nutrient deficiencies. J. Elisha Mitchell Sci. Soc. 92: 144-147.

Cresswell JE (1993). The morphological correlates of prey capture and resource parasitism in pitchers of the carnivorous plant *Sarracenia purpurea*. Am. Midl. Nat. 129: 35-41.

Doran P (2000). Foreign protein production in plant tissue cultures. Curr. Opin. Biotech. 11: 199-204.

Drake PM, Chargelegue DM, Vine ND, Van Dolleweerd CJ, Obregon P, Julian KC (2003). Rhizosecretion of a monoclonal antibody protein complex from transgenic tobacco roots. Plant Mol. Bio. 52: 233-241.

Ellison AM, Buckley HL, Miller TE, Gotelli NJ (2004). Morphological variation in *Sarracenia purpurea* (*Sarraceniaceae*): Geographic, environmental and taxonomic correlates. Am. J. Bot. 91: 1930-1935.

Ellison AM, Gotelli NJ (2002). Nitrogen availability alters the expression of carnivory in the northern pitcher plant, *Sarracenia purpurea*. Proc. Natl. Acad. Sci. 99: 4409-4412.

Ferrante E, Simpson D (2001). A review of the progression of transgenic plants used to produce plantibodies for human usage. J. Young Invest. 4: 1-10. Gallie D, Chang S (1997). Signal transduction in the carnivorous plant *Sarracenia purpurea*. Plant Physiol. 115: 1461-1471.

Gibson TC (1991). Differential escape of insects from carnivorous plant traps. Am. Midl. Nat. 125: 55-62.

Givnish TJ, Burkhardt EL, Happel RE, Weintraub JD (1984). Carnivory in the bromeliad *Brocchinia reducta*, with a cost/benefit model for the general restriction of carnivorous plants to sunny, moist nutrient-poor habitats. Am. Naturalist. 124: 479-497.

Hara-Nishimura I, Inoue K, Nishimura M (1991). A unique vacuolar processing enzyme responsible for conversion of several proprotein precursors into the mature forms. FEBS Lett. 294: 89-93.f

Harvey E, Miller TE (1996). Variance in composition of inquilines communities in leaves of *Sarracenia purpurea L.* on multiple spatial scales. Oecologia 108: 562-566.

Heard SB (1994). Plant Midges and mosquitoes: a processing chain commensalism. Ecol. 75: 1647-1660.

Hellwig S, Drossard J, Twyman R, Fischer R (2004). Plant cell cultures for the production of recombinant proteins. Nat. Biotech. 22: 1415-1422.

Hepburn JS, St. John EQ (1927). A bacteriological study of the pitcher liquor of the *Sarraceniaceae*. Trans. Wagner Free. Inst. Sci. Phila. 11: 75-83.

Jaffe K, Michelangeli F, Gonzalez JM, Miras B, Ruiz MC (1992). Carnivory in pitcher plants of the genus *Heliamphora* (*Sarraceniaceae*). N. Phytol. 122: 733-744.

Kiihnel BK, Holbrook LA, Moloney MM, van Rooijen GI (1996). Oil bodies of transgenic *Brassica napus* as a source of immobilized 13-glucuronidase. J. Am. Oil Chem. Soc. 73: 1533-1538.

Lindquist J (1975). Bacteriological and ecological observations on the northern pitcher plant, *arracenia purpurea*. Masters Thesis, Department of Bacteriology, University of Wisconsin, Madison, WI.

Ma JK, Drake PM, Christou P (2003). The production of recombinant pharmaceutical proteins in plants. Nat. Rev. Genet. 4: 794-805.

Malek L, Bogorad L, Ayers A, Goldberg A (1984). Newly synthesized proteins are degraded by an ATP-stimulated proteolytic process in isolated pea chloroplasts. FEBS Lett. 166: 253-257.

Mandossian AJ (1996). Variations in the leaf of *Sarracenia purpurea* (pitcher plant). Mich. Bot. 5: 26-35.

Newell SJ, Nastase AJ (1998). Efficiency of Insect Capture by *Sarracenia purpurea* (*Sarraceniaceae*), the Northern Pitcher Plant. Am. J. Bot. 85: 88-91.

Nykiforuk CL, Boothe JG, Murray EW, Keon RG, Goren HJ, Markley NA, Moloney MM (2006). Transgenic expression and recovery of biologically active recombinant human insulin from *Arabidopsis thaliana* seeds. Plant Biotech. J. 4: 77-85.

Robins R (1976). The nature of the stimuli causing digestive juice secretion in *Dionaea muscipula* (Venus's flytrap). Planta 128: 263-265.

Schillberg S, Zimmermann S, Voss A, Fischer R (1999). Apoplastic and cytosolic expression of full-size antibodies and antibody fragments in *Nicotiana tabacum*. Transgenic Res. 8: 255-263.

Schnell DE (1979). A critical review of published variants of *Sarracenia purpurea L.* Castanea 44: 47–59.

Schnell DE (2002). Carnivorous plants of the United States and Canada. Timber Press, Portland, Oregon, USA: Timber Press.

Siragusa AJ, Swenson JE, Casamatta DA (2007). Culturable bacteria present in the fluid of the hooded-pitcher plant *Sarracenia minor* based on 16S rDNA gene sequence data. Microbial Ecol. 54: 324-331.

Stephan, P (2008). Insulin: Product bulletin. SemBioSys Genetics Inc. June 1[updated June 1 2008; cited 2009 May 21]. Available from: http://www.sembiosys.com/pdf/SBS-172 Product FS(Insulin).pdf

Stoger E, Sack M, Fischer R, Christou, P (2002). Plantibodies: applications, advantages and bottlenecks. Curr. Opin. Biotech. 13: 161-166.

Voinnet O, Rivas S, Mestre P, Baulcombe D (2003). An enhanced transient expression system in plants based on suppression of gene silencing by the p19 protein of tomato bushy stunt virus. Plant J. 33: 949-956.

Watson J, Koya V, Leppla S, Daniell H (2004). Expression of *Bacillus anthracis* protective antigen in transgenic chloroplasts of tobacco, a non-food/feed crop. Vaccine 22: 4374-4384.

Whitman RL, Byers SE, Shively DA, Ferguson DM, Byappanahalli M (2005). Occurrence and growth characteristics of *Escherichia coli* and *enterococci* within the accumulated fluid of the northern pitcher plant (*Sarracenia purpurea L.*). Can. J. Microbiol. 51: 1027-1037.

3

A review of the pharmacological aspects of *Solanum nigrum* Linn.

F. O. Atanu[1]*, U. G. Ebiloma[1] and E. I. Ajayi[2]

[1]Department of Biochemistry, Kogi State University, Anyigba, Nigeria.
[2]Department of Biochemistry, Osun State University, Osogbo, Nigeria.

This article reviews, bridges the gap between the folkloric use of *Solanum nigrum* linn. (Sn) and the results of evidence based experiments. Although Sn is a rich source of one of plants most dreaded toxins solanine, it has appreciably demonstrated its potential as a reservoir of antioxidants having hepatoprotective, anti-tumor, cytostatic, anti-convulsant, anti-ulcerogenic and anti-inflammatory effects. The review encompasses *in-vitro, in vivo* and clinical studies done on Sn, while examining whether or not correct scientific measures have been taken in generating experimental evidences for its traditional uses. This review would afford research scientist to know how much is known and what is left undone in the investigation of Sn.

Key words: *Solanum nigrum*, folklore medicine, anticancer, solanine.

INTRODUCTION

Solanum nigrum Linn. (Sn) commonly known as Black Nightshade is a dicot weed in the Solanaceae family. It is an African paediatric plant utilised for several ailments that are responsible for to infant mortality especially feverish convulsions. Sn is an Annual branched herb of up to 90 cm high, with dull dark green leaves, juicy, ovate or lanceolate, and toothless to slightly toothed on the margins. Flowers are small and white with a short pedicellate and five widely spread petals. Fruits are small, black when ripe (Cooper and Johnson, 1984). *S. nigrum* is found mainly around waste land, old fields, ditches, and roadsides, fence rows, or edges of woods and cultivated land. It is a common plant found in most parts of Europe and the African continent. Sn is a popular plant in part due to its toxic content of Solanine, a glyco-alkaloid found in most parts of the plant, with the highest concentrations in the unripened berries (Cooper and Johnson, 1984). Although it is considered a rich source of one of the most popular plant poisons, it has proven also to be a reservoir of phytochemicals with phamacological prospects (Lee and Lim, 2006). The aim of this review is to comprehensively put together the literatures consistent with the pharmacological potentials of Sn.

CHEMICAL CONSTITUENTS

Several compounds have been isolated from different fractions of Sn which have shown pharmacological relevance to the observed effects of whole plant preparation of Sn. Sun et al. (2006) reported the variability of the concentration of organic acids between seedlings of Sn and the mature plants. Acetic acid, tartaric acid, malic acid and citric acid were identified as the major organic acids in Sn. Tartaric acid and citric acid however, were said to be most important in adaptive responses by Sn to environmental stresses. High concentrations of solanine, a glycoalkaloid is found in most parts of Sn, but highest levels are found in unripe berries of Sn. However, when ripe, the berries are the least toxic part of the plant and are sometimes eaten without ill effects. Similarly, the solanine increases in the leaves as the plant matures (Cooper and Johnson, 1984). Solanine presented in Figure 1 may be separated by chromatography into six components: Alpha, beta gamma chaconines, and alpha, beta gamma solanines (Merck, 1989). Solanidine ($C_{27} H_{43}$ NO; MW = 397.62) is obtained after hydrolysis of solqnine, solanine and is less toxic. Bhat et al. (2008) also reported the salinity dependent production of a

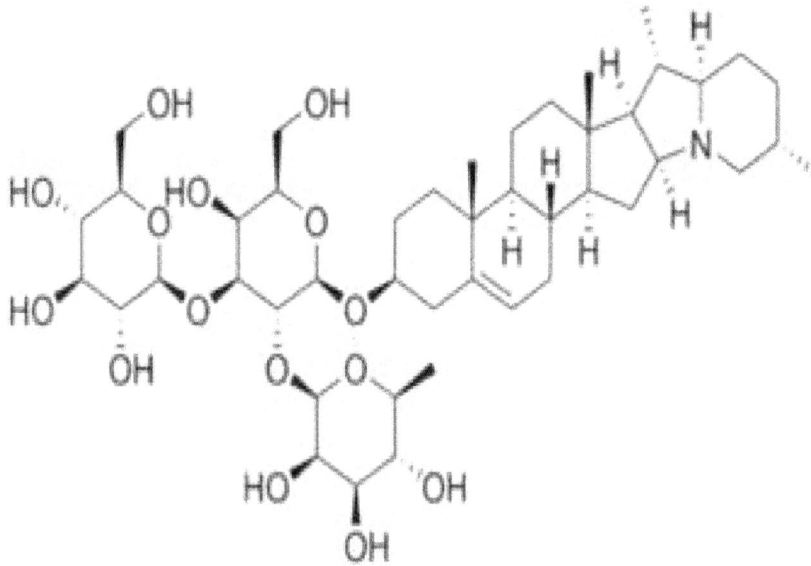

Figure 1. Solanine

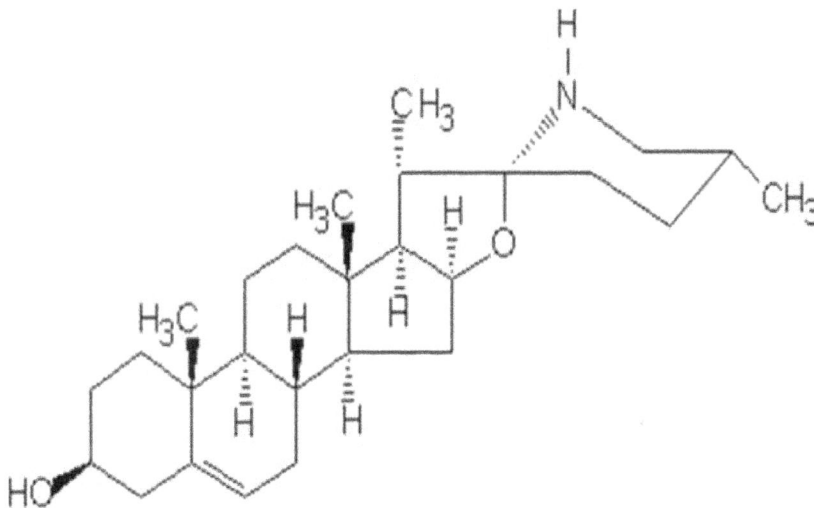

Figure 2. Solasodine

structurally similar steroidal alkaloid, solasodine (Figure 2). Eltayeb et al. (1997) demonstrated that the steroidal alkaloid solasodine was highest in the leaves. However, a somewhat contrasting report by some sources indicate the relative distribution of Solasodine as 9.93 mg g-1 (roots), 6.10 mg g-1 (stems), 4.06 mg g-1 (leaves) and 0.61 mg g-1 (fruits). The absolute amount of alkaloid per leaf increased during leaf development, whereas, the concentration declined. Small unripe fruits of *S. nigrum* had a high concentration of solasodine, but both the concentration and the absolute amount per fruit decreases with fruit maturation. Researches reveal that the alkaloidal content of plant parts changes during development of Sn. Nitrates and nitrites also occur in variable amounts in black nightshade and may contribute to its

toxic effects (Cooper and Johnson, 1984). Hu et al. (1999) isolated three anti-neoplastic steroidal glycosides; beta 2-solamargine, solamargine and degalactotigonin.

Studies on Sn through spectroscopic analysis, chemical degradation and derivitisation led to the identification of six new steroidal saponins collectively called solanigrosides and a one known saponin degalactotigonin (Zhou et al., 2006). Similarly, any set of two steroidal saponin known as nigrumins I and II were characterised from Sn. Nigrumnin I was established as (25R)-5alpha-spirostan-3beta-ol 3-O-betaD-xylopyranosyl-(1-->3)-[alpha-L-arabinopyranosyl-(1 -->2)]- beta-D-glucopyranosyl-(1-->4)-[alpha-L-rhamnopyranosyl(1-->2)] -beta-D- galactopyranoside (1), and nigrumnin II was elucidated as (25R)-3beta, 17alpha-dihydroxy-5alphaspirostan-1 2-one 3-O-beta-D- xylopyranosyl-(1-->3)-[alpha-L-arabinopyranosyl-(1-->2)]-beta-D-glucopyranosyl-(1-->4)-[alpha-L-rhamnopyra-nosyl-(1-->2)l-beta-D-galactopyranoside.

Also, five non-saponin namely 6-methoy-hydroxycoumarin, syringaresinol-4-O-beta-D-glucopyranoside, pinoresinol-4-O-beta-D-glucopyranoside, 3, 4-dihydroxhbenzoic acid (IV), p-hydroxybenzoic acid and 3-methoxy-4-hydroxyienzoic acid were isolated for studies by Wang et al. (2007)

It was discovered by Schmidt and Baldwin (2007) that Sn produces Sysytemin, an 18 amino acid polypeptide similar to systemic wound response protein produced by tomato. Recently, the isolation of a 910 bp cDNA encoding osmotin-like protein with an open reading frame of 744 bp encoding a protein of 247 amino acids (26.8 kDa) was cloned from S. nigrum (SniOLP) (Jami et al., 2007). Phylogenetic analysis revealed the evolutionary conservation of this protein among diverse taxa belonging to a small multigene family and it showed organ-specific expression. The expression of this protein has been discovered to be upregulated by osmotic and oxidative stress inducers.

One spirostanol glycoside and two furostanol glycosides have been isolated from a methanol extract of the stems and roots of S. nigrum (Sharma et al., 1983). Quercitin represents one of the most potent natural antioxidants. Sn contains two quercetin glycosides namely, quercetin 3-O-(2^{Gal}-α-rhamnosyl)-β-glucosyl (1→6)-β-galactoside and quercetin 3-O-α-rhamnosyl(l→2)-β-galactoside. Also, previously known quercetin 3-glucosyl(l→6)galactoside, 3-gentiobioside, 3-galactoside and 3-glucoside, were also found (Nawwar et al., 1989). The most recent phytochemical analysis of S. nigrum has resulted in the isolation of two novel disaccharides. Their structures were determined as ethyl β-d-thevetopyranosyl-(1→4)-β-d-oleandropyranoside and ethyl β-d-thevetopyranosyl-(1→4)-α-d-oleandropyranoside, respectively, by chemical and spectroscopic methods (Chen et al., 2009).

The berries of S. nigrum have been found to contain a saturated steroidal genin, which has been identified as tigogenin by mixed melting point and i.r. spectroscopy (Varshney and Sharma, 1965). 150-kDa glycoprotein was isolated from S. nigrum, which has been used as an antipyretic and anticancer agent in folk medicine. The SNL glycoprotein consists of carbohydrate content (69.74%) and protein content (30.26%), which contains more than 50% hydrophobic amino acids such as glycine and proline (Lee and Lim, 2006).

Although toxic constituents are present in most part of the plants, studies on the nutritional potential of the leaves and seeds revealed that Sn is nutritive despite the presence of some anti-nutritive components like oxalate. Protein content of the leaves and seed was found to be24.90 and 17.63%, respectively. Other findings are ash 10.18 and 8.05%, crude fibre, 6.81 and 6.29% and carbohydrate, 53.51 and 55.85% for the leaves and seed

respectively. Mineral analysis revealed the magnitude of presence in the order Mg>K>Ca>Fe>Na>Mn>Zn in the leaves and Mg>K>Fe>Ca>Na>Mn>Zn in the seeds.

Phosphorus and sulphur levels were 75.22 and 8.55 mg/100 g in the leaves and 62.50 and 14.48, g/100g in the seeds. Vitamin content indicate the order of magnitude as Vit C>Vit B,>Folic acid>Vit E>Vit A in both the leaves and seeds. Phytochemical analysis revealed high oxalate, phenol, but low sterol content in the studied plant materials. Cyanide levels were higher in the leaves compared to the seeds.

ANTICANCER PROPERTIES

The effect of crude polysaccharide isolated from S. nigrum linn. (SNL-P) was examined both in vivo and in vitro on U14 cervical cancer cells. Though exposure to SNL-P had no antiprolifreative effect in vitro at doses up to 1 mg/ml, it decreased the number of ascites tumor cells and survival time of U14 cervical cancer bearing mice which received between 90 - 360mg/kg bw. P.o. FACScan flow cytometer analysis showed that most of the ascites tumor cells were arrested in G2/M phase of cell cycle. This can be considered as the basis for its use as an anticancer agent (Jian et al., 2009). Similarly, in an earlier work by Jian et al. (2007) on the in vivo effect of a 12-day oral administration of SNL-P, showed a significant growth inhibition effect on cervical cancer (U14) of tumor-bearing mice with increased expression of Bax and a decreased expression of Bcl-2 and mutant p53 which had a positive correlation with the number of apoptosising tumor cells. Moreover, SNL-P treatment decreased the level of blood serum TNF-α, this corresponds to triggerring of apoptosis in tumor cells. These findings demonstrated that the SNL-P is a potential antitumor agent (Jian et al., 2007). The review by An Lei et al. (2006) suggests that the anticancer potential of Sn was based on its capacity to interfere with the structure and function of tumor cell membrane, disturb the synthesis of DNA and RNA, change the cell cycle distribution, blocking the anti-apoptotic pathway of NF-kappaB, activating caspase cascades reaction and increasing the production of nitric oxide. The contribution of autophagic cell death in the anticancer pathways of Sn was carefully elucidated through studies utilising LC3-I and LC3-II proteins in Hep G2 cells. Results show a concentration dependent mechanism of Sn in cell autophagy and vacuolisation. This may provide a leverage to treat liver specific cancer.

A case-control study of dietary and social factors was performed for 130 patient/control pairs matched for age, gender, and educational level. Staple diet, consumption of wild vegetables, use of tobacco, and traditional beer consumption were compared between the two groups. S. nigrum contains protease inhibitors capable of oesophagic proliferative and oncogenic drive (Sammon, 1998).

The anti-tumor activity of solanine, a steroid alkaloid isolated from the nightshade has been evaluated by the MTT assay on the three digestive system tumor cell lines namely, HepG(2), SGC-7901, and LS-174. Solanine had a concentration specific IC (50) score for HepG (2), SGC-7901, and LS-174 cell lines (14.47, > 50, and > 50 microg/ml, respectively) and signs for apoptosis were found. These effects are obtainable, although much less, in other cancer cell line, for example, the Chang liver and WRL-68 cells (Lin et al., 2007). Cells in the G(2)/M phases disappeared, while the number of cells in the S phase increased significantly for treated groups, which decreased the expression of Bcl-2 protein. Therefore, the target of solanine in inducing apoptosis in HepG (2) cells seems to be mediated by the inhibition in the expression of Bcl-2 protein (Ji et al., 2008). There seems to be a differential response of exposure to either high or low concentrations SNE on the nature of cell death. While high doses elicited apoptotic cell death with corresponding mitochondria release of cytochrome c, and caspase activation at low concentrations SNE (50-1000 µg/ml), revealed morphological and ultrastructural changes of autophagocytic death. Furthermore, these cells showed increased levels of autophagic vacuoles and LC3-I and LC3-II proteins, and specific markers of autophagy. Taken together, these findings indicate that SNE induced cell death in hepatoma cells via two distinct antineoplastic activities of SNE- the ability to induce apoptosis and autophagocytosis, therefore, suggesting that it may provide leverage to treat liver cancer (Lin et al., 2007).

Also, the aqueous fraction of Sn was tested for its antitumor activity *in vivo*. This closely mimics the preparation of plant in folklore medicine. Aqueous extract of *S. nigrum* (SNL-AE) inhibited U14 cervical carcinoma growth and increased the number of CD4+ T lymphocyte subsets as well as the ratio of CD4+/CD8+ T lymphocyte, decreased the number of CD8+ T lymphocyte subsets of tumor-bearing mice and PCNA positive cells. Furthermore, SNL-AE caused cell cycle arrest in G0/G1 phase and induced apoptosis of more transplanted tumor cells in a dose-dependent manner, suggesting that the anti-tumor activity of SNL-AE is embedded in its immune-modulatory effects (Jian et al., 2008). Results of some studies show that Sn achieves antitumor activity by beefing up the oxidative stress threshold of the neoplastic cell.

A proline and glycine rich glycoprotein (150kDa) isolate of Sn had modulatory effects on transcriptional factors (NF-kappa B and AP-1) and iNO production which in turn enhanced NO production in MCF-7 cells (Heo et al., 2004; Son et al., 2003; Lim, 2005). It is evident that this glycoprotein stimulates the mitochondrial release of cytochrome C, which culminates in caspase activation and the eventual death of tumor cells. Another 43 amino acid, 4.8 kDa peptide Lunasin originally found in soybean was identified in Sn. It elicits anticancer and cell-cycle arrest by inhibiting phosphorylation of retinoblastoma protein (Rb) and acetylation of core histone H3 and H4 (Jeong et al., 2008). Anti-angiogenesis is an established antineoplastic mechanism of many chemotherapeutic agents. Sn displayed anti-angiogenic activity on chick chorioalliantoic membrane (Xu et al., 2008).

IMMUNOMODULATORY EFFECTS

In vivo experiments showed that the ratio of CD4+/CD8+ peripheral blood T-lymphocyte subpopulations were restored following the treatment of SNL-P. Furthermore, treatment with SNL-P also caused a significant increased in IFN-α (p < 0.01, 90, 180 and 360 mg/kg bw) and a remarkable decrease in IL-α (p < 0.01, 90, 180 mg/kg b.w.; p < 0.05, 360 mg/kg b.w.) measured by the method of ELISA.

These data showed that SNL-P possess potent antitumor activity and SNL-P might exert antitumor activity via activation of different immune responses in the host rather than by directly attacking cancer cells on the U14 cervical cancer bearing mice. Thus, SNL-P could be used as an immunomodulator (Jian et al., 2009).

ANTIMICROBIAL, NEMATICIDAL AND MOLLUSCICIDAL PROPERTIES

Root extracts of black nightshade (*S. nigrum*) were analyzed for its activity against isolates ABA-31 and ABA-104 of *Alternaria brassicicola*, the causal agent of black leaf spot of Chinese cabbage (*Brassica pekinensis*). Methanolic extracts of dried root tissues of black nightshade contained antifungal properties wchich act against *A. brassicicola*. Further fractionation and anti-microbial screening of ethyl acetate, n-butanol and water fractions of root extracts showed that n-butanol extracts was the most potent. Saponins were identified as the active principles conferring antimicrobial effects on Sn (Muto et al., 2006). Afaf s' and Soads' (2007) investi-gation on the effect of sub-lethal (LC25) concentration of leaves of Sn on Saudi Arabian mollusc *Biomphalria arabica* revealed that AST, ALT and LDH activities were affected in them and may suggest the mechanism for its molluscicidal activities. Similarly, binary combination of Sn and *Iris pseudacorus* by Amer and Manal (2005) had molluscicidal and cercaricidal efficacy against *Biomphalaria alexandrina* and *Schistosoma mansoni cercariae*, respectively (Ahmed and Ramzy, 1998). The effect of a 30 min pre-treatment of mice with varied concentration (2.5 – 10 mg/ml) of crude water extract of Sn on penetration and infectivity of *S. mansoni cercariae* showed a significant reduction in penetration (p < 0.001) and infectivity (p < 0.01) (Amer and Manal, 2005). Also, a recent work done by Raghavendra et al. (2009) and Ahmed et al., (2002) appraised Sn extracts as a larvicidal agent against five laboratory colonised strains of mosquito species.

ANTIOXIDANT PROPERTIES

Many pathological states encompassing both communicable and non-communicable diseases have been shown to have association with oxidative stress.

Consequently, the need for potent antioxidants in our diet and drug supplements becomes very necessary. A study which utilises six pretreatment methods before cooking on the peroxidase activity, chlorophyll and antioxidant status of S. nigrum L., showed that pretreatment methods have significant effects ($p < 0.05$) on the parameters measured. A sharp difference in the carotenoids, phenolics, flavonoids and tannins contents has been reported, indicating the fragility of this antioxidant present in Sn (Adebooye et al., 2008). SNL glycoprotein showed a dose-dependent radical scavenging activity on radicals, including 1, 1-diphenyl-2-picrylhydrazyl (DPPH) radicals, hydroxyl radical (OH), and superoxide anion (O_2^-).

Although Sn acts as an anti-tumor, the SNL glycoprotein may induce apoptosis through the inhibition of NF-κB activation, induced by oxidative stress in HT-29 cells (Heo et al., 2004). A 50% ethanol extract of the whole plant of S. nigrum also possess hydroxyl radical scavenging potential which is suggested as cytoprotective mechanism (Kumar et al., 2001; Mohamed et al., 2007). Evaluation of the antioxidant potential of Sn leaves on the modulation of a 6 h restraint induced oxidative stress, which suggest that Sn was better as an antioxidant with post-restraint treatment than with pre-restraint administration.

ANTI-CONVULSANT ACTIVITY

Central nervous system-depressant action of Sn was ascertained by measuring the effects of intraperitoneal injection of Sn on various neuropharmacological parameters. Fruit extracts of Sn significantly prolonged pentobarbital-induced sleeping time, produced alteration in the general behaviour pattern, reduced exploratory behaviour pattern, suppressed the aggressive behaviour, affected locomotor activity and reduced spontaneous motility. This buttresses its usage as an anti-convulsant and may concur with its acetylcholine-like activity (Perez et al., 1998). The potency of Sn in combating infant convulsion is widely accepted in African paediatric medicine. Wannang et al. (2008) tested the anti-convulsant effects of leaves of S. nigrum in chicks, mice and rats.

A 30 min pretreatment by intraperitoneal injection of Sn leaf extract protected the animal subjects against different types of proconvulsants. The aqueous leaf extract produced a significantly ($p < 0.05$) dose dependent protection against electrically-induced seizure in chicks and rats, pentylenetetrazole-induced seizure in mice and rats and picrotoxin-induced seizure in mice and rats (Wannang et al., 2008). De Melo et al. (1978) were the first to present experimental data supporting the claims of the acetylcholine-like activity of Sn. They based their conclusion on the basis of the observation of the following effects: Isotonic contraction of the isolated toad rectus abdominis; 2) Negative chronotropic and inotropic action on the isolated toad heart; 3) Isotonic contraction of the isolated guinea pig's ileum; 4) Isotonic contraction of the rat's isolated jejunum; 5) Decrease on the cat's arterial blood pressure; 6) Secretory effects on the rat's submaxillary gland. Fruits of Sn were also found to contain acetylcholine-activity compounds up to 250 micrograms/g of fruit (de Melo et al., 1978).

HEPATOPROTECTIVE EFFECTS

S. nigrum L. (SN) is an herbal plant that has been used as hepatoprotective and anti-inflammatory agent in Chinese medicine. Sprague-Dawley (SD) rats orally fed with SNE (0.2, 0.5, and 1.0 g kg^{-1} bw) along with the administration of CCl_4 (20% CCl_4/corn oil; 0.5 mL kg^{-1} bw) for 6 weeks displayed that Sn had hepatoprotective effects against CCl_4. The test drug significantly lowered the CCl_4-induced elevation of hepatic enzyme markers (GOT, GPT, ALP, and total bilirubin) and decreased superoxide and hydroxyl radical generation (Raju et al., 2003) in comparison with the CCl_4 treatment group. Liver histology showed that SNE reduced the incidence of liver lesions including hepatic cells cloudy swelling, lymphocytes infiltration, hepatic necrosis, and fibrous connective tissue proliferation induced by CCl_4 in rats (Lin et al., 2008). Other studies using other hepatotoxic challenges such as thioacetamide (TAA), a liver fibrosis inducer was attenuated by a 12 days administration of SNE (0.2 or 1.0 g/kg) via gastrogavage throughout the experimental period. SNE reduced the hepatic hydroxyproline and α-smooth muscle actin protein levels of TAA-treated mice. SNE inhibited TAA-induced collagen (α1) (I) and transforming growth factor-β1 (TGF-β1) mRNA levels in the liver (Hsieh et al., 2008; Sultana et al., 1995).

Another study by Hsu et al. (2009) utilised 2-acetylaminoflorene as an inducer of hepatocarcino-genesis. Sn inhibited hepatocarcinogenesis which is consistent with increased expression of glutathione S-transferase-alpha and -mu, the level of transcription factor Nrf2, glutathione peroxidase, superoxide dismutase-1, and catalase (Hsu et al., 2009).

In Africa, Aflatoxin B1 induced liver cancer is a common cause of hepatocarcinogenesis. Activation of several cytochrome p450 systems and depression in the expression of phase II enzymes responsible for AFB1 metabolism precludes its toxicity. Sn increased the activity of uridine diphosphate glucuronyltransferase (UDPGT) and glutathione S-transferase in female rats toxicated with AFB1 (0.2 or 0.4 mg/kg bw) (Moundipa and Domngang, 1991).

ANTIULCEROGENIC AND ANTI-INFLAMMATORY EFFECTS

Sn is recommended in ayurveda for the management of gastric ulcers, it is therefore, essential to find out what the mechanism of anti-ulcerogenic effect is. Rats were exposed to various types of stress (cold restraint stress, indomethacin, pyloric ligation, ethanol and acetic acid) to induce stress ulcers. Sn fruits extract significantly inhibited the gastric lesions induced by 76.6, 73.8, 80.1 and 70.6%, respectively, with equal or higher potency than omeprazole. Sn extracts showed concomitant attenuation of gastric secretory volume, acidity and pepsin secretion in ulcerated rats (Akhtar and Munir, 1989). In addition, SNE (200 and 400 mg/kg b.w.) accelerated the healing of acetic acid induced ulcers after treatment for 7 days. Enzymatic studies on $H^+K^+ATPase$ activity to ascertain the antisecretory action showed that SNE significantly inhibits $H^+K^+ATPase$ activity and decreases the gastrin secretion in EtOH-induced ulcer model. Histological studies revealed a reduction of ulcer size by SNE (Jainu et al., 2006). Data suggesting the anti-inflammatory and analgesic potential of Sn is provided by Zainul et al. (2006). Quite recently, the potency of the 150 kDa glycoprotein of Sn in preventing dextran sodium sulfate-induced colitis in A/J mouse was determined. It was observed that Sn had suppressive effects on the concentrations of nitric oxide production, lactate dehydrogenase release and thiobarbituric acid reactive substances. This is achieved through the regulation of transcription factors such as NF-kappaB (p50) and AP-1 (c-Jun). Also, Sn regulates the expression of iNOS and COX-2, which are principal enzymes in inflammatory response pathways.

HYPOLIDEAMIC, ANTI-HYPERGLYCEMIC AND HYPOTENSIVE POTENTIALS

Hyperlipidemia is a major risk factor in cardiovascular pathologies. Atherosclerosis and other forms of cardiovascular dysfunction are promoted by excessive agitation of the cation pumps on the cell membranes. It is therefore, conceivable that since Sn had inhibitory effects on the H+K+ATPase, it could serve as cardioprotective regimen. Hypolipidemic agents are the first defence against lipid associated pathologies, therefore, the investigation of the effects of 150 kDa glycoprotein isolated from S. nigrum Linn. (SNL), which has been used as a hepatoprotective and anticancer agent in folk medicine is necessary. Mice treated with Sn had decreased levels of the plasma lipoprotein levels (TG, TC and LDL). In addition, SNL glycoprotein inhibits the activity of cholestyramine-induced hepatic HMG-CoA reductase at 40 µg/g head body weight (Lee et al., 2005). Validation of the ethnobotanical use of the leaves of S. nigrum Linn. (Solanaceae), Vitex negundo Linn. (Verbenaceae) and stems of Nopalea cochinellifera

(Linn.) as anti-diabetic agents using the oral glucose tolerance test showed that there was no significant lowering in BGLs by S. nigrum (Villaseñor and Lamadrid, 2006).

Sn which has been used as an antipyretic and anticancer in folk medicine was investigated for its anti-hypertensive properties. A 150 kDa glycoprotein isolated from Sn is made up of carbohydrates (69.74%) and protein (30.26%), which contains more than 50% hydrophobic amino acids such as glycine and proline blocked nuclear factor-kappa B (NF-κB) activation, and reduced inducible nitric oxide (iNO) production in vitro at a concentration of 40 µg/ml (Lee and Lim, 2006).

REFERENCES

Adebooye OC, Ram V, Vasudeva S (2008). Peroxidase activity, chlorophylls and antioxidant profile of two leaf vegetables (Solanum nigrum L. and Amaranthus cruentus L.) under six pretreatment methods before cooking. Int. J. Food Sci. Technol., 43(1): 173-178.

Afaf KE, Soad KA (2007). Effect of sublethal concentration of Solanum nigrum on transaminases and lactate dehydrogenase of Biomphalria arabica, in Saudi Arabia. J. Egyptian Soc. Parasitol., 37(1): 39-50.

Ahmed AH, Kamal IH, Ramzy RM (2002). Studies on the molluscicidal and larvicidal properties of Solanum nigrum L. leaves ethanol extract. J. Egyptian Soc. Parasitol., 31(3): 843-852.

Ahmed AH, Ramzy RM (1999). Seasonal variation in molluscicidal activity of Solanum nigrum L. J. Egyptian Soc. Parasitol., 28(3): 621-629.

Akhtar MS, Munir M (1989). Evaluation of the gastric antiulcerogenic effects of Solanum nigrum, Brassica oleracea and Ocimum basilicum in rats. J. Ethnopharm., 27(1,2):163-176.

Amer HA, Manal MAR (2005). Molluscicidal and cercaricidal efficacy of Acanthus mollis and its binary and tertiary combinations with Solanum nigrum and Iris pseudacorus against Biomphalaria alexandrina. J. Egyptian Soc. Parasitol., 34(3): 1041-1050.

An Lei, Tang Jin-tian, Liu Xin-min, Gao Nan-nan (2006). Review about mechanisms of anti-cancer of Solanum nigrum. China J. Chinese Materia Medica. 31(15):1225-1226.

Bhat MA, Ahmad S, Aslam J, Mujib A, Mahmooduzzfar (2008). Salinity stress enhances production of solasodine in Solanum nigrum L. Chem. Pharm. Bull., 56(1): 17-21.

Chen Rong, Feng Lin, Li Hai-Dao, Zhang Hua, Yang Fei (2009). Two novel oligosaccharides from Solanum nigrum. Carbohydrate Res. 344(13,8):1775-1777.

Cooper MR, Johnson AW (1984). Poisonous Plants in Britain and other effects on Animals and Man. Ministry of Agriculture, Fisheries Food, 161: 219-220.

De Melo Ac, Perec CJ, Rubio MC (1978). Acetylcholine-like activity in the fruit of the black nightshade (Solanaceae) Acta Physiol. Lat. Am., 28(4,5):19-26.

Eltayeb Elsadig A, Al-Ansari Alia S, Roddick James G (1997). Changes in the steroidal alkaloid solasodine during development of Solanum nigrum and Solanum incanum. Phytochemistry, 46(3): 489-494

Heo KS, Lee S, Lim KT (2004). Cytotoxic effect of glycoprotein isolated from Solanum nigrum L. through the inhibition of hydroxyl radical-induced DNA-binding activities of NF-kappa B in HT-29 cells. Environ. Toxicol. Pharmacol., 17(1): 45-54.

Heo KS, Lee SJ, Ko JH, Lim K, Lim KT (2004). Glycoprotein isolated from Solanum nigrum L. inhibits the DNA-binding activities of NF-κB and AP-1, and increases the production of nitric oxide in TPA-stimulated MCF-7 cells. Toxicol. in Vitro, 18(6): 755-763.

Hsieh CC, Fang HL, Lina WC (2008). Inhibitory effect of Solanum nigrum on thioacetamide-induced liver fibrosis in mice. J. Ethnopharmacol., 119(1,2): 117-121.

Hu K, Dong A, Jing Y, Iwasaki S, Yao X (1999). Antineoplastic agents. III: Steroidal glycosides from Solanum nigrum. Planta medica., 65(1): 35-38.

Jainu M, Chennam S, Shyamala D (2006). Antiulcerogenic and ulcer healing effects of *Solanum nigrum* (L.) on experimental ulcer models: Possible mechanism for the inhibition of acid formation. J. Ethnopharmacol., 104(1,2,8): 156-163.

Jami SK, Anuradha TS, Guruprasad L, Kirti PB (2007). Molecular, biochemical and structural characterization of osmotin-like protein from black nightshade (*Solanum nigrum*). J. Plant Physiol., 164(3,7): 238-252.

Jeong JB, Jeong HJ, Park JH, Lee SH, Lee JR, Lee HK, Chung GY, Choi J D, Lumen BO (2008). Cancer-preventive peptide lunasin from *Solanum nigrum* L. inhibits acetylation of core histones H3 and H4 and phosphorylation of retinoblastoma protein (Rb). J. Agric. Food Chem. 55(26): 10707-10713.

Ji YB, Gao SY, Ji CF, Zou X (2008). Induction of apoptosis in HepG2 cells by solanine and Bcl-2 protein. J. Ethnopharmacol., 115(2): 194-202.

Jian L, Qing-Wang L, Da-Wei G, Zeng-Sheng H, Wen-Zong L (2009). Antitumor and immunomodulating effects of polysaccharides isolated from *Solanum nigrum* Linne. Phytother. Res., 23(11): 1524-1530.

Jian L, Qingwang L, Tao F, Kun L (2008). Aqueous extract of *Solanum nigrum* inhibit growth of cervical carcinoma (U14) via modulating immune response of tumor bearing mice and inducing apoptosis of tumor cells Fitoterapia., 79(7,8): 548-556.

Jian L, Qingwang L, Tao F, Tao Z, Kun L, Rui Z, Zengsheng H, Dawei G (2007). Antitumor activity of crude polysaccharides isolated from *Solanum nigrum* Linne on U14 cervical carcinoma bearing mice. Phytother. Res., 21(9): 832-840.

Kumar VP, Shashidhara S, Kumar MM, Sridhara BY (2001). Cytoprotective role of *Solanum nigrum* against gentamicin-induced kidney cell (Vero cells) damage *in vitro*. Fitoterapia. 72(5): 481-486.

Lee SJ, Ko JH, Lim K, Lim KT (2005). 150 kDa glycoprotein isolated from *Solanum nigrum* Linne enhances activities of detoxicant enzymes and lowers plasmic cholesterol in mouse. Pharmacol. Res., 51(5): 399-408.

Lee S-J, Lim K-T (2006). 150 kDa glycoprotein isolated from *Solanum nigrum* Linne stimulates caspase-3 activation and reduces inducible nitric oxide production in HCT-116 cells. Toxicol. *in Vitro*, 20(7): 1088-1097.

Lim K-T (2005). Glycoprotein isolated from *Solanum nigrum* L. kills HT-29 cells through apoptosis. J. Med. Food, 8(2): 215-226.

Lin HM, Tseng HC, Wang CJ, Chyau CC, Liao KK, Peng PL, Chou FP (2007). Induction of autophagy and apoptosis by the extract of *Solanum nigrum* Linn in HepG2 cells. J. Agric. Food Chem., 55(9): 3620-3628.

Lin H-M, Tseng H-C, Wang C-J, Lin J-J, Lo C-W, Chou F-P (2008). Hepatoprotective effects of *Solanum nigrum* Linn extract against CCl₄-induced oxidative damage in rats. Chemico-Biological Interactions, 171(3,15): 283-293.

Lin H-M, Tseng H-C, Wang CJ, Lin JJ, Lo CW, Chou FP (2008). Hepatoprotective effects of *Solanum nigrum* Linn extract against CCl4-induced oxidative damage in rats. Chem. Biol. Interact, 171(3): 283-293.

Merck I (1989). Ed. S. Budavari, 11th Ed., NJ, USA. pp:1371. Watt JM; Breyer-Brandwijk MG (1962). The Medicinal and Poisonous Plants of Southern and Eastern Africa E & S, Livingston Ltd., Edinburgh and London, UK, pp. 996-1000.

Mohamed AF, Martina W, Gudrun Sr, Ulrike L (2007). Antioxidant, antimicrobial and cytotoxic activities of selected medicinal plants from Yemen. J. Ethnopharmacol., 111(3): 657-666.

Moundipa PF, Domngang FM (1991). Effect of the leafy vegetable *Solanum nigrum* on the activities of some liver drug-metabolizing enzymes after aflatoxin B1 treatment in female rats. British J. Nutr., 65(1): 81-91.

Muto M, Mulabagal V, Huang H-C, Takahashi H, Tsay H-S, Huang J-W (2006). Toxicity of black nightshade (*Solanum nigrum*) extracts on (Alternaria brassicicola, causal agent of black leaf spot of Chinese cabbage (Brassica pekinensis). J. Phytopathol., 154 (1): 45-50.

Nawwar MAM, El-Mousallamy AMD, Barakat HH (1989). Quercetin 3-glycosides from the leaves of *Solanum nigrum*. Phytochemistry, 28(6): 1755-1757.

Perez G, Perez L, Garcia D, Sossa M (1998). Neuropharmacological activity of *Solanum nigrum* fruit. J. Ethnopharmacol., 62(1): 43-48.

Raghavendra K, Singh SP, Subbarao SK, Dash AP (2009). Laboratory studies on mosquito larvicidal efficacy of aqueous and hexane extracts of dried fruit of *Solanum nigrum* Linn. Indian J. Med. Res., 130(1): 74-77.

Raju K, Anbuganapathi G, Gokulakrishnan V, Rajkapoor B, Jayakar B, Manian S (2003). Effect of dried fruits of *Solanum nigrum* LINN against CCl4-induced hepatic damage in rats. Biol. Pharm. Bull., 26(11): 1618-1619.

Sammon AM (1998). Protease inhibitors and carcinoma of the esophagus. Cancer, 83(3): 405-408.

Sharma SC, Chand R, Sati OP, Sharma AK (1983). Oligofurostanosides from *Solanum nigrum*. Phytochem., 22(5): 1241-1244.

Son YO, Kim J, Lim JC, Chung Y, Chung GH, Lee JC (2003). Ripe fruits of *Solanum nigrum* L. inhibits cell growth and induces apoptosis in MCF-7 cells. Food Chem. Toxicol., 41(10): 1421-1428.

Sultana S, Perwaiz S, Iqbal M, Athar M (1995). Crude extracts of hepatoprotective plants, *Solanum nigrum* and *Cichorium intybus* inhibit free radical-mediated DNA damage. J. Ethnopharmacol., 45(3): 189-192.

Sun RI, Zhou Q-x, Wang X (2006). Relationships between cadmium accumulation and organic acids in leaves of *Solanum nigrum* L. as a cadmium-hyperaccumulator. Huan jing ke xue., 27(4): 765-769.

Varshney IP, Sharma SC (1965). Saponins and sapogenins-XXIX: The sapogenin of *Solanum nigrum* L. berries. Phytochemistry, 4(6): 967-968.

Villaseñor IM, Lamadrid MRA. (2006). Comparative anti-hyperglycemic potentials of medicinal plants . J. Ethnopharmacol., 104(1,2,8): 129-131.

Villaseñor IM, Lamadrid MR (2006). Comparative anti-hyperglycaemic potentials of medicinal plants. J. Ethnopharmacol., 104(1,2): 129-131.

Wang L-y, Wang N-l, Yao X-s (2007). Non-saponins from *Solanum nigrum* L. J. Chin. Med. Mat. 30(7): 792-794.

Wannang NN, Anuka JA, Kwanashie HO, Gyang SS, Auta A (2008). Anti-seizure activity of the aqueous leaf extract of *Solanum nigrum* linn (solanaceae) in experimental animals. Afr. Health Sci., 8(2): 74-79.

Xu Y, Pan R-L, Chang Q, Qin M, Liu Y, Tang J-T (2008). Experimental study of *Solanum nigrum* on inhibiting angiogenesis in chick chorioallantoic membrane. China J. Chin. Materia Med., 33(5):549-552.

Zainul AZ, Hanan KG, Hairani Z, Nur HMP, Nur AM, Anwariah A, Mohd RS (2006). Antinociceptive, anti-inflammatory and antipyretic effects of *Solanum nigrum* chloroform extract in animal models. J. Pharm. Soc. Japan, 126(11): 1171-1178.

Zhou X, He X, Wang G, Gao H, Zhou G, Ye W, Yao X (2006). Steroidal saponins from *Solanum nigrum*. J. Nat. Prod., 69(8): 1158-1163.

4

An investigation for potential development on biosurfactants

A. Salihu[1*], I. Abdulkadir[2] and M. N. Almustapha[3]

[1*]Department of Biochemistry, Ahmadu Bello University, Zaria, Nigeria.
[2]Department of Chemistry, Ahmadu Bello University, Zaria, Nigeria.
[3]Department of Pure and Applied Chemistry, Usman Danfodio University, Sokoto, Nigeria.

Biosurfactants are surface-active metabolites produced by microorganisms. The applications of these biological compounds in the field of enhanced oil recovery and bioremediation proved effective. Besides their environmental applications; biosurfactants have shown interesting properties in several processes. Thus, this article attempts to organize this Information for ease of reference and further stimulates those that have interests in the area to explore further especially in biodegradation of recalcitrant compounds.

Key words: Biosurfactant, bioremediation, biodegradation.

INTRODUCTION

Biosurfactants have received more and more attention in recent years as surface-active compounds released by microorganisms that have some influence on interfaces, most notably on the surface tension of liquid - vapor interfaces. They are interesting amphiphilic molecules of microbial origin; whose hydrophobic and hydrophilic domains depend on the carbon substrate and the organism strain. They have various biological functions/ properties and have potentials in commercial applications in the food, microbiological, pharmaceutical and therapeutical agents in biological industries, as a bio-control agent in agricultural applications and in health and beauty products for the cosmetic industries (Nayak et al., 2009; Mukherjee et al., 2009; Tugrul and Cansunar, 2005; Benincasa et al., 2004; Volkering et al., 1998; Fiechter, 1992). Many microorganisms have ability to produce a wide range of biosurfactants, as such initial classification was made into two; based on molecular weights, properties and cellular localizations. The low molecular weight biosurfactants e.g. glycolipids, lipope-ptides, flavolipids, corynomycolic acids and phospholipids lower the surface and interfacial tensions at the air/water interfaces while the high molecular weights are called bioemulsans, (such as emulsan, alasan, liposan, poly-saccharides and protein complexes) and are more effec-tive in stabilizing oil-in-water emulsions (Neu, 1996; Franzetti et al., 2009). These high molecular weight biosurfactants are highly efficient emulsifiers that work at low concentrations and exhibit considerable substrate specificity (Dastgheib et al., 2008).

However, general classification based on parent chemi-cal structure and surface properties are represented in the following groups:

1. The glycolipids, in which carbohydrates such as sophorose, trehalose or rhamnose are incorporated to a long-chain aliphatic acid or lipopeptide. This is seen in rhamnolipids produced by *Pseudomonas aeruginosa* which constitute one or two sugar moieties linked to caprilic acid group by a glycosidic bond (Rosenberg and Ron, 1999).
2. Biosurfactant with amino-acid moieties such as sur-factin produced by *Bacillus subtilis* composed of several amino-acid structures linked to a molecule of 3-hydroxy-13-methyl tetradecanoic acid.
3. The emulsan synthesized by *Acinetobacter calcoa-ceticus* RAG-1 is an extracellular polysaccharide - lipid complex (Rosenberg et al., 1979).
4. Protein-like substances (e.g. liposan) constitute small amount of carbohydrates as produced by *Candida lipoltica* (Calvo et al., 2008).

Biosurfactants have unique advantages which include: structural diversity that may lead to unique properties

---
*Corresponding author. E -mail: salihualiyu@yahoo.com.

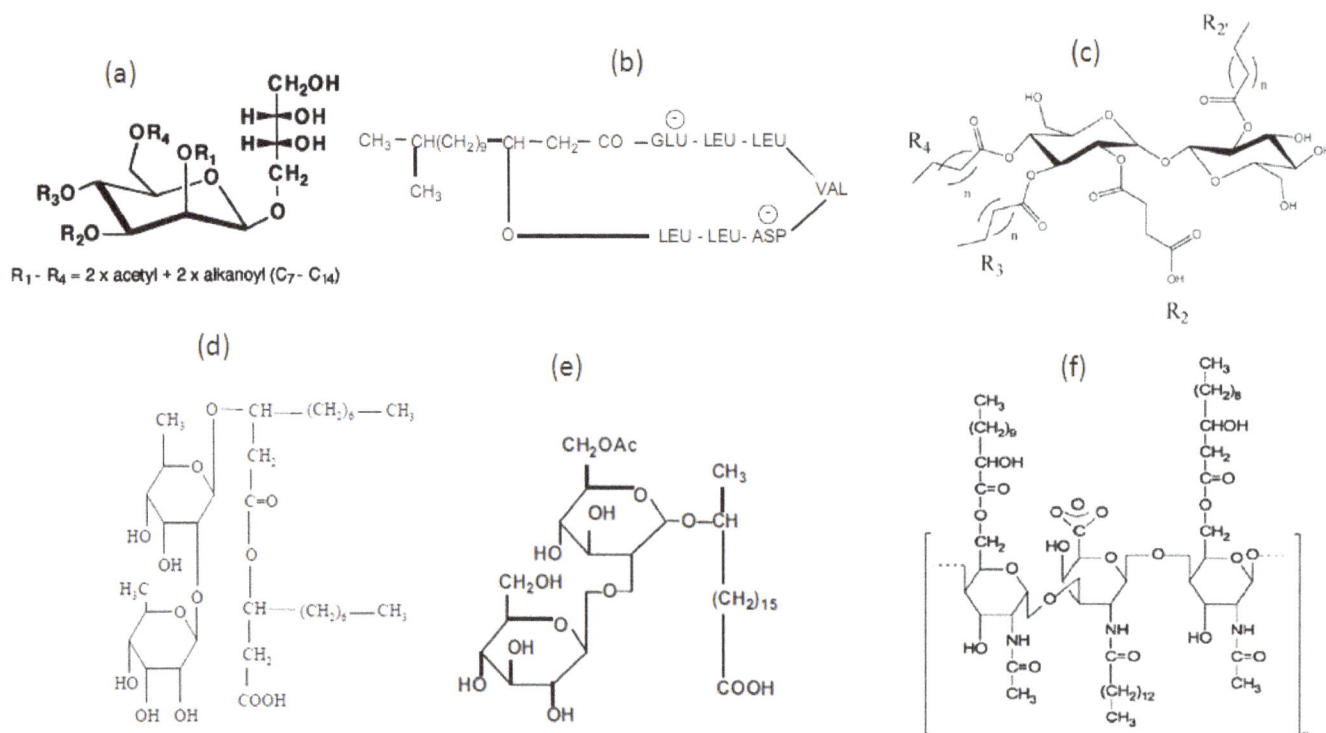

Figure 1. Chemical structures of some common biosurfactants (a) Mannosylerythritol lipid (b) Surfactin (c) trehalose lipid (d) Sophorolipid (e) Rhamnolipid (f) Emulsan.

(chemical structures of commonly studied biosurfactant are shown in Figure 1); the possibility of cost effective production and their biodegradability. These properties make them a promising choice for applications in enhancing hydrocarbon bioremediation (Whang et al., 2008). Biosurfactants are able to retain their properties even under extreme conditions of pH, temperature, salinity (Ron and Rosenberg, 2002) and have low irritancy and compatibility with human skin (Pornsunthorntawee et al., 2009).

Other important advantages, such as bioavailability, activity under a variety of conditions, ecological acceptability, low toxicity, their capacity to be modified by biotechnology and genetic engineering and their capability of increasing the bioavailability of poorly soluble organic compounds, such as polycyclic aromatics are among the unique properties of these agents (Tugrul and Cansunar, 2005). Thus, their use could offer some solutions in bioremediation of contaminated soil and subsurface environments (Lai et al., 2009). Also, biosurfactants could easily be produced from renewable resources via microbial fermentation, making them have an additional advantage over chemically synthetic surfactants. The important challenges for the competitive production of biosurfactants include high yields, alternative low-cost substrates and cost-effective bioprocesses (Pornsunthorntawee et al., 2009). Some of the biosurfactants that have been studied using alternative low-cost

substrates, such as molasses (Makkar and Cameotra, 1997) agro-industrial wastes,(Makkar and Cameotra, 1999), soapstock and a by-product of the vegetable oil refining processes (Benincasa et al., 2004), are surfactin produced by *B. subtilis* and rhamnolipids produced by *P. aeruginosa*. Others produced by different species of microorganisms are presented in Table 1. The foci for reduction of biosurfactant production costs are the microbes (selected, adapted, or engineered for high yields of product), the process (selected, designed and engineered for low capital and operating costs), the microbial growth substrate and/or the process by-products (minimized or managed as saleable products rather than treated and discarded as wastes).

Synthetic surfactants

Conventional chemical surfactants have been extensively used and their derivatives are costly and of serious environmental concern; they are potential threats to the environment due to their recalcitrant and persistent nature. Therefore, with current advances in biotechnology, attentions have been paid to the alternative environmental friendly processes for production of different types of biosurfactants from microorganisms (Lotfabad et al., 2009).

Also, as part of the problems; most manufacturers of

Table 1. Some of the common biosurfactants with their producing microorganisms.

Organisms	Biosurfactants	Sources	References
Pseudomonas aeruginosa	Rhamnolipids	Petrochemical wastewater, soap stock, Petroleum contaminated soil, etc.	Whang, et al., 2008; Haba, et al., 2003; Wei, et al., 2008.
Pseudoxanthomonas spp PNK-04			Nayak, et al., 2009.
Pseudomonas alcaligenes			Oliveira, et al., 2009.
Rhodococcus erythropolis 51T7 *Micrococcus luteus* BN56	Trehalose lipids	Oil contaminated soil, Hydrocarbon gas station soil, etc.	Marquez, et al., 2009; Tuleva, et al., 2009.
Bacillus subtilis ATCC 21332	Surfactin	Petroleum sludge, Waste soybean oil, etc.	Pornsunthorntawee, et al., 2008; Lee, et al., 2008.
Candida antarctica	Mannosylerythritol lipids	Vegetable oil and soybean oil waste	Kim, et al., 2002; Kitamoto, et al., 2001.
Gordonia spp BS29	Bioemulsan	Diesel contaminated soil	Franzetti, et al., 2009.

chemical surfactants set the recommended dispersal ratios for their products on the basis of the economics and effectiveness of the dispersant with minimal consideration for the potential harm that can be caused in the receiving ecosystem (Laux et al., 2000; Putheti and Patil, 2009). However, concern still exists on the possible toxic effects of these surfactants on aquatic organisms, especially if they are used in near shore waters (Otitoloju and Popoola, 2009; Venosa and Holder, 2007).

The effects of chemical surfactants on biostimulation of indigenous microorganisms in enhancing the removal of organic pollutants yielded inconsistent results. Decrease in the rate of biodegradation of organic pollutants especially at higher concentrations as suggested by Sun et al. (2008) may be linked to the interaction of surfactant with the lipid membrane and their effects on enzymes and other cellular proteins necessary for basic functions of the microorganisms. Thus, the use of biosurfactants in place of chemical surfactants can minimize all sets of threats posed by the latter.

Despite all these effects; chemical surfactants find applications in different operations, such as drilling, cement production, slurries, fracturing, acidization, demulsification, corrosion inhibition, transportation, cleaning, water flooding, chemical, foam and steam flooding and environment protection as oil spill dispersants (Atta et al., 2006).

Factors affecting biosurfactant production

Composition and emulsifying activity of the biosurfactant depend not only on the producer strain but also on the culture conditions. Thus, the nature of the carbon source,

the nitrogen source as well as the C: N ratio, nutritional limitations and chemical and physical parameters such as temperature, aeration, divalent cations and pH influence not only the amount but also the type of polymer produced (Toledo et al., 2008; Lotfabad et al., 2009). The commonly used carbon sources include carbohydrates, hydrocarbons and vegetable oils. It has been concluded from a number of studies that different carbon sources can influence the composition of biosurfactant formation. *Arthrobacter* produces 75% extracellular biosurfactant when grown on acetate or ethanol but it is totally extracellular when grown on hydrocarbon (Mulligan and Gibbs, 1993; Putheti and Patil, 2009). The nitrogen source in the medium plays a significant role in production and contributes to pH control. Several organic and inorganic sources proved effective. Whang et al. (2009) showed that ammonium concentration and pH enhanced biosurfactant efficiency and increase emulsification ability on diesel oil, thus establishing the optimum pH for the production to be towards neutrality. Environmental factors such as temperature, pH, agitation and oxygen availability also affect the production of biosurfactant (Banat, 1995).

METHODS USED FOR IDENTIFICATION

Several methods for screening and estimation of biosurfactants have been developed.

Drop-collapse method is one of the qualitative methods used to determine the presence of biosurfactant. Tugrul and Cansunar (2005) conducted experiments to confirm the reliability of the method using polystyrene microwell plate; oil-coated wells collapse was observed when the

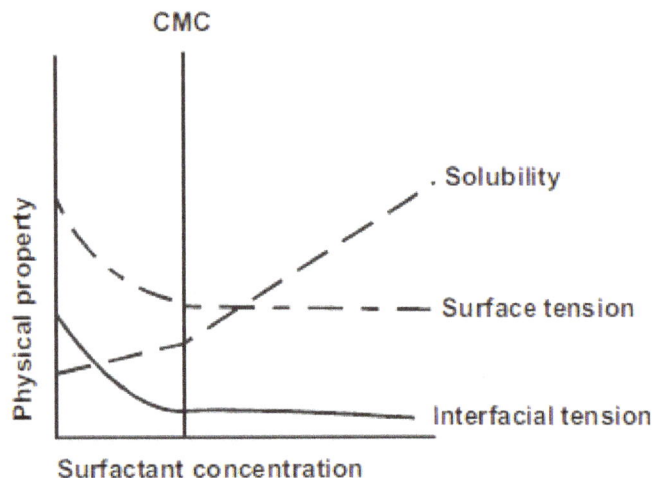

Figure 2. Relationship of surface tension, interfacial tension and the CMC with surfactant concentration (Mulligan, 2005).

culture broth contained biosurfactant and there was no change in the shape of the droplets in the absence of biosurfactant.

Thin layer chromatography (TLC) is also used in preliminary characterization of the biosurfactant where the cell free extract containing biosurfactant is separated on a silica gel plate using chloroform: methanol: water (70:10:0.5, v/v/v); this is then followed by using colour developing reagents. Lipopeptide biosurfactant showed red spots in the presence of ninhydrin reagent, while glycolipid biosurfactant is detected as yellow spots when anthrone is used as the colour reagent (Yin et al., 2009).

Additionally, blood agar hemolysis tests is another method used; where the organisms with biosurfactant ability are streaked on blood agar plates and incubated at 40°C. The plates are visually monitored for the presence of clearing zone around the colonies which is indicative of surfactant biosynthesis. The diameter of the clear zones depends on the concentration of the biosurfactant produced (Youssef et al., 2004; Ghojavand et al., 2008).

Surface tension measurement by a du Nöuy ring-type tensiometer (Krüss, K10T) is one of the simplest techniques used. The surface tension measurement is carried out at room temperature after dipping the platinum ring in the solution for a while in order to attain equilibrium conditions. A higher biosurfactant concentration in the test sample provides a lower surface tension until the critical micelle concentration (CMC) is reached. The CMC is obtained where the surface tension remains steady despite the changes in concentration as represented in Figure 2 (Desai and Banat, 1997; Crosman et al., 2002). This is widely used because of its accuracy, eases to use and provides a fairly rapid measurement of surface and interfacial tension. In general, biosurfactants produced by *P. aeruginosa* strains have the ability to reduce the surface tension of water, when measured with a tension-metre from 72 to 30 mN/m (Pornsunthorntawee et al.,

2008).

In contrast to the methods described so far, high performance liquid chromatography (HPLC) is not only appropriate for the complete separation of different biosurfactants, but can also be coupled with various detection devices (UV, MS, evaporative light scattering detection, ELSD) for identification and quantification of biosurfactants (Heyd et al., 2008).

Das and Mukherjee (2005, 2007) described the acidification of the solution containing biosurfactant in the presence of HCl and equal volume of ethyl acetate in a separatory funnel. Organic layer is collected following phase separation and further dehydrated in rotatory evaporator. The crude sample is then dissolved in sodium Bicarbonate and filtered; adjusting the pH with HCl; precipitation occurs overnight at 4°C. The pure product of biosurfactant is quantitatively obtained by centrifugation and freeze-drying.

Biosurfactants enhanced biodegradation of hydrocarbon compounds

Itoh and Suzuki (1972) were the first to show that hydrocarbon culture media stimulated the growth of a rhamnolipid producing strain of *P. aeruginosa*. Recent researches confirmed the biosurfactant effects on hydrocarbon biodegradation by increasing microbial accessibility to insoluble substrates and thus enhance their biodegradation (Zhang and Miller, 1992; Hunt et al., 1994).

Various experiments have been conducted that showed the effects of biosurfactants on hydrocarbons; enhancing their water solubility and increasing the displacement of oily substances from soil particles. Thus, biosurfactants increase the apparent solubility of these organic compounds at concentrations above the critical micelle concentration (CMC), which enhance their availability for microbial uptake (Chang et al., 2008). For these reasons, inclusion of biosurfactants in a bioremediation treatment of a hydrocarbon polluted environment could be really promising, facilitating their assimilation by microorganisms (Calvo et al., 2008).

Many of the biosurfactants known today have been studied to examine their possible technical applications (Nayak et al., 2009). Most of these applications involve their efficiency in bioremediation, dispersion of oil spills and enhanced oil recovery. *Alcanivorax* and *Cycloclasticus* genera are highly specialized hydrocarbon degraders in marine environments. *Alcanivorax borkumensis* utilizes aliphatic hydrocarbons as its main carbon source for growth and produces an anionic glucose lipid biosurfactant and thus potentials of *Alcanivorax* strains during bioremediation of hydrocarbon pollution in marine habitats have been studied (Olivera et al., 2009); thus, this property needs to be studied extensively in soil to ensure its effectiveness.

Several species of *P. aeruginosa* and *B. subtilis* produce rhamnolipid, a commonly isolated glycolipid biosurfactant and surfactin, a lipoprotein type biosur-factant respectively; these two biosurfactants have been shown by Whang et al. (2008), to increase solubility and bioavailability of a petrochemical mixture and also stimulate indigenous microorganisms for enhanced biodegradation of diesel contaminated soil. *Gordonia* species BS29 grows on aliphatic hydrocarbons as sole carbon source has been found to produce Bioemulsan, which effectively degrade crude oil, PAHs and other recalcitrant branched hydrocarbons from contaminated soils. The rate of biodegradation is dependent on the chemico-physical properties of the biosurfactants and not by the effects on microbial metabolism (Franzetti et al., 2008; Franzetti et al., 2009).

Corynebacterium alkanolyticum produces a phosphorlipid biosurfactant with a relatively low yield; however, the use of self-cycling fermentation processes resulted in three fold increase in the biosufactant production. Also, the yield could be further increased to five fold by addition high amount of the limiting substrates (Crosman et al., 2002)

Persistent organic pollutants found in oil containing wastewater and sediments, such as PAHs (phenanthrene, crysene), are also hydrophobic in nature and thus water solubility of PAHs normally decrease with the increasing number of rings in molecular structure. This property induces the low bioavailability of these organic compounds that is a crucial factor in the biodegradation of PAHs. The water solubility of some PAHs can be improved by addition of biosurfactants owing to their amphipathic structure by several folds (Yin et al., 2009). In addition, most hydrocarbons exist in strongly adsorbed forms when they are introduced into soils. Thus, their removal efficiency can be limited in low mass transfer phases. However, additions of solubilization agents, such as biosurfactants to the system enhance the bioavailability of low solubility and highly sorptive compounds (Shin et al., 2004).

Conclusion

The biosurfactant family constitutes an interesting group of microbial secondary products. Production of biosurfactant is related to the utilization of available hydrophobic substrates by the producing microbes from their natural habitat. Selection of suitable alternative substrates and the design of feasible processes for cost-effective production which involves media and process optimization are the main research focus. The potential use of some hyper producing microbial strains in addition to novel cost-effective bioprocesses throws real challenges and offers tremendous opportunities for making industrial production of biosurfactants a success story.

However, on the basis of their biological and physico-chemical properties, biosurfactants can be used as an antimicrobial, emulsifier and conservative agent in agrochemical, food industries, as well as in bioremediation applications.

These surface-active compounds enhanced the bioavailability of recalcitrant and hydrophobic organic pollutants through different mechanisms as suggested by Volkering et al. (1998):

1. Decrease in interfacial tension between an aqueous and a non-aqueous phase resulting in formation of emulsions that lead to improved mass transfer of pollutants to the aqueous phase.
2. Formation of micelles results in solubilization of hydrophobic organic compounds; thus the hydrophobic compounds remain in the core of the micelles whereas the hydrophilic molecules interact with the exterior part (Volkering et al., 1995); however, opinion is divided whether the solubilized hydrocarbons are directly available to the degrading micro-organisms.
3. Surfactants' interaction with hydrocarbons in the soil through mobilization lowers the surface tension of the pore water and hydrocarbons in soil particles (Deitsch and Smith, 1995).

Finally, more insight is required for large scale use of biosurfactants, this involves laboratory and field experiments, especially *in situ* studies with proper control processes.

REFERENCES

Atta AM, Abdel-Rauf ME, Maysour NE, Abdul-Rahiem AM, Abdel-Azim AA (2006). Surfactants from Recycled Poly (ethylene terephthalate) Waste as Water Based Oil Spill Dispersants, J. Polymer Res. 13: 39-52.

Banat I (1995). Biosurfactants production and possible uses in microbial enhanced oil recovery and oil pollution remediation: a review, Bioresour. Technol. 51: 1-12.

Benincasa M, Abalos A, Oliveira I, Manresa A (2004). Chemical structure, surface properties and biological activities of the biosurfactant produced by *Pseudomonas aeruginosa* LBI from soapstock, Antonie van Leeuwenhoek 85: 1-8.

Calvo C, Manzanera M, Silva-Castro GA, Uad I, González-López J (2008). Application of bioemulsifiers in soil oil bioremediation processes. Future prospects, Sci. Total Environ., doi:10.1016/j.scitotenv.2008.07.008.

Chang MW, Holoman TP, Yi, H (2008). Molecular characterization of surfactant-driven microbial community changes in anaerobic phenanthrene-degrading cultures under methanogenic conditions, Biotechnol. Letter, 30:1595-1601.

Crosman JT, Pinchuk RJ, Cooper DG (2002). Enhanced Biosurfactant Production by *Corynebacterium alkanolyticum* ATCC 21511 using Self-Cycling Fermentation, JAOCS 79(5): 467-472.

Das K, Mukherjee AK (2005). Characterization of biochemical properties and biological activities of biosurfactants produced by *Pseudomonas aeruginosa* mucoid and non-mucoid strains isolated from hydrocarbon-contaminated soil samples. Appl. Microbiol. Biotechnol. 69: 192-199.

Das K, Mukherjee AK (2007). Crude petroleum-oil biodegradation efficiency of *Bacillus subtilis* and *Pseudomonas aeruginosa* strains isolated from a petroleum-oil contaminated soil from North-East India, Bioresour. Technol. 98: 1339-1345.

Dastgheib SMM, Amoozegar MA, Elahi E, Asad S, Banat IM (2008). Bioemulsifier production by a halothermophilic Bacillus strain with potential applications in microbially enhanced oil recovery, Biotechnol.

Lett. 30: 263-270.

Deitsch JJ, Smith JA (1995). Effect of Triton X-100 on the rate of trichloroethene desorption from soil to water. Environ. Sci. Technol. 29: 1069-1080.

Desai JD, Banat IM (1997). Microbial production of surfactants and their commercial potential. Microbiol. Mol. Biol. Rev. 61: 47-64.

Fiechter A (1992). Biosurfactants: moving towards industrial application. Trends Biotechnol. 10: 208-217.

Franzetti A, Bestetti G, Caredda P, La Colla P, Tamburini E (2008). Surface-active compounds and their role in bacterial access to hydrocarbons in Gordonia strains. FEMS Microbiol. Ecol. 63: 238-248.

Franzetti A, Caredda P, Ruggeri C, La Colla P, Tamburini E, Papacchini M, Bestetti G (2009). Potential applications of surface active compounds by Gordonia sp. strain BS29 in soil remediation technologies, Chemosphere, doi:10.1016/j.chemosphere.2008.12.052 (Article accepted in Press).

Ghojavand H, Vahabzadeh F, Mehranian M, Radmehr M, Shahraki KA, Zolfagharian F, Emadi MA, Roayae E (2008). Isolation of thermotolerant, halotolerant, facultative biosurfactant-producing bacteria, Appl. Microbiol. Biotechnol. 80: 1073-1085.

Haba E, Abalos A, Jáuregui O, Espuny MJ, Manresa A (2003). Use of Liquid Chromatography–Mass Spectroscopy for Studying the Composition and Properties of Rhamnolipids Produced by Different Strains of Pseudomonas aeruginosa, J. Surfactants Detergents 6(2): 155-161.

Heyd M, Kohnert A, Tan TH, Nusser M, Kirschhöfer F, Brenner-Weiss G, Franzreb M, Berensmeier S (2008). Development and trends of biosurfactant analysis and purification using rhamnolipids as an example, Anal. Bioanal. Chem. 391: 1579-1590.

Hunt WP, Robinson K, Ghosh MM (1994). The role of biosurfactant in biotic degradation of hydrophobic organic compounds. In: Hinchee RE, Alleman BC, Hoeppel RE, Miller RN (Eds.), Hydrocarbon Bioremediation. Lewis Publishers, Boca Raton, Florida pp. 318-322.

Itoh S, Suzuki T (1972). Effect of rhamnolipids on growth of Pseudomonas aeruginosa mutant deficient in n-paraffin-utilizing ability. Agric. Biol. Chem. 36: 2233-2235.

Kim H, Jeon J, Lee H, Park Y, Seo W, Oh H, Katsuragi T, Tani Y, Yoon B (2002). Extracellular production of a glycolipid biosurfactant, mannosylerythritol lipid, from Candida Antarctica, Biotechnol. Lett. 24: 225-229.

Kitamoto D, Ikegami T, Suzuki GT, Sasaki A, Takeyama Y, Idemoto Y, Koura N, Yanagishita H (2001). Microbial conversion of n-alkanes into glycolipid biosurfactants, mannosylerythritol lipids, by Pseudozyma (Candida antarctica), Biotechnol. Lett. 23: 1709-1714.

Lai C, Huang Y, Wei Y, Chang J (2009). Biosurfactant-enhanced removal of total petroleum hydrocarbons from contaminated soil, J. Hazard. Mater. doi:10.1016/j.jhazmat.2009.01.017.

Laux H, Rahimian I, Butz T (2000). Theoretical and practical approach to the selection of asphaltene dispersing agents. Fuel Process Technol. 67: 79-89.

Lee S, Lee S-J, Kim S, Park I, Lee Y, Chung S, Choi Y (2008). Characterization of new biosurfactant produced by Klebsiella sp. Y6-1 isolated from waste soybean oil, Bioresour. Technol. 99: 2288-2292.

Lotfabad TB, Shourian M, Roostaazad R, Najafabadi AR, Adelzadeh MR, Noghabi KA (2009). An efficient biosurfactant-producing bacterium Pseudomonas aeruginosa MR01, isolated from oil excavation areas in south of Iran, Colloids Surf. B: Biointerfaces 69(2): 183-193.

Makkar S, Cameotra SS (1997). Utilization of Molasses for Biosurfactant Production by Two Bacillus Strains at Thermophilic Conditions, J. Am. Oil Chem. Soc. 74(7): 887-889.

Makkar S, Cameotra SS (1999). Biosurfactant Production by Microorganisms on Unconventional Carbon Sources, J. Surfactants Detergents 2(2): 237-241.

Marquez AM, Pinazo A, Farfan M, Aranda FJ, Teruel JA, Ortiz A, Manresa A, Espuny MJ (2009). The physicochemical properties and chemical composition of trehalose lipids produced by Rhodococcus erythropolis 51T7, Chem. Phys. Lipids, doi:10.1016/j.chemphyslip.2009.01.001.

Mukherjee S, Das P, Sen R (2009). Rapid quantification of a microbial surfactant by a simple turbidometric method, J. Microbiol. Methods, 76: 38-42.

Mulligan CN (2005). Environmental applications for biosurfactants, Environ. Pollut. 133:183-198.

Mulligan CN, Gibbs BF (1993). Factors influencing the economics of biosurfactant. In: Kosaric, N. (Ed). Biosurfactant production, properties and application, New York; Mercel Decker pp. 329-371.

Nayak AS, Vijaykumar MH, Karegoudar T B (2009). Characterization of biosurfactant produced by Pseudoxanthomonas sp. PNK-04 and its application in bioremediation, Int. Biodeterior. Biodegrad. 63: 73-79.

Neu T (1996). Significance of bacterial surface-active compounds in interaction of bacteria with interfaces. Microbiol. Rev. 60: 151-166.

Olivera FJS, Vazquez L, de Campos NP, de Franca FP (2009). Production of rhamnolipids by a Pseudomonas alcaligenes strain, Process Biochem, doi:10.1016/j.procbio.2008.11.014

Olivera ND, Nievas ML, Lozada M, del Prado G, Dionisi HM, Si~neriz F (2009). Isolation and characterization of biosurfactant-producing Alcanivorax strains: hydrocarbon accession strategies and alkane hydroxylase gene analysis, Res. Microbiol. 160: 19-26.

Otitoloju AA, Popoola TO (2009). Estimation of "environmentally sensitive" dispersal ratios for chemical dispersants used in crude oil spill control, The Environmentalist, DOI 10.1007/s10669-008-9212-2.

Pornsunthorntawee O, Arttaweeporn N, Paisanjit S, Somboonthanate P, Abe M, Rujiravanit R, Chavadej S (2008).Isolation and comparison of biosurfactants produced by Bacillus subtilis PT2 and Pseudomonas aeruginosa SP4 for microbial surfactant-enhanced oil recovery, Biochem. Eng. J. 42: 172-179.

Pornsunthorntawee O, Maksung S, Huayyai O, Rujiravanit R, Chavadej S (2009). Biosurfactant production by Pseudomonas aeruginosa SP4 using sequencing batch reactors: Effects of oil loading rate and cycle time, Bioresour. Technol., 100: 812-818.

Putheti RR, Patil MC (2009). Pharmaceutical formulation development of floating and swellable sustained drug delivery systems: a review, E-J. Sci. Technol. 4(2): 1-12.

Ron E, Rosenberg E (2002). Biosurfactants and oil bioremediation, Current Opinion in Biotechnol. 13: 249-252.

Rosenberg E, Ron EZ (1999). High- and low-molecular-mass microbial surfactants. Appl. Microbiol. Biotechnol. 2: 154-162.

Rosenberg E, Zuckerberg A, Rubinovitz C, Gutnick DL (1979). Emulsifier of Arthrobacter RAG-1: isolation and emulsifying properties. Appl. Environ. Microbiol. 37: 402-408.

Shin KH, Kim KW, Seagren EA (2004). Combined effects of pH and biosurfactant addition on solubilization and biodegradation of phenanthrene, Appl. Microbiol. Biotechnol. 65: 336-343.

Sun N, Wang H, Chen Y, Lu S, Xiong Y (2008). Effect of Surfactant SDS, Tween 80, Triton X-100 and Rhamnolipid on Biodegradation of Hydrophobic Organic Pollutants, presented at 2nd International conference of Bioinformatics and Biomedical Engineering pp. 4730-4734.

Toledo FL, Gonzalez-Lopez J, Calvo C (2008). Production of bioemulsifier by Bacillus subtilis, Alcaligenes faecalis and Enterobacter species in liquid culture, Bioresour. Technol. 99: 8470-8475.

Tugrul T, Cansunar E (2005). Detecting surfactant-producing microorganisms by the drop-collapse test, World J. Microbiol. Biotechnol. 21: 851-853.

Tuleva B, Christova N, Cohen R, Antonova D, Todorov T, Stoineva I (2009). Isolation and characterization of trehalose tetraester biosurfactants from a soil strain Micrococcus luteus BN56, Process Biochemistry, 44: 135-141.

Venosa AD, Holder EL (2007). Biodegradability of dispersed crude oil at two different temperatures, Marine Pollution Bulletin, 54: 545-553.

Volkering F, Breure AM, Rulkens WH (1998). Microbiological aspects of surfactant use for biological soil remediation, Biodegradation 8: 401-417.

Volkering F, Breure AM, Van Andel JG, Rulkens WH (1995). Influence of nonionic surfactants on bioavailability and biodegradation of polycyclic aromatic hydrocarbons. Appl. Environ. Microbiol. 61: 1699-1705.

Wei Y, Cheng C, Chien C, Wan H (2008). Enhanced di-rhamnolipid production with an indigenous isolate Pseudomonas aeruginosa J16, Process Biochemistry, 43: 769-774.

Whang L, Liu PG, Ma C, Cheng S (2008). Application of biosurfactants, rhamnolipid and surfactin, for enhanced biodegradation of diesel-contaminated water and soil, J. Hazardous Materials, 151: 155-163.

Whang L, Liu PG, Ma C, Cheng S (2009). Application of rhamnolipid and surfactin for enhanced diesel biodegradation—Effects of pH and ammonium, J. Hazard. Mater. 164(2-3): 1045-1050.

Yin H, Qiang J, Jia Y, Ye J, Peng H, Qin H, Zhang N, He B (2009). Characteristics of biosurfactant produced by *Pseudomonas aeruginosa* S6 isolated from oil-containing wastewater, Process Biochem., 44: 302-308.

Youssef NH, Duncan KE, Nagle DP, Savage KN, Knapp RM, McInerney MJ (2004). Comparison of methods to detect biosurfactant production by diverse microorganisms, J. Microbiol. Methods 56: 339-347.

Zhang Y, Miller RM (1992). Enhanced octadecane dispersion and biodegradation by a Pseudomonas rhamnolipid surfactant (biosurfactant). Appl. Environ. Microbiol. 58: 3276-3282.

5

The retinoblastoma binding protein 6 is a potential target for therapeutic drugs

Monde Ntwasa

School of Molecular and Cell Biology, University of the Witwatersrand, Wits, 2050. South Africa. E-mail: monde.ntwasa@wits.ac.za.

The retinoblastoma binding protein 6 (RBBP6) proteins (also called P-53 Associated Cell Testis Derived (PACT)) are highly upregulated in esophageal cancer and enhance the activity of MDM2, a p53 inhibitor with ubiquitin ligase activity that is overexpressed in many human cancers. Mammalian RBBP6 binds the tumour suppressor proteins p53 and the retinoblastoma protein (Rb). The invertebrate orthologues, on the other hand, have not been shown experimentally to have these properties and they have no obvious sequence features that are similar to the mammalian p53- and Rb-binding domains. General features of RBBP6 proteins such as a highly conserved N-terminal ubiquitin-like domain and a RING-finger indicate that they may be involved in proteolytic degradation of substrate proteins via the proteasome pathway. They have recently been found to act downstream hedgehog in certain normal developmental processes. This may implicate RBBP6 proteins in a wider range of human cancers. These data imply that antagonists of RBBP6 can be used as effective antitumour agents to treat tumours that have functional p53.

Key words: p53, RBBP6, PACT, SNAMA, cell cycle, apoptosis, cancer.

INTRODUCTION

RBBP6 proteins are found only in eukaryotes. The mammalian RBBP6 protein binds to the tumour suppressor proteins p53 and Rb and promotes p53 degradation by enhancing the activity of Mdm2 , the key p53 negative regulator (Li et al., 2007; Scott et al., 2003; Simons et al., 1997). The *Drosophila* orthologue called SNAMA has not been shown to bind p53 but is involved in apoptosis and is essential for embryonic development (Mather et al., 2005) while the yeast one, Mpe1 is involved in pre-mRNA processing (Vo et al., 2001).

p53 and Rb are key regulators of the cell cycle and p53 also plays an important role in maintaining genome integrity through its role in nucleotide excision repair systems (Wang et al., 2003). In normal cells p53 is kept at low levels by MDM2, a RING finger protein that me-

mediates its ubiquitination and proteasome degradation. *Drosophila* is peculiar in this regard because it seems to lack a MDM2 homologue. RBBP6 proteins have a characteristic highly conserved ubiquitin-like N-terminal domain called the Domain With No Name (DWNN) ((Mather et al., 2005; Pugh et al., 2006)). Overall, the vertebrate and invertebrate proteins have generally low homology but show high level of homology at the N-terminal domain. When compared with one another, the mammalian sequences have high identity in the p53 and Rb binding regions. These were experimentally delineated in the mouse P2P-R (Figure 3) (Witte and Scott, 1997) (see also Table 1).

In addition to the highly conserved DWNN, RBBP6 proteins show various combinations of other sequence features such as the CCHC zinc finger, a RING-finger-like (RFL) motif, lysine-rich, and proline-rich regions, coiled-coils and RS regions. This suggests that these proteins may interact with a number of proteins and indeed that they could have multiple functions. The RS

Abbreviations: RBBP6: retinoblastoma binding protein 6; Rb: retinoblastoma protein; P2P-R: proliferation potential protein-related; MDM2: mouse double mutant 2.

Table 1. RBBP6 orthologues that have been experimentally characterized.

Organism	Isoform	Name	Accession number	Length	Reference
Mouse (Chr 7)	One	PACT/P2P-R	NP_035377	1560	(Witte and Scott, 1997) (Simons et al., 1997)
	Two	RBBP6	P97868	1790	Predicted
	Three	RBBP6	XP_145621	786	Predicted
	Four	DWNN	NP_778188, NM_175023	123	(Mather et al., 2005; Pugh et al., 2006)
Human (Chr. 16)	One	RBBP6	NP_008841.	1792	(Pugh et al., 2006)
	Two	RBBP6/RBQ-1	NP_061173/X85133	1758	(Sakai et al., 1995)
	Three	DWNN	NP_116015.	118	(Mather et al., 2005; Pugh et al., 2006)
Fruitfly (Chr 2R)		SNAMA/Mnm^p	NP_611884, CG3231-PA	1231	(Jones et al., 2006; Mather et al., 2005)

region is often found in proteins that are involved in RNA processing and indeed Mpe1 has been shown to be involved in the 3' mRNA cleavage complex (Vo et al., 2001).

Interestingly, the mammalian RBBP6 N-terminal DWNN can exist as an independent splice isoform in the same manner as ubiquitin. Furthermore, the mammalian proteins have a conserved di-glycine peptide closer to and downstream the final conserved proline in DWNN. This feature is noteworthy because in ubiquitin the di-peptide is crucial for conjugation of ubiquitin to other proteins. It can be speculated that DWNN represents another form of protein modification that is similar to ubiquitination.

Evidence found in *Drosophila* indicates that the RBBP6 protein, SNAMA/Mnm^p, plays an important role in cell proliferation and cell survival and is directly implicated in nucleic acid metabolism and apoptosis during development. Furthermore, SNAMA/Mnm^p acts downstream hedgehog in the morphogenetic furrow during the development of the compound eye of the fly (Jones et al., 2006; Mather et al., 2005). The involvement of RBBP6 protein in the hedgehog pathway may widen the number of pathological conditions in which RBBP6 proteins are involved as hedgehog is implicated in a number of human tumors and developmental abnormalities.

The role that RBBP6 proteins play in regulation of the cell cycle and more especially the influence they have on the prototypical tumor suppressors, p53 and Rb, underscores their importance as targets for anticancer therapy. Inhibitors of these proteins should prevent p53 degradation and increase apoptosis in tumour cells. Indeed small molecule inhibitors of the E3 ubiquitin called nutlins have been tried in retinoblastoma cells and found to induce p53-mediated cell death (Laurie et al., 2007). Antisense oligonucleotides have also been used to inhibit expression of the Mdm2 gene (Bianco et al., 2005; Wang et al., 2001; Zhang et al., 2005). Other approaches could target the p53-RBBP6 interface. Again such approaches have been explored in attempts to design molecules that interfere with p53-Mdm2 interaction (Justin K. Murray, 2007).

Evolution of structural features of vertebrate RBBP6 proteins

Invertebrate and vertebrate RBBP6 family members have acquired a number of structural features in their sequence through evolution. Even the DWNN has acquired interesting structural features that appear late in evolution (Figure 1). For instance the di-glycine peptide that follows the conserved proline (asterisk) is found in mammals and birds but is absent in plants, arthropods and fish indicating that this may be a late evolutionary event.

The new protein modules acquired through evolution, probably confer new functions to the RBBP6 proteins. For instance more recent organisms in evolution such as insects, mammals and birds have acquired the lysine-rich region, arginine–rich region, and the RS region (Figure 2.). Other late evolutionary features are the p53 and Rb-binding domains (Figures 3 and 4). These were mapped in the mouse isoform, PACT/P2P-R (Witte and Scott, 1997). It is therefore possible that through evolution this family acquired new functions such as apoptosis and DNA repair. This region of the mammalian sequence also interacts with MDM2, a ubiquitin-ligase that negatively regulates p53 (Li et al., 2007). The role of RBBP6 proteins in invertebrates is likely to be somewhat different with respect to p53 because in addition to the lack of an obvious MDM2 homologue in the *Drosophila* genome, an MDM2 binding site seems to be absent in *Drosophila* p53.

RBBP6 proteins and the tumor suppressors

The role of RBBP6 proteins in cell cycle processes is in-

```
arabidopsis  ..MAIYVKFKSARDYDTIAMDGPFISVGILKDKIFETKHIGTGKDIDIVVSNAQTNEEYLPEAMLPKNTSVLIRRVPGRPRITV....ITTQEPRIQNK   94
chick        .MSCVHYKFKSKLNYDTVTFDGLHISLCDLRKQIMAREKLKAADC.DLQITNACTKERYTDDNALIPKNSS.VIVRRIPTGGVKA....TSKTYVISRTE   93
dog          .MSCVHYKFKSKLNYDTVTFDGLHTSLCDLRKQIMGREKLKAADC.DLQITNACTKERYTDDNALIPKNSS.VIVRRIPTGGVRS....TSKTYVISRTE   93
elephant     .MSCVHYKFKSKLNYDTVTFDGLHISLCDLRKQIMGREKLKAADC.DLQITNACTKERYTDDNALIPKNSS.VIVRRIPTGGVRS....TSKTYVISRTE   93
fly          ..MSVFHYKPTLNFDTITFDGLHISVSDLKREIVQOKRLGKIIDFDLQITNAQSKEEYKPDGFIPKNIT..LIISRIPIAHPTK...KGWEPPAAENA    93
fugu         .MAHIYKFKAKRSPDTVLIFDGPHITLQRLKRLIMEREKLRSGDC.DLQITNSQTKPBFTLRGEGLIAKCSS.VIVRRIPVVGARS....SSNSKIRNIQR  93
human        .MSCVHYKFKSKLNYDTVTFDGLHISLCDLRKQIMGREKLKAADC.DLQITNACTKERYTDDNALIPKNSS.VIVRRIPTGGVKS....TSKTYVISRTE   93
mouse        .MSCVHYKFKSKLNYDTVTFDGLHTSLCDLRKQIMGREKLKAADS.DLQITNACTKERYTDDNALIPKNSS.VIVRRIPTGGVKS....TSKTYVISRTE   93
yeast        MSSTIHYKFKSQRNTSRILFDGTGLTVHDLKREIIQENKLGLGIDFQLYNPDTEEEYLPDDAIVPRSTSVIVKRSPAIKSFSVHSRLKGNVGAAAIGN  100
zebrafish    .MSCVHYKFRSQSRLTYDSLQFEGLNITVGELRQIMRRERLKFCQLKISK....AQTDEEYLPDDALIPKNTS.VIIRRVPAAGLKS....SNRRFVGHQAG  91
Consensus          v  i  s       d     lk i  l          ee   d   l*                                     ΔΔ

arabidopsis  VEDVQAETTNFPVADPSAAE.......FPEDEYDEFGTDLYSIPDTQDAQHI                                                 139
chick        PVSGTSKAIDDSSASTSLAQ.......LTKTANL..................                                                120
dog          PVMGTSKAIDDSSASTSLAQ.......LTKTANL..................                                                120
elephant     PVMGTSKAIDDSSASTSLAQ.......LTKTANL..................                                                120
fly          FSAAPAKQDNFNMDLSKMQG.......TEEDKIQ..................                                                120
fugu         SDGHLQOPFGAFRAMTNLAF.......VDASEED..................                                                120
human        PAMATTKAIDDSSASTSLAQ.......LTKTANL..................                                                120
mouse        PVMGTTK...........................................                                                 100
yeast        ATRYVTGRPRVLQKROHTATTTANVSGTTEEERIASMFAT...........                                                140
zebrafish    PSAKSSQTG.DSSLL.SLEQ.......LLKTENLAEAK..............                                                120
Consensus
```

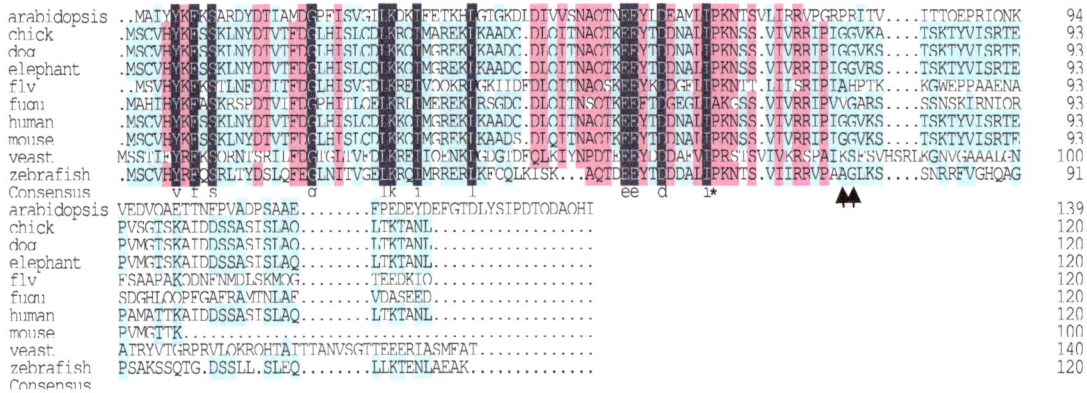

Figure 1. Alignment of DWNN of plant, vertebrate and invertebrate RBBP6 proteins. The asterisk refers to the conserved proline. The arrows indicate the di-glycine peptides that are conserved in mammals. In ubiquitin, a similar conserved di-glycine at position 75 and 76 is crucial for conjugation of the ubiquitin moiety to itself or to other proteins that are targeted for proteasome degradation.

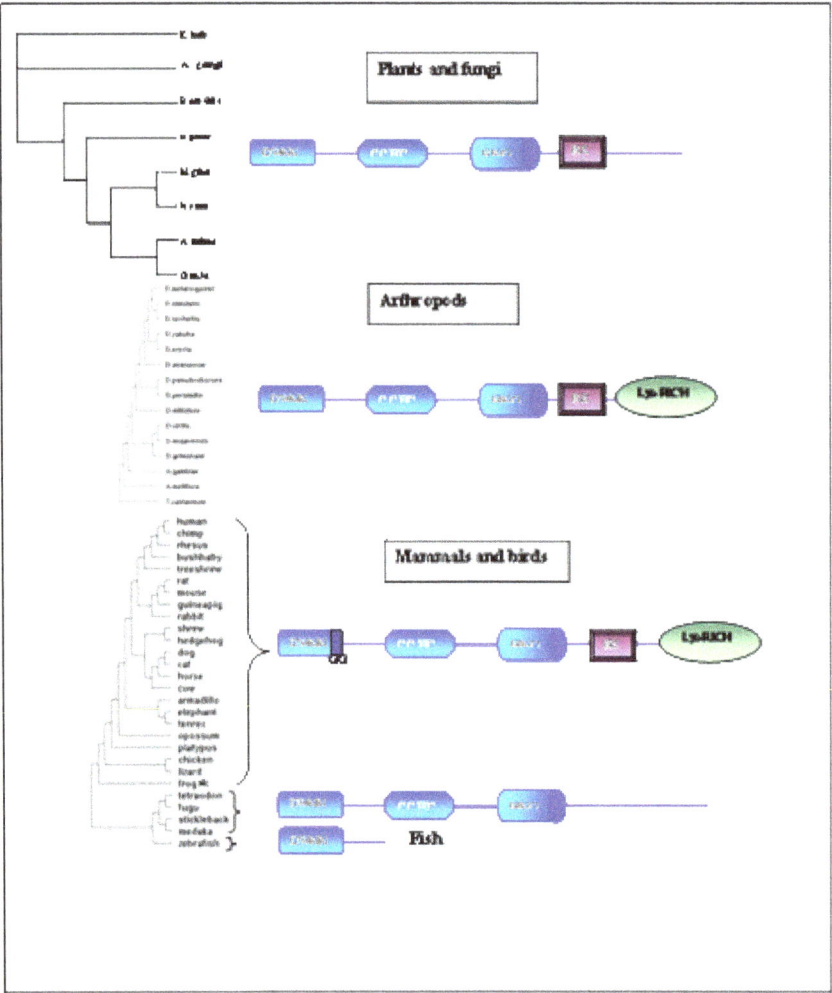

Figure 2. Domains structure of RBBP proteins. The phylogenetic trees show how these structures have evolved. Note that the fish orthologues lack the zinc finger and the RING finger motif – but exist as an independent DWNN with a short C-terminal extension.

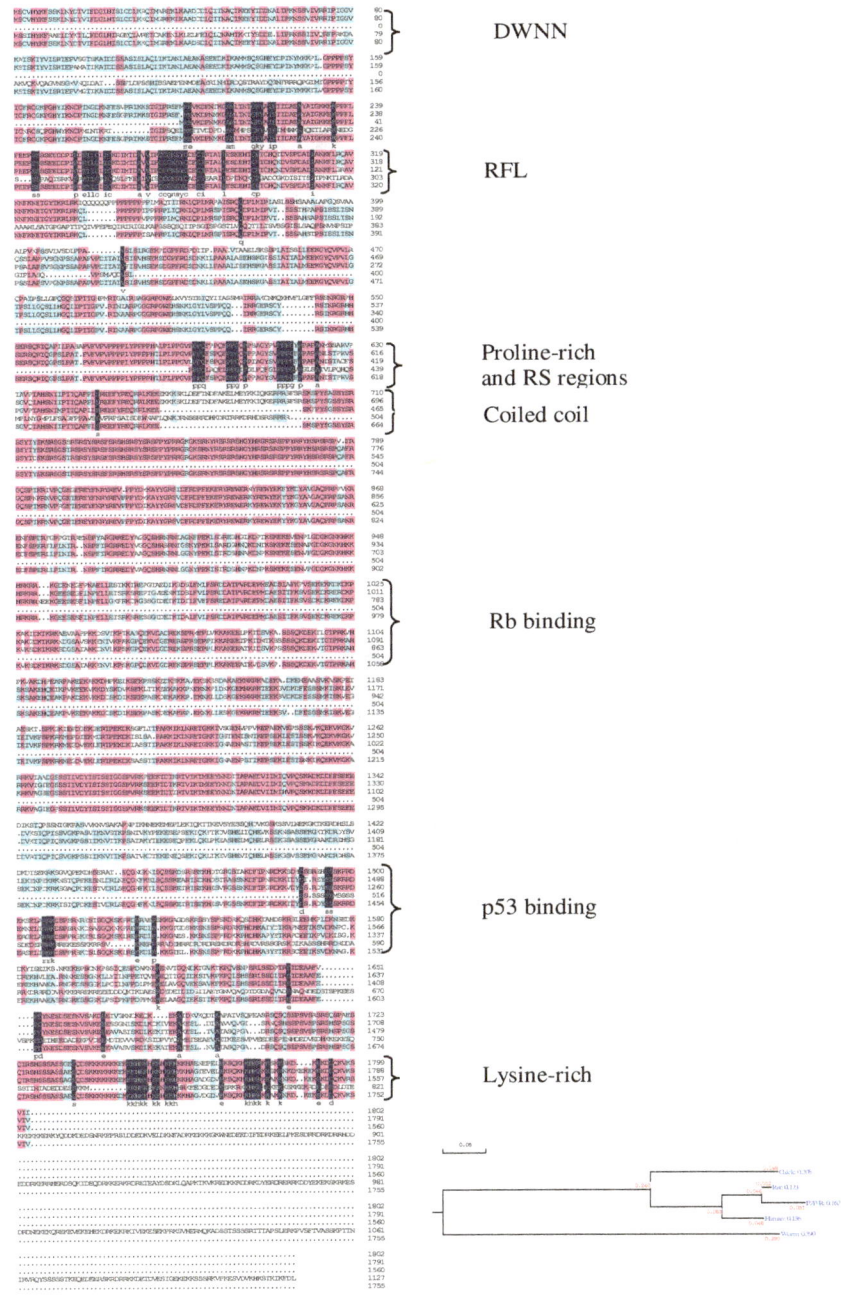

Figure 3. Sequence alignment of vertebrate RBBP6 proteins sequences. Alignment human (*Homo sapiens*) chick (*G. gallus*), Rat (*R. norvegicus*), mouse (*M. musculus*) and worm (*C. elegans*) protein sequences. This alignment and Phylogenetic three (insert) was produced by using DNAMAN shows conserved regions of the proteins.

dicated by subcellular localization in nuclear spec-kles, expression in mitotically active cells, by aberrant apoptosis when perturbed and by association with cellular differentiation (Gao and Scott, 2002; Gao and Scott, 2003; Gao et al., 2002; Robert et al., 2003; Scott et al., 2003; Scott et al., 2005; Witte and Scott, 1997). RBBP6 proteins are widely expressed in many tumor cell lines and are upregulated in esophageal cancer (Yoshitake et al., 2004). They are normally expressed in the heart, lung, liver skeletal muscle and most prominently in the testis (Witte and Scott, 1997).

Many cancers are caused by alterations in tumor suppressor proteins such as breast cancer 1 and breast cancer 2 (BRCA1 and BRCA2) (Greenberg, 2008), patched (Chidambaram et al., 1996), E2F (Du and Dyson, 1999; Du et al., 1996) and many others, resulting in aberrant proliferation of cells. Tumor suppressor proteins have therefore become important targets for anticancer therapy. Controlling the activity of tumor suppressor regulatory proteins is also a growing area of drug discovery.

The role of RBBP6 proteins as negative regulators of p53 was elucidated in the mouse system where RBBP6/PACT was shown to negatively affect p53 levels by enhancing the activity of MDM2. In these experiments the essential part that RBBP6 plays is emphasized by the fact that embryos lacking a functional RBBP6/PACT ($Pact^{-/-}$) had a reduced size, were developmentally retarded and died before E7.5. Moreover, lethality caused by the disruption of the $PACT$ gene was partially rescued by a $p53$ null mutation (Li et al., 2007). Notably, the $Pact^{-/-}$ phenotype is similar to that of $mdm2^{-/-}$ mice which also die during embryogenesis. This phenotype is also rescued by the concomitant absence of p53 indicating that MDM2 and RBBP6/PACT are critical for maintaining optimum p53 levels (Luna et al., 1995). These results are a significant contribution to the understanding of the relationship between p53 and RBBP6 proteins.

MDM2 interacts with and negatively regulates two key tumor suppressor proteins, p53 (Jones et al., 1995; Luna et al., 1995; Michael and Oren, 2003) and Rb (Xiao et al., 1995) and is amplified in a number of human tumors. In addition to this, MDM2 promotes cell cycle progression by stimulating the S-phase transcription factors E2F/DP1 (Martin et al., 1995). MDM2 is associated with aberrant p53 gene expression and with invasiveness of hepatocellular carcinoma (Qiu et al., 1997). MDM2 splice isoforms that lack the p53 binding domain are associated with advanced malignancy in ovarian tumors, in bladder and breast cancers and in human astrocytic neoplasms indicating that MDM2 can promote malignant cell proliferation independently of p53 (Matsumoto et al., 1998; Sigalas et al., 1996). p53 was also found to be stable in glioblastoma cells despite the amplification of MDM2 splice isoforms (Kraus et al., 1999).

Because it is rare for proteins to bind both p53 and Rb it is speculated that RBBP6 proteins may act as scaffold for the assembly of tumor suppressor proteins (Li et al., 2007). This view is consistent with an earlier view that proposed a formation of a complex that comprises an RBBP6 protein in matrix attachment regions (MARS) to influence gene transcription and chromatin organization (Scott et al., 2003). In addition the mammalian proteins occur in isoforms including one which has a coiled coil domain that is encoded by a separate exon (Figure 4).

Coiled coils in proteins often control oligomerisation and are associated with the cytoskeleton, the Golgi, centromeres, centrosomes, the nuclear matrix, and chromatin. This feature indicates that this isoform may dimerize and perform a unique role.

The structural and functional features of RBBP6 proteins, namely, involvement in the cell cycle and association with key tumor suppressor proteins, p53 and RB make them attractive candidate targets for anticancer therapy. The RBBP6 functional relative, MDM2, is already a promising target of anticancer therapeutic agents. For example, small molecule MDM2 inhibitors have been developed as anticancer agents by exploiting the p53-MDM2 interface. These are either non-genotoxic molecules that bind to the p53 binding pocket in MDM2 without interfering with normal p53 activity or mimetic peptides (Sakurai et al., 2004; Secchiero et al., 2007).

RBBP6 proteins and hedgehog signaling

Hedgehog signaling is involved in many human congenital diseases and in many cancers. A catalogue of pathological conditions that involve the hedgehog pathway lists abnormalities in the central and peripheral nervous systems, the circulatory system, the gut, the kidney and many bone related abnormalities (McMahon et al., 2003). Moreover cyclins D and E, which are regulators of the Rb/E2F pathway, acting in some cases downstream of hedgehog and its receptor patched (Ptc) (Figure 5), are implicated in many human cancers (Donnellan and Chetty, 1999). This is a highly conserved pathway in both vertebrates and invertebrates.

Recent experiments demonstrate that the $Drosophila$ RBBP6 protein, SNAMA/Mnm[p] acts downstream hedgehog during development of the $Drosophila$ eye (Jones et al., 2006). This was the first report that links RBBP6 proteins with the hedgehog pathway. Hedgehog signaling is known to control cell cycle exit via the Dpp-dependent pathway and cell cycle reentry via a Notch dependent pathway in the $Drosophila$ retina. RBF, the $Drosophila$ homologue of Rb, acts downstream hedgehog in the $Drosophila$ retina during cell differentiation when it mediates cell cycle exit in a Notch-dependent pathway. During cell cycle reentry RBF is antagonized by a Notch-dependent mechanism resulting in the release of the transcription factor, dE2F1 into the nucleus (reviewed by (Neumann, 2005). E2F is physiologically inhibited in a complex with Rb and is released, upon phosphorylation of Rb, into the nucleus to activate or repress genes that are involved in various cellular functions, such as cell cycle phase transitions, DNA synthesis, mitosis, apoptosis, DNA repair and differentiation depending on the context and source of the signal (DeGregori, 2005).

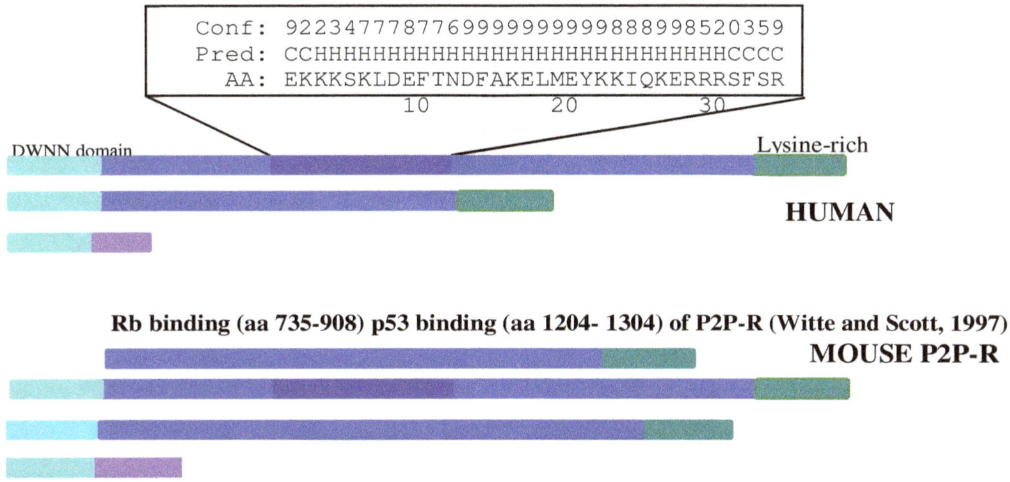

```
Conf: 9223477787769999999999888998520359
Pred: CCHHHHHHHHHHHHHHHHHHHHHHHHHHHHHHCCCC
  AA: EKKKSKLDEFTNDFAKELMEYKKIQKERRRSFSR
              10        20        30
```

DWNN domain   Lysine-rich

HUMAN

Rb binding (aa 735-908) p53 binding (aa 1204- 1304) of P2P-R (Witte and Scott, 1997)
MOUSE P2P-R

Figure 4. Schematic representation of mouse and human RBBP6 proteins. The Rb and p53-binding domains of the mouse P2P-R proteins were delineated experimentally. The dark blue block shows the coiled coil region that is missing in other isoforms and in invertebrate proteins.

The role of the *Drosophila* RBBP6 protein, SNAMA/-Mnm[p], in the hedgehog pathway probably entails the control of dE2F inhibition by interacting with RBF. It could then influence cell cycle exit or reentry in a context dependent manner.

Taken together all these results suggest a potential for RBBP6 proteins to catalyze the degradation of either Rb or p53 in a proteasome-dependent manner. This scenario would be consistent with an earlier proposal that the *Drosophila* protein SNAMA/Mnm[p] suppresses apoptosis directly or negatively regulates a proapoptotic molecule (Mather et al., 2005). RBBP6 family members therefore provide another level of intervention in the fight against cancer and probably other diseases that are caused by aberration in hedgehog signaling. These include several human con-genital malformations such as brachydactyly and some limb defects. Hedgehog signaling is involved in several other processes affecting development of the central nervous tissue, the gut, the circulatory system, the respi-ratory system, the neural crest, and others (McMahon et al., 2003). It is, however, not clear yet how RBBP6 proteins influence the hedgehog pathway. Further work is required to dissect this pathway before therapeutic interventions targeting its components can be designed. The signaling pathway depicted in Figure 5 could provide basis for speculation about the influence of RBBP6 proteins in the hedgehog pathway, but more experimental evidence is required.

Conclusions

The regulatory effect that RBBP6 proteins have on key tumor suppressors illustrates that they represent a noteworthy class of potential targets for anticancer therapy. Inhibition of these proteins is likely to elevate the levels of p53 in tumor cells thus promoting apoptosis. These strategies involve small molecule antagonists that perturb protein-protein interactions between the tumor suppressors and the E3 ligases. Antisense oligonucleo-tides are also being explored to prevent expression of these enzymes.

Further work needs to be done to refine these strate-gies because there are still formidable challenges. For example, specific and targeted inhibitors are required because the E3 ligases have diverse substrates and are required for normal cell function. However, local (subcon-juctival) delivery of the small molecule, nutlin-3, seems to have been successful in preclinical retinoblas-toma models reinforcing the feasibility of this approach (Laurie et al., 2007). Delivery techniques where the inhibitor is confined to the diseased tissue may help avoid extensive damage to normal tissue.

Future prospects

It has to be confirmed experimentally that RBBP6 pro-teins have ubiquitin-like activities and their function in different cell types must be elucidated. Protein-protein interaction studies to further delineate the RBBP6-p53 and RBBP6-Rb interfaces should contribute significantly in drug design. Structural studies of these proteins will also help design antagonists that could be useful in the-rapy. Cell based assays and animal models that over express RBBP6 proteins will help screen for molecules

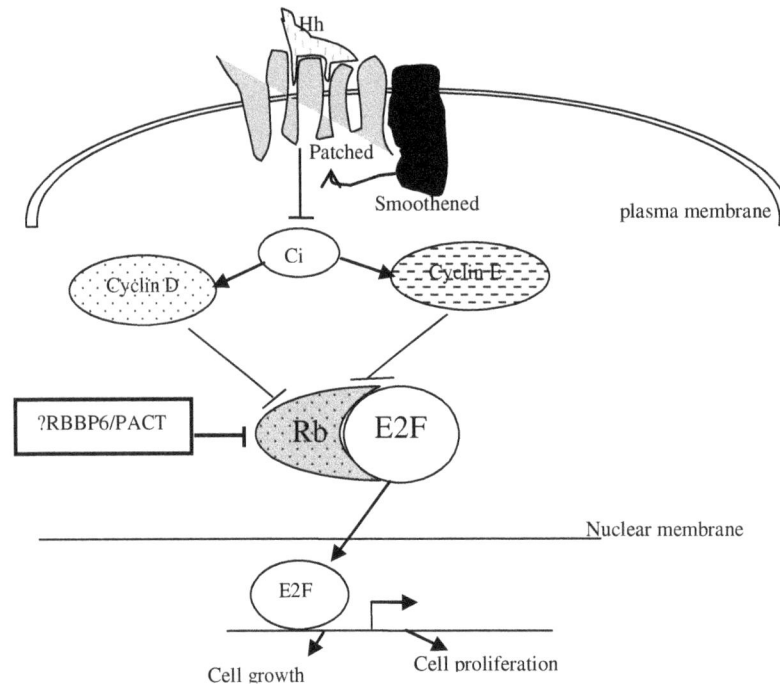

Figure 5. Simplified hedgehog signaling pathway. RBBP6 proteins probably promote cell proliferation and growth by negatively regulating Rb. Binding of hedgehog to patched, its receptor, leads to the release of smoothened (Smo) and to the subsequent activation of downstream molecules. This leads to phosphorylation of Rb by cyclin D or cyclin E and its dissociation from E2F. E2F is a transcription factor that activates or represses transcription of genes that are involved in cell proliferation and cell growth.

that could be used in cancer therapy or in managing developmental defects.

REFERENCES

Bianco R, Ciardiello F, Tortora G (2005). Chemosensitization by antisense oligonucleotides targeting MDM2. Curr. Cancer Drug Targets, 5: 51-56.

Chidambaram A, Goldstein AM, Gailani MR, Gerrard B, Bale SJ, DiGiovanna JJ, Bale A E, Dean M (1996). Mutations in the Human Homologue of the Drosophila patched Gene in Caucasian and African-American Nevoid Basal Cell Carcinoma Syndrome Patients. Cancer Res. 56: 4599-4601.

DeGregori J (2005). E2F and cell survival: Context really is key. Developmental Cell 9, 442-444.

Donnellan R, Chetty R (1999). Cyclin E in human cancers. Faseb. J. 13: 773–780.

Du W, Dyson N (1999). The role of RBF in the introduction of G1 regulation during Drosophila embryogenesis. EMBO. J. 18: 916-925.

Du W, Vidal M, Xie J, Dyson N (1996). RBF, a novel RB-related gene that regulates E2F activity and interacts with cyclin E in Drosophila. Genes & Dev. Vol 10: 1206-1218.

Gao S, Scott RE (2002). P2P-R protein overexpression restricts mitotic progression at prometaphase and promotes mitotic apoptosis. J. Cell. Physiol. 193: 199-207.

Gao S, Scott RE (2003). Stable overexpression of specific segments of the P2P-R protein in human MCF-7 cells promotes camptothecin-induced apoptosis. J. Cell Physiol. p. 197.

Gao S, White MM, Scott RE (2002). P2P-R protein localizes to the nucleus of interphase cells and the periphery of chromosomes in mitotic cells which show maximum P2P-R immunoreactivity. J. Cell Physiol. 191: 145-154.

Greenberg R (2008). Recognition of DNA double strand breaks by the BRCA1 tumor suppressor network. Chromosoma Epub ahead of print.

Jones C, Reifegerste R, Moses K (2006). Characterization of Drosophila mini-me, a gene required for cell proliferation and survival. Genetics 173: 793-808.

Jones SN, Roe AE, Donehower LA, Bradley A (1995). Rescue of embryonic lethality in Mdm2-deficient mice by absence of p53. Nature 378: 206-208.

Justin K, Murray SHG (2007). Targeting protein-protein interactions: Lessons from p53/MDM2. Peptide Science 88: 657-686.

Kraus A, Neff F, Behn M, Schuermann M, Muenkel K, Schlegel J (1999). Expression of alternatively sliced mdm2 transcripts correlates with stabilized wild-type p53 protein in human gliblastoma cells. Int J Cancer: 80: 930–934.

Laurie N, Schin-Shih C, MAyer D (2007). Targeting MDM2 and MDMX in retinoblastoma. Curr Cancer Drug Targets 7: 689-695.

Li L, Deng B, Xing G, Teng Y, Tian C, Cheng X, Yin X, Yang J, Gao X, Zhu Y, et al provide all other name (2007). PACT is a negative regulator of p53 and essential for cell growth and embryonic development. Proc. Natl. Acad. Sci. USA 104: 7951-7956.

Luna RM, d O, Wagner DS, Lozano G (1995). Rescue of early embryonic lethality in mdm2-deficient mice by deletion of p. 53.

Nature 378: 203-206.

Martin K, Trouche D, Hagemeier C, Sorensen TS, Thangue NBL, Kouzarides T (1995). Stimulation of E2F/DP1 transcriptional activity my MDM2 oncoporotein. Nature 375: 691-694.

Mather A, Rakghotho M, Ntwasa M (2005). SNAMA, a novel protein with a DWNN domain and a RING finger-like motif:A possible role in apoptosis. Biochim. Biophys. Acta. 1727: 169-176.

Matsumoto R, Tada M, Nozaki M, Zhang C-L, Sawamura Y, Abe H (1998). Short alternative splice transcripts of the mdm2 oncogene correlate to malignancy in human astrocytic neoplasms. Cancer Res 58: 609-613.

McMahon A, Ingham P, Tabin C (2003). Developmental roles and clinical significance of hedgehog signaling. Curr. Top Dev. Biol. 53: 1-114.

Michael D, Oren M (2003). The p53-Mdm2 module and the ubiquitin system. Seminars in Cancer Biology 13: 49-58.

Neumann CJ (2005). Hedgehogs as negative regulators of the cell cycle. Cell Cycle 4: 1139-1140.

Pugh DJR, Eiso AB, Faro A, Lutya PT, Hoffmann E, Rees DJG (2006). DWNN, a novel ubiquitin-like domain, implicates RBBP6 in mRNA processing and ubiquitin-like pathways. BMC Struct. Biol. 6: 1-12.

Qiu S-J, Ye S-L, Wu Z-Q, Tang Z-Y, Liu Y-K (1997). The expression of the mdm2 gene may be related to the aberration of the p53 gene in human hepatocellular carcinoma. J. Cancer Res. Clin. Oncol. 124: 253-258.

Robert SE, Giannakouros T, Gao S, Peidis P (2003). Functional Potential of P2P-R: Role in the Cell Cycle and Cell Differentiation Related to its interactions With Proteins That Bind to Matrix associated Regions of DNA. J. Cell. Biochem. 90: 6-12.

Sakai Y, Saijo M, Coelho K, Kishino T, Niikawa N, Taya Y (1995). cDNA sequence and chromosomal localisation of a novel human protein, RBQ-1 (RBBP6), that binds to the retinoblastoma gene product. Genomics 30: 98-101.

Sakurai K, Chung HS, Kahne D (2004). Use of a retroinverso p53 peptide as an inhibitor of MDM2. J. Am. Chem. Soc. 126: 16288-16289.

Scott RE, Giannakouros T, Gao S, Peidis P (2003). Functional potential of P2P-R: A role in the cell cycle and cell differentiation related to its interactions with proteins that bind to matrix associated regions of DNA? J. Cell. Biochem. 90: 6-12.

Scott RE, White-Grindley E, Ruley HE, Chesler EJ, Williams RW (2005). P2P-R expression is genetically coregulated with components of the translation machinery and with PUM2, a translational repressor that associates with the P2P-R mRNA. J. Cell Physiol. 204: 99-105.

Secchiero P, Corallini F, Gonelli A, Dell'Eva R, Vitale M, Capitani S, Albini A, Zauli G (2007). Antiangiogenic activity of the MDM2 antagonist Nutlin-3. Circ. Res. 100: 61-69.

Sigalas I, Calvert A, Anderson J, Neal D, Lunec J (1996). Alternatively spliced mdm2 transcripts with loss of p53 binding domain sequences: transforming ability and frequent detection in human cancer. Nat Med 2: 912-917.

Simons A, Melamed-Bessudo C, Wolkowicz R, Sperling J, Sperling R, Eisenbach L, Rotter V (1997). PACT: cloning and characterization of a cellular p53 binding protein that interacts with Rb. Oncogene 14: 145-155.

Vo LTA, Minet M, Schmitter J, Lacroute F, Wyers F (2001). Mpe1, a zinc knuckle protein, is an essential component of yeast cleavage and polyadenylation factor required for the cleavage and polyadenylation of mRNA. Mol. Cell. Biol. 21: 8346-8356.

Wang H, Nan L, Yu D, Agrawal S, Zhang R (2001). Antisense Anti-MDM2 oligonucleotides as a novel therapeutic approach to human breast cancer: In vitro and in vivo activities and mechanisms. Clin. Cancer Res. 7: 3613-3624.

Witte MM, Scott RE (1997). The proliferation potential protein-related (P2P-R) gene with domains encoding heterogeneous nuclear ribonucleoprotein association and Rb1 binding shows repressed expression during terminal differentiation. Proc. Natl. Acad. Sci. USA. 94: 1212-1217.

Xiao Z.-X, Chen J, Levine AJ, Modjtahedi N, Xing J, Sellers WR, Livingstone DM (1995). Interaction between the retinoblastoma protein and the oncoprotein MDM2. Nature. 375: 694-698.

Yoshitake Y, Nakatsura T, Monji M, Senju S, Matsuyoshi H, Tsukamoto H, Hosaka S, Komori H, Fukuma D, Ikuta Y, et al provide all other name (2004). Proliferation potential-related protein, an ideal esophageal cancer antigen for imunotherapy, identified using complimentary DNA microarray analysis. Clin. Cancer Res. 10: 6437-6448.

Zhang R, Wang H, Agrawal S (2005). Novel antisense anti-MDM2 mixed-backbone oligonucleotides: Proof of principle, in vitro and in vivo activities, and mechanisms. Curr. Cancer Drug Targets, 5: 43-49.

6

Plant terpenoids: applications and future potentials

Sam Zwenger and Chhandak Basu*

University of Northern Colorado, School of Biological Sciences, Greeley, Colorado, 80639, USA.

The importance of terpenes in both nature and human application is difficult to overstate. Basic knowledge of terpene and isoprene biosynthesis and chemistry has accelerated the pace at which scientists have come to understand many plant biochemical and metabolic processes. The abundance and diversity of terpene compounds in nature can have ecosystem-wide influences. Although terpenes have permeated human civilization since the Egyptians, terpene synthesis pathways are only now being understood in great detail. The use of bioinformatics and molecular databases has largely contributed to analyzing exactly how and when terpenes are synthesized. Additionally, terpene synthesis is beginning to be understood in respect to the various stages of plant development. Much of this knowledge has been contributed by the plant model, *Arabidopsis thaliana*. Considering the advances in plant terpene knowledge and potential uses, it is conceivable that they may soon be used in agrobiotechnology.

Key words: Terpenes, terpene synthase, secondary metabolites, transgenic plants

TABLE OF CONTENT

INTRODUCTION

Plants produce primary and secondary metabolites which encompass a wide array of functions (Croteau et al., 2000). Primary metabolites, which include amino acids, simple sugars, nucleic acids, and lipids, are compounds that are necessary for cellular processes. Secondary metabolites include compounds produced in response to stress, such as the case when acting as a deterrent against herbivores (Keeling, 2006). Plants can manufacture many different types of secondary metabolites, which have been subsequently exploited by humans for their beneficial role in a diverse array of applications (Balandrin et al., 1985). Often, plant secondary metabolites may be referred to as plant natural products, in which case they illicit effects on other organisms. Although this review focuses on plant terpenes, it should be realized that other organisms are able to synthesize terpenes. For example, the endophytic fungus isolated from St. John's Wort (*Hypericum perforatum*) was recently shown to produce hypericin and emodin, two types of terpene lactones (Kusari et al., 2008). There are three broad categories of plant secondary metabolites as natural products; terpenes and terpenoids (~25,000 types), alkaloids (~12,000 types), and phenolic compounds (~8,000 types) (Croteau et al., 2000).

Terpene chemistry and biosynthesis

Ever since techniques such as low-temperature chromatography, were used to separate plant terpenes nearly a half of a century ago (Clements, 1958), great strides have been made to explore molecular details of terpenes. For instance, subjecting plant vegetation to pyrolysis techniques and gas chromatography has allowed for identifi-

* Corresponding author. E-mail: chhandak.basu@unco.edu.

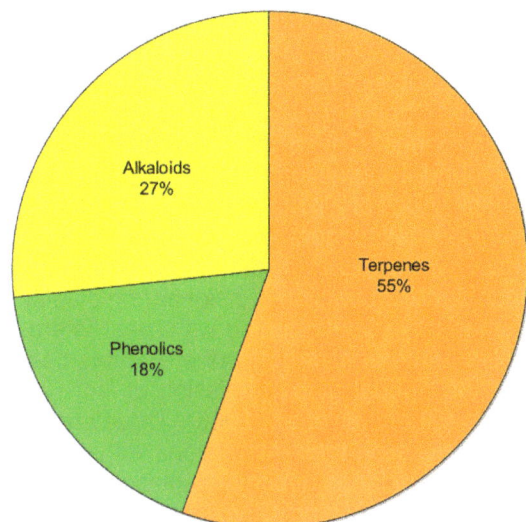

Figure 1. Pie chart representing the major groups of plant secondary metabolites according to Croteau et al. (2000). Based on their numbers and diversity, terpenes offer much potential in an array of industrial and medicinal applications.

cation of different volatile organic compounds (VOCs) (Greenberg et al., 2006). Some extraction techniques have relied on using ultra pure water, also dubbed subcritical water. Although this method works relatively well, it has been pointed out that increasing the temperature of the water decreases the stability of the terpenes (Yang et al., 2007). Lai et al. (2005) performed a crude extraction terpene trilactones from the leaves of *Ginko biloba* by refluxing the leaves in ethanol and then dissolving the extracts in water. They subsequently isolated trilactones with either column chromatography or a liquid-liquid extraction using ethyl acetate as a solvent.

Ma et al. (2007) used two-dimensional gas chromatography time-of-flight mass spectrometry (GC x GC-TOF MS) to analyze volatile oils in the leaves and flowers of *Artemisia annua*. The authors concluded that the number of components was close to 700 and the majority was terpenes. In a comparative analysis they used the same extraction techniques but instead of GC x GC-TOF MS they used GC-MS, which resulted in a much lower sensitivity for molecular diversity. This illustrates the fact that newer isolation and identification methods are helping with terpene analysis.

It has long been known that the basic unit of most secondary plant metabolites, including terpenes, consists of isoprene, a simple hydrocarbon molecule. The term terpene usually refers to a hydrocarbon molecule while terpenoid refers to a terpene that has been modified, such as by the addition of oxygen. Isoprenoids are, therefore, the building blocks of other metabolites such as plant hormones, sterols, carotenoids, rubber, the phytol tail of chlorophyll, and turpentine.

Terpenes are the most numerous and structurally diverse plant natural products (Figure 1). For this reason, a system of nomenclature has been established. The nomenclature of terpene compounds is ostensibly complex, yet can be quickly elucidated upon closer examination. The isoprene unit, which can build upon itself in various ways, is a five-carbon molecule. The single isoprene unit, therefore, represents the most basic class of terpenes, the hemiterpenes. An isoprene unit bonded with a second isoprene is the defining characteristic of terpene, which is also a monoterpene (C_{10}). While sesquiterpenes contain three isoprene units (C_{15}), diterpenes (C_{20}) and triterpenes (C_{30}) contain two and three terpene units, respectively. Tetraterpenes consist of four terpene units and polyterpenes are those terpenes containing more than four terpene units (i.e., more than eight isoprene units).

McGarvey and Croteau (1995) reviewed terpene biosynthesis and suggested that a more detailed study of the terpene synthases were needed and that this in turn would increase the role of terpenes in, perhaps, commercial uses such as flavor enhancers. Since their publication more than a decade ago, terpene biosynthesis enzymes have been studied in detail. For instance, Greenhagen et al. (2006) used mapping strategies to determine the variance and composition of amino acids within terpene synthase active sites. This is, arguably, very useful in determining the evolutionary divergence of the terpene synthases and elucidating relationships among plants. Trapp and Croteau (2001) reviewed the genomic organization of terpene synthase genes across different species. They suggest that terpene synthase genes may impact phylogenetic organization of some plants. For example, some terpene genes are more closely related in certain plant species, in which the species themselves were previously thought to be distantly related.

In a more recent review of terpene synthase genes, Zwenger and Basu (2007) performed *in silico* analysis of publicly available microarray data using Genevesitgator software (Zimmerman et al., 2004). Such software allows for assaying an organism of choice, in this case, *Arabidopsis thaliana*. In their study, more than 2,500 microarrays were simultaneously compared for expression of terpene synthase genes. Multiple biotic and abiotic factors, which may or may not induce expression, were also considered in respect to terpene synthase gene expression. Possibly even more important is the fact that expression of terpene synthases were examined across the life cycle of *Arabidopsis*, which countered some wet lab experimental data previously published. In addition to expanding the comprehension of terpene synthase genes across the life cycle of *Arabidopsis*, the authors determined that five terpene synthase genes, which appeared in many microarray analyses, were lacking in experimental studies. As discussed by the authors, further experiments may lead to better understanding the roles of previously uncharacterized genes.

Terpenes in nature

The distribution of terpenes in nature has been studied extensively. Indeed, the distribution of terpenes within species has received attention. To better understand terpene and other volatile organic compound emissions, from loblolly pine (*Pinus taeda*), Thompson et al. (2006) analyzed tree core samples. They found the highest concentrations of terpenes in heartwood, lowest in outer sapwood, and moderate levels in the inner sapwood. In a less invasive study by Martin et al. (2003) methyl jasmonate was applied onto foliage of Norway spruce (*Picea abies*) trees which led to a two fold increase of terpenes within the needles. In another investigation, the amounts of different terpenes in Scots pine (*Pinus sylvestris*) needles varied across Finnish and Turkish regions, showing the diversity of terpene distribution can vary within a species (Semiz et al., 2007).

Although more commonly associated with coniferous species, terpenes have been detected in other plant phyla, including angiosperms. Aside from terpenes manufactured by plants in response to herbivory or stress factors, it has also been shown that flowers can emit terpenoids to attract pollinating insects (Maimone and Baran 2007). Interestingly, terpenoids have also been shown to attract beneficial mites, which feed on the herbivorous insects (Kappers et al., 2005). Terpene emissions and subsequent attracting mechanisms have been shown to play an indirect role in plant defense mechanisms in other studies as well. Kessler and Baldwin (2001) have shown that herbivorous insects can induce terpene release from a plant, and also cause the plant to release signals which attracts predatory species. These experiments provide not only powerful evidence for the role of terpenes for plant defense, but also give an exemplary model for co-evolution between plants, mites, and insects. Chen et al. (2003) have shown that many different volatiles, including terpenes, may be emitted from flowers of *Arabidopsis*. They propose that the role as insect attractants of at least some emitted terpenes seems inconclusive, but still strongly suggest they might play a role in reproduction.

Of course, other studies have extended the understanding of plant terpenes and insects. Johnson et al. (2007) examined fragrance mixtures including terpenes and found scent chemistry of the emitted fragrance played a role in beetles and wasps pollinating an orchid species (*Satyrium microrrhynchum*). After performing GC-MS to identify fragrance compounds, they manipulated antennae to determine electrophysiological responses. Molecules which elicited effects included monoterpenes and sesquiterpenes. While the beetles were generalists in pollination, the wasps were more specific. However, Urzúa et al. (2007) studied terpenoids from an Asteraceae (*Haplopappus berterii*) and suggested little or no correlation between fragrance molecules and insect preference.

Ecological roles of terpenes extend beyond plant-insect coevolution. Cheng et al. (2007) discuss ecological impacts of terpenes. These include their roles above and below ground in attracting predatory species upon herbivory attack. Additionally, they point out terpenes may act as chemical messengers which influence the expression of genes involved in plant defense mechanisms or even influence gene expression of neighboring plants.

Terpenes have been studied with great interest, due to their roles in the earth's atmosphere. It has been estimated that the annual global emission of isoprenes is 500 teragrams (Guenther et al., 2006). Therefore, it is tempting to speculate on their interactions with solar radiation. Due to the abundance of citrus plantations in the mediterranean area, Thunis and Cuvelier (2000) helped identify the influence and composition of VOCs on ozone formation in this region and found some of the biogenic VOCs included α-pinene and *d*-limonene. In a study by VanReken et al. (2006) a biogenic emissions chamber was used to measure terpenoids released from Holm oak (*Quercus ilex*), loblolly pine and a dilute mixture of α-pinene. They suggested a large majority of emissions are chemically oxidized or otherwise transformed into different aerosol compounds. Llusiá and Penñuelas (2000) have examined stomatal conductance to better understand how plants interact with abiotic atmospheric conditions such as temperature, water availability, and irradiance to alter the diffusive resistance of terpenes from plant leaves. They also describe the seasonal fluctuation of terpenes.

Since many plants contribute to the earth's atmospheric composition by releasing volatile organic compounds, which include terpenes, they should arguably be studied more extensively. Future research may therefore help pave the road to understanding the global influence of terpenes.

Society and terpenes

There have been many applications of terpenes in human societies. Pharmaceutical and food industries have exploited them for their potentials and effectiveness as medicines and flavor enhancers. Perhaps the most widely known terpene is rubber, which has been used extensively by humans. Rubber is a polyterpene, composed of repeating subunits of isoprene. The addition of sulfur to rubber by Charles Goodyear led to vulcanized rubber, which yields various degrees of pliability depending on the mixture ratio (Stiehler and Wakelin, 1947). Other important terpenes include camphor, menthol, pyrethrins (insecticides), cleaners, antiallergenic agents, and solvents. Rosin (a diterpene), limonene, carvone, nepetalactone (in catnip), hecogenin (a detergent), and digitoxigenin are also important terpenes (Croteau et al., 2000).

Agriculture has also shown an increasing interest in

terpenes. In a study by Villalba et al. (2006) sheep were suggested to have increased tolerance for terpene consumption if they consumed more grains. They also showed terpenes can influence ungulate herbivory on other plants. This may help agronomists balance diets of ruminants if they consume plants such as sagebrush (Artemesia sp.). Terpenes have also shown antimicrobial activities (Islam et al., 2003). This is important due to the increase in antibiotic resistant bacteria, which is occurring globally and at an alarming rate. Addition of terpenes into livestock feed may replace conventional antibiotic addition, which in turn would slow the rate of antibiotic resistance in bacteria.

The effect of some terpenes on microorganisms has been seriously studied since at least the 1980's (Andrews et al., 1980). Plant oils, which contain terpenes, have shown increasing promise in vivo, inhibiting multiple species of bacteria. For example, cinnamon oil has shown broad-spectrum activity against Pseudomonas aeruginosa (Prabuseenivasan et al., 2006). The various compositions of terpenes can be markedly different from one species to another. For example, John et al. (2007) found plant oils from Neolitsea foliosa, which also showed some antibacterial properties, included sesquiterpenes such as β-caryophyllene but lacked monoterpenes.

Other microbes have also shown inhibition by terpenes. Murata et al. (2008) extracted numerous compounds from stem bark of the cape ash (Ekebergia capensis) growing in Kenya. Ten of these were triterpenes, whose structures were determined using spectroscopic analysis such as NMR (nuclear magnetic resonance). Determining the precise molecular activities of these triterpenes may be an important step towards finding newer and more effective drugs against Plasmodium falciparum, the causative agent of malaria. Susceptibility to terpenes has been tested by Morales et al. (2003), in which extracts from Artemisia copa showed inhibitory effects against yeast (Candida albicans). They also showed that some plant extracts containing terpenes tested showed biotoxicity effects against brine shrimp (Artemia salina).

Cumene (isopropylbenzene) is a terpene that has been used in bioremediation studies. In an experiment carried out by Suttinun et al. (2004), bacteria used in bioremediation of trichloroethylene (TCE) showed an increased capability to uptake TCE in the presence of cumene. In their study, 75% of the TCE present was successfully metabolized, allowing for a more robust degradation and bioremediation. Additional terpenes included in the study were limonene, carvone, and pinene. However, cumene showed the most beneficial effects. Without the knowledge and application of cumene, such success in bioremediation studies might not have been possible.

Because terpenes have been incorporated into much antibacterial soaps, cosmetics and household products, descriptive studies have been published on absorption and penetration into skin. Due to their properties of lipid organization disruption, Cal et al., (2006) studied the ab-

sorption kinetics of four cyclic terpenes; α-pinene, β-pinene, eucalyptol, and terpinen-4-ol. Each terpene varied in accumulation and elimination time with terpinen-4-ol showing the fastest penetration. Matura et al. (2005) investigated the role some terpenes play as causative agents of contact dermatitis and fragrance allergies. Out of approximately 1500 patients tests, just over 1% had reactions to oxidized linalool.

To better understand the vast array of terpenes, genetically modified organisms have been used. For example, the biosynthesis of terpenes has been studied in transformed E. coli (Adam et al., 2002). As described by Adam et al. (2002), modification of organisms is important to help understand the various pathways of terpene synthesis for the purpose of producing antimicrobial and antiparasitic drugs (Goulart et al., 2004).

Transgenic plants and future research

Plant tissue culture is an in vitro technique that allows clonal propagation of transformed clones. A review of tissue culture methods and applications by Vanisree et al. (2004) discusses the importance of inserting genes for plant secondary metabolites, including taxol (a diterpene alkaloid), a well-known anticancer agent. They point out that in vitro cell culture methods provides systematic advantages such as the ability to manipulate plant environment, control of cell growth, and regulation and extraction of metabolic products.

In contrast to producing terpenes in the laboratory, others have suggested extending the methods to create transgenic crops for terpene sythesis and production. Genetic modification of Arabidopsis has been performed to study the production of different terpenoids by up-regulating terpene synthase genes (Aharoni et al., 2003), which has led to an increase in understanding of how terpenes might function. For example, it has been shown that genetic engineering of Arabidopsis plants has allowed for an increase in pest resistance (Kappers et al., 2005). Others who have shown terpenes to influence insect behavior have suggested the use of terpene expression as a possible control mechanism for aphid infestations (Harmel et al., 2007). A study by Lweinsohn et al. (2001) determined that genetically modified tomatoes could be produced, which had enhanced levels of linalool and thus enhanced flavor and aroma. Degenhardt et al. (2003) discuss how monoterpenes and sesquiterpenes are the two most common terpenes emitted from plants post-herbivory. They suggest that finding the proper mixture and timing of terpene release from crop plants is key to creating an adequte transgenic plant. Additionally, the properties of terpene emission should be tightly regulated by an herbivore-responsive promoter.

Genetic transformation of tobacco (Nicotiana tobacum) was carried out by Lücker et al. (2004) (Table 1). After inserting monoterpene synthase genes the plants showed

Table 1. Partial representation of organisms which have been genetically transformed with at least one terpene synthase gene

Species	Organism	Citation
Escherichia coli	bacteria	Adam et al., 2002
Candida albicans	yeast	Jackson et al., 2006
Arabidopsis thaliana	thale cress	Aharoni et al., 2003
Lycpersicon esculentum	tomato	Lweinsohn et al., 2001
Nicotiana tobacum	tobacco	Lücker et al., 2004
Lactuca sativa	lettuce	Wook et al., 2005
Mentha piperita	mint	Wildung et al., 2005

an increase of terpene emission from leaves. To better understand terpene biosynthetic pathways Pateraki et al. (2007) isolated multiple cDNAs from *Cistus creticus.* They used polymerase chain reaction (PCR) techniques to amplify sequences from the plant and found additional terpene synthase genes after searching expressed sequence tag (EST) libraries. Expression of genes is at least partly dependent on their promoters. Davidovich-Rikanati et al. (2007) used a ripening-specific promoter to modify the aroma and flavor of tomatoes. Unfortunately, although levels of flavor-causing monoterpenes increased, the lycopene decreased.

The medicinal value of terpenes has not been ignored. Canter et al. (2005) discuss some areas within biotechnology to improve medicinal plant cultivation. These include incorporating agronomic traits into medicinal plants, pathway engineering and exploring additional transformation systems. In a more recent examination, Tyo et al. (2007) describe methods for studying and engineering cells such as using 'omics' technologies, screening libraries, and synthetic and computational systems biology. As they have mentioned, some of these technologies are currently being used to extend comprehension of metabolic pathways in plants. In a study by Yao et al. (2008) bioinformatics helped characterize a terpene synthase pathway after comparative analysis of isolated cDNA from a cultured callus line of an endangered medicinal plant (*Camptotheca acuminata*) native to China. Similar to many other bioinformatic-based approaches, they benefited from NCBI's (National Center for Biotechnology Information) BLAST (Basic Local Alignment Search Tool), which can help understand phylogenetic relationships among nucleotide sequences (Altschul et al., 1990).

Melvin Calvin, the Nobel laureate known for his contribution to the scientific understanding of the carbon fixation pathways in plant chloroplasts, studied the tropical copiaba (*Copaifera langsdorfii*) for its natural biofuel production (Calvin, 1980). Although it has not yet been examined, diesel from *C. langsdorfii* is largely composed of terpenes. The current interest in biofuels is not only in the United States but also other countries such as Brazil and the European Union. This has sparked new interest in finding renewable sources, or plants which may contribute to biofuels. Considering this global interest in

biofuels, research describing the up-regulation of terpene synthase genes in *C. langsdorfii* may prove very beneficial. This research would be very useful, for instance, in providing a more cost effective extraction and by-pass typical conversion of biomass (corn ethanol) to more contemporary biofuels (Demirbas and Balat, 2006). Therefore, future studies may include terpene production in the diesel tree or related biofuel plants.

Conclusions

Many terpenes remain to be discovered so they will undoubtedly intrigue scientists for years, as their applications are only beginning to be fully realized. Arguably, society has benefited tremendously from terpenes. In addition, understanding the function of genes in terpene production could lead to discovering novel compounds or pathways, which might reveal new important aspects for many human applications. For instance, the ability to up-regulate terpene synthesis in *C. langsdorfii* could result in an increase in the diesel-like resin harvested from this tree, might prove beneficial to the global market of biofules. As we continue into the agrobiotechnology age, it is highly likely the applications and potentials of terpenes will be further explored.

REFERENCES

Adam P, Hecht S, Eisenreich W, Kaiser J, Grawert T, Arigoni D, Adelbert B, Rohdich F (2002). Biosynthesis of terpenes: Studies on 1-hydroxy-2methyl-2-(*E*)-butenyl 4-diphosphate reductase. Proc. Nat. Acad. Sci. 99: 12108-12113.

Aharoni A, Giri AP, Deuerlein S, Griepink F, de Kogel W, Verstappen F, Verhoeven HA, Jongsma MA, Schwab W, Bouwmeester HJ (2003). Terpenoid metabolism in wild-type and transgenic *Arabidopsis* plants. Plant Cell. 15: 2866-2884.

Altschul SF, Gish W, Miller W, Myers EW, Lipman DJ (1990). Basic local alignment search tool. J. Mol. Biol. 215: 403–410.

Andrews RE, Parks LW, Spence KD (1980). Some effects of Douglas fir terpenes on certain microorganisms. App. Environ. Microbiol. 40: 301-304.

Balandrin MF, Klocke JA, Wurtele ES, W.H Bollinger (1985). Natural plant chemicals: Sources of industrial and medicinal materials. Science. 228:1154-1160.

Cal K, Kupiec K, Sznitowska M (2005). Effect of physicochemical properties of cyclic terpenes on their ex vivo skin absorption and elimination kinetics. J. Dermatol. Sci. 41:137-142.

Calvin M (1980). Hydrocarbons from plants: Analytical methods and observations. Naturwissenschaften. 67: 525-533.

Canter PH, Thomas H, Ernst E (2005). Bringing medicinal plants into cultivation: Opportunities and challenges for biotechnology. Trends in Biotechnol. 23: 180-185.

Chen F, Dorothea T, D'Auria JC, Farooq A, Pichersky E, Gershenzon J. (2003). Biosynthesis and emission of terpenoid volatiles from Arabidopsis flowers. The Plant Cell. 15: 481-494.

Cheng A, Lou Y, Mao Y, Lu S, Wang L, Chen X (2007). Plant terpenoids: Biosythesis and ecological functions. J. Integrative Plant Biol. 49: 179-186.

Cho DW, Park YD, Chung KH (2005). Agrobacterium-mediated transformation of lettuce with a terpene synthase gene. J. Korean Soc. Hortic. Sci. 46: 169-175.

Clements RL (1958). Low-temperature chromatography as a means for separating terpene hydrocarbons. Science. 128: 899-900.

Croteau R, Kutchan, TM, Lewis NG, (2000). Natural products (secondary metabolites). In Buchanan B, Gruissem W, Jones R (Eds.), Biochemistry and molecular biology of plants. Rockville, MD: American Society of Plant Physiologists. pp. 1250-1318.

Davdidovich-Rikanati R, Sitrit Y, Tadmor Y, Iijima Y, Bilenko N, bar E, Carmona B, Fallik E, Dudai N, Simon J, Pichersky E, Lewinsohn E (2007). Enrichment of tomato flavor by diversion of the early plastidial terpenoid pathway. Nat. Biotechnol. 25: 899-901.

Degenhardt J, Gershenzon J, Baldwin IT, Kessler A (2003). Attracting friends to feast on foes: Engineering terpene emission to make crop plants more attractive to herbivore enemies. Curr. Opin. Biotechnol. 14: 169-176.

Demirbas MF, Balat M (2006) Advances on the production and utilization trends of bio-fuels: A global perspective. Energy Convers. Manage. 47: 2371-2381.

Goulart HR, Kimura EA, Peres VJ, Couto AS, Duarte FA, Katzin AM (2004). Terpenes arrest parasite development and inhibit biosynthesis of isoprenoids in Plasmodium falciparum. Antimicrob. Agents and Chemother. 48: 2502-2509.

Greenberg JP, Friedli H, Guenther AB, Hanson D, Harley P, Karl T (2006). Volatile organic emissions from the distillation and pyrolysis of vegetation. Atmos. Chem. and Phys. 6: 81-91.

Greenhagen BT, O'Maille PE, Noel JP, Chappell J (2006). Identifying and manipulating structural determinates linking catalytic specificities in terpene synthases. Proc. Nat. Acad. Sci. 103: 9826-9831.

Guenther A, Karl T, Harley P, Wiedinmyer C, Palmer PI, Geron C (2006). Estimates of global terrestrial isoprene emissions using MEGAN (model of emissions of gases and aerosols from nature). Atmos. Chem. Phys. 5: 715–737.

Harmel N, Almohamad R, Fauconnier M, Du Jardin P, Verheggen F, Marlier N, Haubruge E, Francis F (2007). Role of terpenes from aphid-infested potato on searching and oviposition behavior of Episyrphus balteatus. Insect Sci. 14: 57-63.

Islam AK, Ali MA, Sayeed A, Salam SM, Islam A, Rahman M, Khan GR, Khatun S (2003). An antimircrobial terpenoid from Caesalpinia pulcherrima Swartz.: Its characterization, antimicrobial and cytotoxic activities. Asian J. Plant Sci. 2: 17-24.

Jackson BE, Hart-Wells EA, Matsuda SPT (2003). Metabolically engineering yeast to produce sesquiterpenes in yeast. 5: 1629-1632.

John AJ, Karunakran VP, George V (2007). Chemical composition an antibacterial activity of Neolitsea foliosa (Nees) Gamble var. caesia (Meisner) Gamble. J. Essent. Oil Res. 19: 498-500.

Johnson SD, Ellis A, Dötterl B (2007). Specialization for pollination by beetles and wasps: The role of lollipop hairs and fragrance in Satyrium microrrhynchum (Orchidaceae). A. J. Bot. 94: 47-55.

Kappers IF, Aharoni A, Van Herpen T, Luckerhoff L, Dicke M, Bouwmeester HJ (2005). Genetic engineering of terpenoid metabolism attracts bodyguards to Arabidopsis. Science. 309: 2070-2072.

Keeling CI, Bohlmann J (2006). Genes, enzymes, and chemicals of terpenoid diversity in the constitutive and induced defence of conifers against insects and pathogens. New. Phytol. 170: 657-675.

Kessler A, Baldwin T (2001). Defensive function of herbivore-induced plant volatile emission in nature. Science. 291: 2141-2144.

Kusari S, Lamshöft M, Zühlke S, Spittelleractivity M (2008). An endophytic fungus from Hypericum perforatum that produces hypericin. J.

Nat. Prod. In press.

Lai S, Chen I, Tsai M (2005). Preparative isolation of terpene trilactones from Ginkgo biloba leaves. J. Chromatgr. A .1092: 125-134.

Lewinsohn E, Schalechet F, Wilkinson J, Matsui K, Tadmor Y, Nam K, Amar O, Lastochkin E, Larkov O, Ravid U, Hiatt W, Gepstein S, Pichersky E (2001). Enhanced levels of the aroma and flavor compound S-linalool by metabolic engineering of the terpenoid pathway in tomato fruits. Plant Physiol. 127: 1256-1265.

Llusiá J, Peññuelas J (2000). Seasonal patterns of terpene content and emission from seven Mediterranean woody species in field conditions. Am. J. Bot. 87:133-140.

Lücker J, Schwab W, van Hautum B, Blaas J, van der Plas L, Bouwmeester HJ, Verhoeven HA (2004). Increased and altered fragrance of tobacco plants after metabolic engineering using three monoterpene synthases from lemon. Plant Physiol. 134: 510-519.

Ma C, Want H, Lu X, Li H, Liu B, Xu G (2007). Analysis of Artemisia annua L. volatile oil by comprehensive two-dimensional gas chromatography time-of-flight mass spectrometry. J. Chromatogr. A. 1150: 50-53.

Martin D, Tholl D, Gershenzon J, Bohlmann J (2003). Induction of volatile terpene biosynthesis and diurnal emission by methyl jasmonate in foliage of Norway spruce. Plant Phys. 132: 1586-1599.

Matura M, Sköld M, Börje A, Andersen KE, Bruze M, Frosch P, Goossens A, Johansen JD, Svedman C, White IR, Karlberg A (2005). Selected oxidized fragrance terpenes are common contact allergens. Contact Dermat. 52: 320-328.

McGarvey DJ, Croteau R (1995). Terpenoid metabolism. Plant Cell. 7: 1015-1026.

Morales G, Sierra P, Mancilla A, Paredes A, Loyola LA, Gallardo O, Borquez J (2002). Secondary metabolites from four medicinal plants from northern Chile: Antimicrobial activity and biotoxicity against Artemia salina. J. Chil. Chem. Soc. 48:13-18.

Murata T, Miyase T, Muregi FW, Naoshima-Ishibashi Y,| Umehara K, Warashina T, Kanou S, Mkoji GM, Terada M, Ishih A (2008) Antiplasmodial triterpenoid from Ekebergia capensis. J. Plant Nat. Prod. In press

Pateraki I, Falara V, Kanellis A (2007). Isolation and expression profile of Cistus creticus ssp. creticus genes involved in terpenoid biosynthesis. J. Biotechnol. 131: S15 ECB 13.

Prabuseenivasan S, Jayakumar M, Ignacimuthu S (2006). In vitro antibacterial activity of some plant essential oils. BMC Complement. Altern. Med. 6:39.

Semiz G, Heijari J, Isik K, Holopainen JK (2007). Variation in needle terpenoids among Pinus sylvestris L (Pinaceae) provenances from Turkey. Biochem. Syst. Ecol. 35: 652-661.

Stiehler, R.D. and J.H. Wakelin. 1947. Mechanism and theory of vulcanization. Ind. Eng. Chem. 39:1647-1654.

Suttinun O, Lederman PB, Luepromachai E (2004). Application of terpene-induced cell for enhancing biodegradation of TCE contaminated soil. Songklanakarin J. Sci. Technol. 26: 131-142.

Thompson A, Cooper J, Ingram LL (2006). Distribution of terpenes in heartwood and sapwood of loblolly pine. Forest Prod. J. 56:7-8.

Thunis P, Cuvelier C (2000). Impact of biogenic emissions on ozone formation in the Mediterranean area – a BEMA modelling study. Atmos. Environ. 34: 467-481.

Trapp SC, Croteau RB (2001). Genomic organization of plant terpene synthases and molecular evolutionary implications. Genetics. 158: 811-832.

Tyo KE, Hal SA, Stehpanopoulos GN (2007). Expanding the metabolic engineering toolbox: More options to engineering cells. Trends in Biotechnol. 25: 132-137.

Urzúa A, Santander R, Echeverría J, Rezende MC (2007). Secondary metabolites in the flower heads of Haplopappus berterii (Asteraceae) and its relation with insect-attracting mechanisms. J. Chil. Chem. 52: 1142-1144.

Vanisree M, Lee C, Lo S, Nalawade SM, Lin CY Tsay H (2004). Studies on the production of some important secondary metabolites from medicinal plants by plant tissue cultures. Bot. Bull. Acad. Sin. 45: 1-22.

VanReken TM, Greenberg JP, Harley PC, Guenther AB, Smith JN (2006). Direct measurement of particle formation and growth from the oxidation of biogenic emissions. Atm. Chem. Phys. 6:4403-4413.

Villalba JJ, Provenza FD, Olson KC (2006). Terpenes and carbohydrate source influence rumen fermentation, digestibility, intake, and preference in sheep. J. Anim. Sci. 84: 2463-2473.

Wildung MR, Croteau R (2005). Genetic engineering of peppermint for improved essential oil composition and yield. Transgenic Resear. 14: 365-372.

Yang Y, Kayan B, Bozer N, Pate B, Baker C, Gizir AM (2007). Terpene degradation and extraction from basil and oregano leaves using subcritical water. Journal of Chromatography A. 1152: 262-267.

Yao H, Gong Y, Zuo K, Ling H, Qiu C, Zhang F, Wang Y, Pi Y, Liu X, Sun X, Tang K (2008). Molecular cloning, expression profiling and functional analysis of a *DXR* gene encoding 1-deoxy-D-xylulose 5-phosphate reductoisomerase from *Camptotheca acuminata*. Plant Physiol. 165: 203-213.

Zimmermann P, Hirsch-Hoffmann M, Hennig L, Gruissem W (2004). GENEVESTIGATOR. *Arabidopsis* microarray database and analysis toolbox. Plant Physiol. 136: 2621-2632.

Zwenger S, Basu C (2007). *In Silico* analysis of terpene synthase genes in *Arabidopsis thaliana*. EXCLI Journal. 6: 203-211.

7

Key aspects of the mesenchymal stem cells (MSCs) in tissue engineering for *in vitro* skeletal muscle regeneration

Biswadeep Chaudhuri and Krishna Pramanik*

Department of Biotechnology and Medical Engineering, National Institute of Technology – Rourkela - 769008, India.

Tissue engineering, directly associated with Biotechnology and Biomedical Sciences, is an emerging field of research and development. The main issue of tissue engineering is to precisely and safely regenerate or reconstruct injured tissues of skeletal muscle, bone, teeth, neural, cardiac, cartilage etc. One of the primary requirements for tissue engineering development is a constant source of supplementary stem cells which have the ability to be differentiated into various tissue types such as condroblast, osteoblast or myoblast cells. In modern tissue engineering, mesenchymal stem cells (MSCs) take the most important part for *in vitro* growth or regeneration of the required tissues. Selective growth factors are also needed to optimize the growth process. In the preset review, an attempt has been made to focus on the crucial beneficial issues of mesenchymal stem cells for the skeletal muscle regeneration and repair. Though the detailed processes on how dystrophic muscles are replaced by fibrotic tissues inside living organs is still not very clearly understood, we have briefly discussed the overall ideas and future prospects of skeletal muscle regeneration (*in vitro*) using MSCs on 3D scaffold with optimum experimental conditions (use of various medias, growth factors etc.).

Key words: Mesenchymal stem cell, skeletal muscle, tissue engineering, tissue regeneration, growth factors, medical implant, biomaterials.

INTRODUCTION

Recently, considerable interest has been paid on the skeletal muscle regeneration by tissue engineering (Ground, 1999; Kagami et al., 2011; Grefte et al., 2007; Charge and Runiki, 2004; Jin et al., 2008). Tissue engineering research merges different branches of bioscience, engineering and medicine (Kagami et al., 2011; Mauro, 1961; Muir et al., 1965; Chen and Goldhamer, 2003; Shi and Garry, 2006). In case of a minor muscle injury, some growth factors based therapy seems to improve the muscle healing. The effects of growth factors on the activation, proliferation and differentiation of satellite cells have already been discussed elsewhere (Collins et al., 2005; Wagers and Conboy, 2005). Growth factors with stimulatory effects act *in vivo* to enhance the regeneration of the muscle tissue. However, in case of major muscle injuries, scaffold-based tissue engineering therapy (TET) is generally implemented to fill up the large defects. Discovery of different sources (like cord blood etc.) of multifunctional mesenchymal stem cells and rapid technological progress in biodegradable scaffold designing have encouraged tissue engineers and biotechnologists to apply their obtained results in therapeutics (Sadat et al., 2007; Schulze et al., 2005).

In the entire tissue regeneration process, a vital role is played by the Mesenchymal stem cell (MSC) which can differentiate into various tissue types such as bone, cartilage or skeletal muscle (Deans and Moseley, 2000; Harris et al., 2007; Hruba et al., 2008; Labarge and Blau,

---

*Corresponding author. E-mail: kpr@nitrkl.ac.in.

2002). Prockop (1997). Friedenstein and his research group (Friedenstein et al., 1966; 1974; 1976) first defined the bone marrow (BM) derived fibroblast-colony-forming-cells that adhered to cell culture surfaces. The fibroblast-colony forming cells were termed mesenchymal stem cells (MSC) or BM stromal cells (BMSC) (Pittenger et al., 1999). The MSCs might conventionally be defined as the adherent non-hematopoietic cells expressing positive markers such as CD90, CD105, CD73, and negative markers for CD14, CD34, and CD45 (Campioni et al., 2009; Jin et al., 2008). It has become a conventional and important method to derive MSCs from bone marrow. Subsequently, MSCs were also obtained from many other unconventional sources such as human placenta, adipose tissue (Zannettino et al., 2008), heart (Hoogduijin et al., 2007), Wharton's Jelly (Chao et al., 2008), dental pulp (Jo et al., 2007), peripheral blood (He et al., 2007), cord blood (Oh et al., 2008), menstrual blood (Patel et al., 2008; Hida et al., 2008) etc.

Recent researchers working with stem cells have realized the important functions of the mesenchymal stem cells (MSCs) in therapeutics and tissue regeneration. For example, much attention has recently been paid to overcome one of the major problems of tissue engineering associate with immune rejection (Martinez et al., 2011; Ichim et al., 2010) where MSCs take a vital role to overcome this problem along with the usual tissue regeneration process. In the present review, we have attempted to highlight some of the key functions of MSCs that trigger the *in vitro* tissue regeneration process without immune rejection. We have presented an overview of the progress and prospects of next generation skeletal muscle regeneration. For the sake of completeness, the importance of scaffold structure and growth factors in the tissue regeneration process has also been briefly discussed.

COMPARISON AMONG DIFFERENT SOURCES OF MESENCHYMAL STEM CELLS

The masenchymal stem cells for tissue engineering are available from different sources. Among the various available sources of MSCs, bone marrow (BM) has been widely accepted as the conventional source of MSCs. BM derived MSCs (abbreviated as BM-MSC) could be cultured rapidly and the process is also comparatively quicker. But other sources initially contain only mononuclear cell (MNCs), which have to be further differentiated into MSCs. Researchers have proposed that adult bone marrow-derived cells can gradually contribute to muscle cells (Jin et al., 2008). Even the transplanted bone marrow-derived cells have the potential to become satellite cells (mononucleated myogenic cells) that are found in muscle fibres and have the tendency to differentiate into specific type of cells

(muscle cells) to repair muscle injury (Musaro et al., 2007). The main problems associated with bone marrow derived stem cells are their unavailability in large number, it needs suitable donors and the process of SC collection from bone marrow by surgery is rather painful. It has, however, been observed that adipose tissues (AT) have a higher capacity of muscle differentiation ability than that of the bone marrow derived MSCs (Hoogduijin et al., 2007). But it takes relatively much longer time to regenerate skeletal muscle from adipose tissue derived MSCs. To obtain MSCs from other sources like cord blood, adipose tissue or placenta, it takes more time even over a month (Semenov et al., 2009; Miao et al., 2006; Green et al., 2010; Kang et al., 2006). Compared to those obtained from other sources, above mentioned adipose tissue having more number of MNC has better quality in terms of ability to differentiate into MSCs. Some research reports also claimed highest concentration of MSCs found from adipose tissue (Green et al., 2010; Im et al., 2005; Puissant et al., 2005). However, the most easily available sources of stem cells are considered to be cord blood, placenta, and adipose tissues. No ethical problem is involved in the procurement of MSCs from such sources. Importantly, it is possible to obtain MSCs from cryo-preserved human cord blood which is an additional advantage. The obtained cells have also multi-differentiation capacity similarly to that of bone the marrow derived MSC (Lee et al., 2004a; Robinson et al., 2011).

Recently, umbilical cord blood (UCB) is being recognized as an alternative source of hematopoietic stem cells (HSC) and MSCs can be well used for transplantation and in regenerative medicine (Lee et al ., 2004b; Zhou et al., 2003). The SCs derived from the UCB (UCB-SCs) are younger and there is little or no problem of variation of number of UCB derived MSCs with age as encountered in the case of BM derived SCs (Gang et al., 2004). UCB-SCs have already been successfully used *in vitro* to differentiate into insulin and c-peptide-producing cells (Denner et al., 2007; Oh et al., 2008). Most interestingly, UCB-MSCs showed no adipogenic differentiation capacity, in contrast to BM- and adipose tissue (AT)-MSCs. Both UCB and AT are attractive alternative to BM in isolating MSCs. AT contains MSC at the highest frequency, while UCB seems to be expanded to higher numbers (Kern et al., 2066). There are also some other additional advantages of using UCB derived stem cells in tissue engineering as mentioned below: (i) The CB-SCs are available in large numbers. (ii) Lower risk of virus infection. Infectious agents such as cytomegalo virus are rare exceptions (Rubinstein et al., 1993). (iii) They are more potent in application in allograft (same patient) transplantation. The cord blood derived SC, compared to BM-SC, has a bigger telomerase length and demonstrates higher proliferation potential (Vaziri et al., 1994). (iv) The Human placenta derived MSCs can even be combined with HSCs from UCB to reduce the

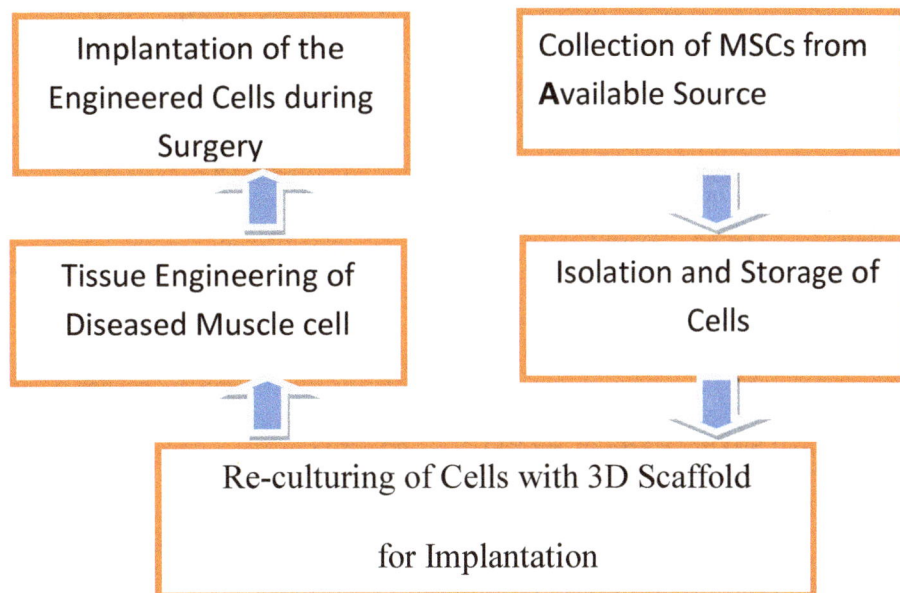

Figure 1. Stem cells collection from various sources, storage of stem cells, culturing on appropriate scaffold and its implantation for tissue regeneration in the body when required.

potential graft-versus-host disease (GVHD) in recipients (Magro et al., 2006). Usage of CB overcomes considerable problems encountered with other sources of CMs, such as allergenic, ethical, and tumorigenic issues. UCB-SC has been reported to repair myocardial hepatocytes (Yamada et al., 2007), muscles and neural tissues (Ikeda et al., 2004). Moreover, recently attempts are being made for the use of both autologous and allogeneic SCs as potential sources of safe and effective immunomodulation (Limbert et al., 2006). So when properly expanded in culture, UCB-MSCs are expected to be the most important practical units for skeletal, cardiac (Chen et al., 2004) or other muscle regeneration in the near future. However, to work with UCB-MSCs, the main drawbacks are the prolonged time to cell recovery, the early mortality associated with CB transplant and the overall lack of knowledge of working with UCB transplantation. Intensive research is going on to overcome these problems. It is to be noted that no significant differences concerning the morphology and immune phenotype of the MSCs derived from different sources were observed (Kern et al., 2006).

MESENCHYMAL STEM CELLS AND TISSUE ENGINEERING PROCESS WITH 3D SCAFFOLD

In the tissue regeneration process, as mentioned earlier, MSCs play the vital role due to the fact that it can differentiate into various tissue types such as bone, cartilage or skeletal muscle (Deans and Moseley, 2000). The first idea of the possibility of tissue engineering/tissue regeneration was put forwarded by the paediatric orthopaedic surgeon Dr. W T Green of Boston's Children Hospital. Dr. Green tried to implant chondrocytes on mouse to regenerate new cartilage (Green, 1977). Though this first experiment was only partially successful, but the outcome of this research paved the way for the use of stem cells in tissue regeneration. Dr Green's innovative idea about the use of stem cells for implantation and tissue regeneration later gave birth to the tissue engineering. Along with this, a new technique of biocompatible sophisticated scaffold designing was also subsequently developed for providing the growing tissues a desirable structure, which could be loaded with stem cells for implantation (Tan et al., 2003; Zein et al., 2002). In tissue engineering, the application of a three-dimensional (3D) scaffold is used for filling up the defect and to induce the formation of new muscles. In general, for skeletal muscle regeneration, the first step is to regenerate skeletal muscle cells in vitro on appropriate scaffold by using optimum culture conditions (growth factors etc.) and then to implanting it into the desired part of the body (Charge and Rudnicki, 2004). Such externally developed (i.e. bio-engineered) tissues are allowed to grow inside the body and reconstruct the affected muscle of the patient (Zammit et al., 2002; Lanza et al., 2007; Khademhosseini et al., 2009). A schematic diagram describing the tissue engineering process is shown in the self explanatory Figure 1.

In this context, it is to be noted that after any muscle injury, several series of well-coordinated optimum and compromising events take place inside the body, necessary for the proper development of the damaged tissue (Zammit et al., 2002; Lazarus et al., 1996). The said optimum conditions triggering the cells to regenerate are initiated immediately after injury by the release of several growth factors and cytokines from various cells of the injured blood vessels as well as inflammatory cells. Interestingly, the entire tissue regeneration happens in a specific site and the tissue growth is optimally controlled (Jones et al, 1986). Though the actual mechanism of this phenomenon has not yet been conclusively understood, it is believed that the MSCs have immense potentiality to revolutionize the conventional therapeutic practices (Tedesco et al., 2010; Mauro, 1961).

SOME IMPORTANT ASPECTS OF MASENCHYMAL STEM CELLS IN TISSUE REGENERATION: IMMUNE RESPONSE, INFLAMATION AND HEALING

As mentioned earlier, one of the main problems of skeletal muscle tissue regeneration or myoblasts therapy is associated with chronic immune rejection (Fan et al., 1996; Qu et al., 1998). Various model experiments have been carried out clinically to reduce immune rejection, but these reconstructive strategies do not always yield satisfactory results (Ianni et al., 2008). Interestingly, it is found that MSCs can not only help prevent this problem, but also help to regenerate selective tissues without immune rejection. It was observed that MSCs prevent inflammation and simultaneously accelerates the healing process (Ripoll et al., 2011). This means that MSCs are capable of giving protection against autoimmune pathogenesis (Zhou et al., 2008). Furthermore, it has also been reported that human MSCs suppress *in vitro* allogeneic T cell responses. T cells play a central role in immunity and according to current research reports, T cells appear to mediate muscle damage through secretion of osteopontin (Vetrone et al., 2009). This indicates that T cells have the ability to directly promote fibrosis, as well as direct perforin-mediated cytotoxicity (Spencer et al., 1997).

It is to be noted that immediately after injury, there appears a phase of myofiber degeneration, which is initiated by the release of proteases at the place of damaged tissue (Hurme and Kalimo, 1992; Cantini and Carraro, 1995; Dipietro, 1995). Proteases automatically digest myofibers that result in tissue debris at the zone of injury. Along with this process, there is chemotaxis of neutrophils and macrophages related to this area. Due to macrophage activity, local debris is phagocytosed and proceeds to induce a local inflammatory response (Dipietro, 1995; Tidball et al., 1999; Robertson et al., 1993). So it appears that macrophages take part to induce inflammation. Some experimental studies also suggest that macrophages secrete several growth factors that enhance tissue regeneration process (Robertson et al, 1993; Summan et al., 2006).

It has been reported that MSCs posses various strong anti-inflammatory properties. For instances, MSCs suppress NK (natural killer) and T cytotoxic cell function (Selmani et al., 2008), reduce macrophage activities (Spaggiari et al., 2009; Yang et al., 2008) inducing generation of Treg cells (Casiraghi et al, 2008), inhibit Th1, Th17 cell generations (Batten et al., 2006), suppress Dendritic Cell (DC) maturation etc. Though the exact mechanism how MSCs regulate such immune suppression is not very clear, there are some important experimental results that demonstrate that different immune functions are suppressed by MSCs through the release of immune suppressive cytokines such as TGF-b, LIF etc., expression of T and NK inhibitory enzyme indolamine 2,3-deoxygenase, expressing contact-dependent inhibitory molecules such as PD-1L and via production of soluble HLA-G (Campioni et al., 2009; Bishopric et al., 2008; Nasef et al., 2008). These important immune modulatory properties are found to induce active immune response (Renner et al., 2009; Ryan et al., 2007; Opitz et al. 2009; Rizzo et al., 2008). As active immunity is cell mediated, it might be concluded that MSCs play very important role in this process. In bone marrow too, one of the main functions of MSC is to protect the hematopoietic precursor from inflammatory damage as MSCs have some control over immune system (Riordan et al., 2007). This also confirms anti-inflammatory effects of MSCs, which could reduce the chance of immune rejection and hence optimise the overall growth of tissue (Jones et al., 2007). Several other inhibitions of chronic inflammatory processes, such as models of autoimmune arthritis, diabetes, multiple sclerosis and lupus, have also been well documented which are optimised by the MSCs. It is also recognized that MSCs have the properties that allow transplantation across major histocompatibility complex (MHC) barriers (Blanck and Ringde, 2007). So, after allogeneic haematopoietic stem cell transplantation (HSCT), these beneficial immunomodulatory effects of MSCs could be utilized to prevent rejection of organ transplants and also to repair tissue damage caused by autoimmune-induced inflammatory diseases like Crohn's disease, ulcerous colitis, graft-versus-host disease (GVHD) of the gut etc. These are some of the advanced indications of enormous possibility of using multifunctional MSCs to regenerate selective or desired tissues free from immune rejection and also to protective against several common diseases. Currently intensive research work is going on in these directions.

SITE SPECIFICITY OF MESENCHYMAL STEM CELLS

Site specificity is another important factor of tissue growth. Regeneration of new tissue has to be highly précised and should also be at the required injured site where replacement of old tissue with the new one is only needed. Here some crucial roles are performed by the Stromal cell-derived factor-1 (SDF-1). It has been observed that factor-1 (SDF-1) stimulates stem cell propagation in the desired sites. For example, SDF-1 has been demonstrated to be associated with the mobilization of masenchymal stem cells into the periphery and homing only to the site of injury (Penn, 2009). So it becomes possible to efficiently control the tissue growth as well as selection of the specific site. In addition, this property allows MSCs to differentiate into various specific tissues and to complement dystrophic deficiency. This also indicates the therapeutic aspects of MSCs for Duchenne muscular dystrophy (DMD), which is a lethal X-linked musculodegenerative condition and also a genetic defect whose manifestation is augmented by inflammatory mechanisms. MSCs produce paracrine factors that directly help to inhibit apoptosis, stimulate endogenous cell proliferation and activate tissue resident stem cells only at the site of injury (Tidball, 1995).

It has also been observed that under proper circumstances, MSCs can differentiate to specific tissues that had already been injured and urgently needed to be regenerated to support normal muscle function. Most importantly, the site specificity and efficiency are highly desirable to regenerate the damaged part without disturbing the rest where no further regeneration is required. Recently Tao et al. (2009) have precisely demonstrated that MSCs might differentiate selectively into tissue types that have only been injured. They have systemically administered the growth of MSCs to clone into immune deficient mice after subsequent carbon tetrachloride hepatic injury. Further to add, differentiation of MSCs only into albumin expressing hepatocyte-like cells was also observed in those mice. All these are strong evidences that MSCs possess some unique properties that are specific to the site of injury.

POTENTIAL RELATION OF MESENCHYMAL STEM CELLS WITH CHEMOKINE SIGNALLING

In the tissue regeneration process, chemokine signalling has a major role to support the entire immune response. MSCs continuously support the entire immune system even inside the body. This provides evidences that MSCs might also have the properties to influence cell signalling procedure especially for chemokines. Ichim and his group have shown that chemokines are directly related to major immune system (Ichim et al., 2010; Charge and Rudnicki, 2004). Based on cytokine production and arginine metabolism, two types of macrophages viz. M1 and M2 have been distinguished. The M1 macrophage is primarily antiangiogenic and shows some properties that can inhibit the tissue growth. For example, stimulation of M1 macrophage inhibits tumour growth in cancerous cells

(Eriksson et al., 2009), whereas M2 macrophage shows more constructive role in the tissue regeneration process. Moreover, M2 macrophages are anti-inflammatory, support angiogenesis, and they are also associated with tissue repair via regeneration (Mantovani et al., 2004; Sica et al., 2008). Recent findings also show that regulatory interactions between cytotoxic M1 macrophages in dystrophic muscle and anti-inflammatory M2 macrophages are important in regulating the overall balance between the death of dystrophic muscle and regenerative processes. Manipulation of the balancing between the functions of M1 and M2 macrophages can, therefore, affect the severity of muscular dystrophy. This suggests that manipulation of macrophage phenotype *in vivo* may have potential therapeutic values for the treatment of various diseases (Tidball, 2002; Tdball and Villalta, 2010). In summary, MSCs, cytokine production, M1 and M2 macrophages must jointly play important part in the regenerative tissue engineering process. M1 might be used for the inhibitory function of tissue growth while M2 for the tissue regeneration process. This phenomenon was also described (Wang et al., 2007) as constructive in case of M1 and destructive in case of M2. So, attempts are being made to utilize the dual nature and activity of macrophages to control over tissue regeneration process by accelerating or controlling tissue growth according to our need.

One might find direct link between MSCs and several chemokine signals associated with MSC migration into specific injured tissues for the formation of new ones. The relevance of cytokines for the development of protective immune system has already been studied and well established. It is now quite clear that cytokines might regulate cell immunity during cell regeneration. However, at the present stage, it is not very clear whether cytokines can establish and maintain immunological memory in those stem cells. Further intensive research is indeed necessary for a deeper understanding of the relation between chemokine signalling and MSC migration.

STRUCTURE DESIGN OF NEO TISSUES

Stem cell research also shed light on the regulation of tissue structure determination. When muscle stem cells, present beneath the muscle basal lamina, are activated by massive proliferation and differentiation of myoblasts at the edge of injury, formation of new muscle fibres begins at that site to regenerate tissue. After that, fusion of myoblasts occurs themselves to the damaged tissue to regenerate new myofibers. But, our present knowledge to predict the structure of those muscles that accelerate regenerative medicine is not well developed. According to the recent experimental observations, understanding the control of cell-matrix interaction could revolutionize the idea of determining the tissue architecture to obtain the desired shape. In this connection, extracellular matrix (ECM) has the potential to guide and support the differentiation of MSCs. The ECM, a part of animal tissue, is well known for providing the structural support to the animal cells (Lanfer et al., 2009). It is to be noted here that there might also have some link between MSCs and ECM and combined study of both could reveal the optimum control over the mechanism of tissue structure prediction. At the same time, morphogenesis of various tissue types could also be predictable for controlling the proper shape.

During skeletal muscle repair, muscle stem cells work on necrotic fibres that might be called basement membranes on which the tissue is to be grown to make sure that the newly formed tissue has the proper shape and position. Again, as there are strong relation between the basement membrane and extracellular membrane, there are also some new important aspects that help the regeneration process. The ECM proteins, for examples, fibronectin and tenascin-C are secreted to the wound surface, before the cell migration, to support the regeneration process (Tanaka et al., 1999; Tervo et al., 1991). Thereafter, the wound area is covered by an adjacent epithelial monolayer followed by cell proliferation covering the entire wound area.

3D SCAFFOLD FOR SKELETAL MUSCLE GENERATION

For the development of tissue engineering and regenerative medicine, the importance of biodegradable scaffold for controlling over the activities of stem cells is unimaginable   (Lanza et al., 2007; Khademhosseini et al., 2009). In stem cell therapy, scaffold and its design contribute a lot to determine the desired shape and structure of the neo tissue (Sun and Lal, 2002). Recently computer aided scaffold designing has become more popular (Mulder de et al., 2009). Scientists working with Biomaterials are trying to control over pore geometry and architecture that would be most suitable for the cell growth (Moroni et al., 2006; Yan and Gu, 1996). Bioengineered scaffold made up of porous polyvinyl alcohol (PVA), silk, polycaprolactone (PCL), chitosan, polyhydroxylbutyrate, collagen, heparin, hybrinogen, elating etc. could be efficient choice of scaffold preparations (Mondrinos et al 2006; Miot et al., 2005; Benya and Shaffer, 1982; Yeong et al., 2004). Leong et al., 2003 used non-toxic polyvinyl alcohol (PVA) as it tends to dissolve quickly after implantation. There are several modern techniques used for scaffold preparation viz. Solvent casting (Mikos et al., 1994), Polymerization (Mooney et al, 1997; Bryant and Anseth, 2001), Melt quenching and moulding (Hsu et al., 1997), Phase separation (Thomson et al., 1995; Hua et al., 2000), Freeze drying (Lo et al., 1995) etc. Regeneration of neo

muscle and degradation of scaffold should take place simultaneously. Ultimately the scaffold would disappear and that space would gradually be occupied by the neo muscles (Oh, Kang and Lee, 2006; Hutmacher, Goh and Teoh, 2001). This phenomenon might be compared with the phagocytosis of artificial (scaffold) basement membrane by neo muscles.

GROWTH FACTORS IN TISSUE OF SKELETAL MUSCLE

Skeletal or other muscle regeneration needs collective action of cells, scaffolds, signalling molecules and growth factors (Lee et al., 2011). Growth factors are soluble-secreted signalling polypeptides which instruct specific cellular responses in a biological environment. Under various circumstances, for instance, to regenerate affected tissues, cells secreted growth factors (GFs) protein perform various cellular actions viz. control over migration, differentiation or proliferation of a specific subset of cells and cell survival. Localised delivery of GFs is believed to be therapeutically effective for replication of cellular components directly involved in tissue regeneration and healing process (Chen et al., 2010; Vasita and Katti, 2006). Some essential GFs for tissue regeneration (Lee et al., 2011) are shown in Table 1.

Though all growth factors are important, some have more specific importance over the others. One important GF is Transforming growth factor (TGF)-beta1 which is very effective for fibroblast tissue regeneration. Immunohistochemical results predict TFG-beta as local stimulators for the tissue repairing process (Bourque et al., 1993). It has been shown that TFG-beta1 is one of the best fibrogenic mediators and it is over expressed in human dystrophic muscle (Leask and Abraham, 2004). With increased TFG-beta1, mRNA levels are directly associated with initial stage (Bernasconi et al., 1995) of tissue fibrosis which could be a positive indicator that the starting point of muscular tissue regeneration occurs through the TFG-beta1. It has also been shown that plasma TGF-beta1 level is elevated in patients with DMD and congenital muscular dystrophy (Ishitobi et al., 2000). TFG-beta also shows positive effect on reorganization of extracellular matrix and basement membrane surrounding the damaged myofibers. By stimulating the synthesis of collagens, fibronectin and novel matrix proteins, TFG-beta directly induces angiogenesis to regenerate new blood vessels (Husmann et al., 1996). For example, it has been examined that TFG-beta is expressed by regenerating skeletal muscle within a few days after trauma. So, TFG-beta is undoubtedly one of the major multifunctional growth factors that can motivate the entire skeletal muscle regeneration process. Moreover, TFG-beta also stimulates the production of Platelet Derived Growth Factor (PDGF) that is well known

Table 1. Some essential GFs for skeletal muscle regeneration.

Symbolic name	Name of GFs
Ang	Angiopoietin GF
bFGF	basic fibroblast GF
BMP	Bone morphogenetic GF
HGF	Hepatocyte GF
TGF-betas	Trans Growth Factor
IGF	Insulin- like Growth Factor
PDGF	Platelet-derived GF
TFG	Transforming GF
VEGF	Vascular endothelial GF
NGF	Nerve GF
CSF	Colony stimulating Growth or
FGF-R	Fibroblast Growth Factor Receptor
LIF	Leukaemia Inhibitory Factor
EGF	Epidermal Growth Facto

to cause cell migration to the injured tissues to accelerate regeneration (Canalis et al., 1989). PDGF also acts as a potent stimulator of cell division in fibroblast-like cells. So PDGF is likely to accelerate fracture repair in early stages.

Similarly, after tissue disruption, Fibroblast Growth Factor (FGF) released during inflammation, induces the satellite cells to further proliferate and hence accelerate the regeneration process (van den Boset et al., 1997). Like TFG-beta, it has also been found that FGFs are angiogenic in nature. So, they can also be involved in the growth process of new blood vessel from pre-existing vessels, which gives us another new aspect that TFG-beta not only regenerates new tissues but also helps in the formation of new blood vessels (Grounds, 1991; Baird and Ling, 1987). So it is quite evident that Transforming growth factor TGF-beta1 has some potentiality to directly influence to enhance the fibrotic process of human muscular dystrophy. Leukaemia inhibitory factor (LIF) has also been well examined and found to have some most important role in the regeneration of injured muscle (Husmann et al., 1996). LIF is also addressed as multifunctional cytokine that directly stimulates the growth of skeletal muscle after damage. Finally, growth factors have a great influence in proper growth and development of skeletal muscle regeneration.

SUMMARY AND CONCLUSION

We have discussed the unique differentiation potential of MSCs both in vitro and in vivo along with their ability to secrete various strophic factors and to modulate the immune system. All these make MSCs a promising nature gifted component for the development of next generation regenerative medicine. However, more research is needed to understand the full potentiality of MSCs in tissue engineering .The possibility of using MSCs from nonconventional sources provides an insight into the general processes involved in regeneration of the muscle which opens the perspectives of novel therapies. Though BM or adipose tissue derived MSCs are very important for the skeletal muscle regeneration, easily and abundantly available sources like cord blood, placenta etc. have great potentiality for their use in tissue engineering. During the last decade most of the researches on skeletal muscle regeneration have been done focusing on the characterization of MSCs and understanding their differentiation potential functions. In vivo tissue generation has mostly been traced in animal models. It has been established that the secretion of bioactive materials by MSCs in response to injury mitigates the inflammatory response leading to cure injury and promote repair. The basic mechanisms of tissue generation at the injury sites by the MSCs and their ability to repair are associated with the secretion of various chemo-tactic factors. However, the complex mechanisms and the pathways with which MSCs supports repair are yet to be understood. So, starting from wound healing to tissue regeneration at specific site of injury, giving proper structure (with the help of biocompatible bioengineered scaffolds) to the newly growing tissues to match the actual biological structure of body, and also to eliminate the chances of immune rejection, the success of tissue engineering research significantly depends on the proper use and functioning of the multifunctional MSCs. More successful clinical trials on human are to be made. There is immense scope of research and development in this field of tissue engineering. Finally, proper understanding and utilization of the various novel aspects of MSCs will lead to enormous change of the conventional medicine to the next generation regenerative medicine for curing not only skeletal muscle but also many other acute diseases.

ACKNOWLEDGEMENT

The authors are thankful to the Department of Biotechnology (DBT), Government of India, New Delhi, for financial support to carry out the present work.

REFERENCES

Baird A, Ling N (1987). Fibroblast growth factors are present in the extracellular matrix produced by endothelial cells in vitro: Implication for a role of heparinase-like enzymes in the neovascular response. Biochem. Biophys. Res. Commun., 142: 428-435.

Batten PP, Sarathchandra J, Antoniw W, Tay SS, Lowdell MW, Taylor PM, Yacoub MH (2006). Human mesenchymal stem cells induce T cell energy and down regulate T cell allo-response via the TH2 pathway: relevance to tissue engineering human heart valves. Tissue Eng., 12: 2263-2273.

Bryant SJ, Anseth KS (2001). The effects of scaffold thickness on tissue engineered cartilage in photocrosslinked poly (ethylene oxide) hydrogels. Biomat., 22:619-626.

Bernasconi P, Torchiana E, Confalonieri P, Brugnoni R, Barresi R , Mora M, Cornelio F , Morandi L, Mantegazza R (1995). Expression of transforming growth factor-beta 1 in dystrophic patient muscles correlates with fibrosis. Pathogenetic role of a fibrogenic cytokine. J. Clin. Invest., 96: 1137–1144.

Bishopric NH (2008). Mesenchymal stem cell-derived IL-10 and recovery from infarction: a third pitch for cord blood. Circul. Res., 103: 125-127.

Blanck KL, Ringde O (2007). Immunomodulation by masenchymal stem Cellsa and clinical experience. J. Int. Med., 262:509-525.

Bourque WT, Gross M, Hall BK (1993). Expression of four growth fracture Repair. Int. J. Dev. Biol., 37: 573-579.

Campioni D, Rizzo R, Stignano M , Rizzo R, Stignani M, Melchiorri M, Ferrari L, Moretti S, Russo A, Bagnara GP, Bonsi L, Alviano F, Lanzoni G, Cuneo A, Baricordi OR, Lanza F (2009). A decreased positivity for CD90 on human Mesenchymal stromal cells (MSCs) is associated with a loss of immunosuppressive activity by MSCs. Clinical. Cytomat., 76: 225–230.

Cantini M, Carraro U (1995). Macrophage-released factor stimulates selectively myogenic cells in primary muscle culture. J. Neuropathol. Exp. Neurol., 54:121–128.

Canalis E, McCarthy TL, Centrella M (1989). Effects of platelet-derived growth factor on bone formation in vitro. J. Cell. Physiol., 140: 530-537.

Casiraghi F, Azzollini N, Cassis P, Imberti B, Morigi M, Cugini D, Cavinato RA, Todeschini M, Solini S, Sonzogni A, Perico N, Remuzzi G, Noris M (2008). Pretransplant infusion of mesenchymal stem cells prolongs the survival of a semiallogeneic heart transplant through the generation of regulatory T cells. J. Immunol., 181: 3933–3946.

Charge SB, Rudnicki MA (2004). Cellular and Molecular regulation of muscle regeneration. Physiol. Rev., 84:209-238.

Chao KC, Chao KF, Fu YS, Liu SH (2008). Islet-like clusters derived from mesenchymal stem cells in Wharton's Jelly of the human umbilical cord for transplantation to control type 1 diabetes. PLoS ONE. 3: 1451.

Chen CJ, Goldhamer DJ (2003). Skeletal muscle stem cells. Reprot Biol. Endocrinol., 1:101–107.

Chen FM, Zhang ZF, Wu ZF(2010) Toward delivery of multiple growth factors in tissue engineering. Biomat., 31: 6279-6308.

Chen SL, Fang ZFM Ye F, Liu YH, Qian J, Shan SL, Zhang JJ, Chunhua RZ, Liao LM, Lin S, Sun JP (2004). Effect on left ventricular function of intracoronary transplantation of autologous bone marrow MSC in patients with acute myocardial infection. Am. J. Cardiol., 94(1):92-95.

Collins CA, Olsen I, Zammit PS, Heslop L, Petrie A, Partridgel B (2005). Stem cell function, self renewal, and behavioral heterogeneity of cells from the adult muscle satellite cell niche. Cell, 122:289– 301.

Deans RJ, Moseley AB (2000). Mesenchymal stem cells: Biology and potential clinical uses. Exp. Haematol., 28:875–884.

Denner L, Bodenburg Y, Zhao JG, Howe M, Cappo J, Tilton RG, Copland JA, Forraz N, McGukin C, Urban R (2007). Directed engineering of umbilical cord blood stem cells to produce C-peptide and insulin. Cell Polif., 40(3): 367-380.

Dipietro LA (1995). Wound healing: the role of the macrophage and other immune cells. Shock, 4: 233–240.

Eriksson F, Tsagozis P, Lindberg K, Parsa R, Mangsbo MS, Persson MA, Harris RA, Pisa P (2009). Tumor-specific bacteriophage induce tumour destruction through activation of tumor-associated macrophages. J. Immunol., 182: 3105-3111.

Fan Y, Maley M, Beilharz M, Grounds M (1996). Rapid death of injected myoblasts in myoblast transfer therapy, Mus. Nerv., 19: 853-860.

Friedenstein AJ, Shapiro P, Petrakova KV (1966). Exper. Haematol., 4: 267-674.

Friedenstein AJ, Chailakhyan RK, Latsinik NV, Panasyuk AF (1974). Keiliss-Borok IV. Stromal cells responsible for transferring the microenvironment of the hemopoietic tissues. Cloning in vitro and retransplantation in vivo. Transplant, 17:331-40.

Friedenstein AJ, Gorskaja JF, Kulagina NN (1976) Fibroblast precursors in normal and irradiated mouse hematopoietic organs nonhematopoietic tissues. Science, 76: 71–74.

Gang EJ, Hong SH, Jeong JA, Hwang SH, Kim SW, Yang IH, Ahn C, Han H, Kim H (2004). In vitro mesogenic potential of human UCB-derived MSCs, Biochem. Biophys. Res. Commun., 231(1): 102-208.

Grefte S, Jagtman AMK, Torensma R, Von Den Hoff JW (2007). Skeletal muscle development and regeneration. Stem cells Devt., 16:857–868.

Green WT (1977). Behavior of articular chondrocytes in cell culture. Clinical. Orthopaed. Rel. Res., 124: 237–250.

Green AC, Amorn De NFG, Pinaguy I (2010). Influence of decantation, washing and centrifugation on adipocyte and mesenchymal stem cells content of aspirated adipose tissue: A comparative study. J.Plastic Reconst. Aesthet. Surg., 63: 1375-1381.

Grounds MD (1991). Towards understanding skeletal muscle regeneration. Pathol. Res. Practice, 187: 1-22.

Ground MD (1999). Muscle regeneration: molecular aspects and therapeutic implication. Current Op. Neurol., 12: 535-443.

Harris DT, Badowski M, Ahmad N, Gaballa MA (2007). The potential of cord blood stem cells for use in regenerative medicine. Expert. Opin. Biol. Ther., 7: 1311-1322,

He Q, Wan C, Li G (2007). Concise review: multipotent mesenchymal stromal cells in blood. Stem Cells, 25: 69–77.

Hida N, Nishiyama N, Miyoshi S, Kira S, Segawa K, Uyama T , Mori T, Miyado K, Ikegami Y, Cui CH, Kiyono T, Kyo S, Shimizu T, Okano , Sakamoto S, Ogawa, Umezawa A (2008). Novel cardiac precursor-like cells from human menstrual blood-derived mesenchymal cells. Stem Cells, 26: 1695–1704.

Hoogduijn MJ, Crop MJ, Peeters AM, Van Osch GJ, Balk AH, Ijzermans JN, Weimar W, Baan CC (2007). Human heart, spleen, and perirenal fat-derived mesenchymal stem cells have immunomodulatory capacities. Stem Cells Dev., 16: 597–604.

Hruba A, Velebny V, Kubala L (2008). Isolation and characterization of mesenchymal stem cell population entrapped in bone marrow collection sets. Cell Bio. Int., 32:1116-1125.

Hsu YY, Greaser JD, Trantolo DJ, Lyons CM (1997). Effect of polymer foam morphology and density on kinetics of in vitro controlled release of isoniazid from compressed foam matrices. J. Miomed. Mater. Res., 35: 107-116.

Husmann I, Soulet L, Gautron J, Martelly I, Barritault D (1996). Growth factors in skeletal muscle regeneration. Cytokine Growth Factor Rev., 7: 249-258.

Hua FJ, Kim GE, Lee JD, Son YK, Lee DS (2000). Macroporous scaffold by liquid liquid phase separation of a PLLA dioxane water system. J. Biomed. Mat. Res., 63: 161-167.

Hurme T, Kalimo H (1992). Activation of myogenic precursor cells after muscle injury. Med. Sc. Sports Exercise, 24: 197–205.

Hutmacher DW, Goh JC, Teoh SH (2001). Biodegradable Materials for Tissue Engineering Application. Ann. Acad. Med. Singapore, 30: 183-191.

Ianni MD, Papa BD, Ioanni MD, Moretti L, Bonifacio E, Cecchini D, Sportoletti P, Falzetti F,Tabilio A (2008). Mesenchymal cells recruit and regulate T regulatory cells. Expt. Hematol. 36: 309–318.

Ichim TE, Alexandrescu DT, Solano F, Lara F, Campion RN, Paris E, Woods JE, Murphy MP, Dasanu CA, Patenl AN, Marleau AN, Leal A, Raiordan NH (2010). Mesenchymal stem cells as anti-inflammatories: Implications for treatment of Duchenne muscular dystrophy. Cell Immunol., 260: 75–82.

Ikeda Y, Noboru F, Wada M, Matsumoto T, Satomi A, Yokoyama SI, Saito S, Masumoto K, Katsuo K, Mugishima H (2004). Development of angiogenic cell and gene therapy by transplantation of umbilical cord blood with vascular endothelial growth factor gene. Hyperten. Res., 27(2):119-128.

Im G, Shin YW, Lee KB (2005). Do adipose tissue –derived mesenchymal stem cells have the same osteogenic and chondrogenic potential as bone marrow-derived cells? Osteoarthri. Cartil., 13: 845-853.

Ishitobi M, Haginoya K, Zhao Y, Ohnuma A, Minato J, Yanagisawa T, Tanabu M, Kikuchi M, Iinuma K (2000). Elevated plasma levels of transforming growth factor beta1 in patients with muscular dystrophy. Neuroreport, 11: 4033–4035.

Jin JD, Wang HX, Xiao FJ, Wang JS, Lou X, Hu LD, Wang LS, Guo ZK (2008). A novel rich source of human machenchymal stem cell from debris of bone marrow samples. Biochem. Biochips. Res. Commun., 376: 191-1995.

Jo YY, Lee HJ, Kook SY, Choung HW, Park JY, Chung JH, Choung YH, Kim ES, Yang HC, Choung PH (2007). Isolation and characterization of postnatal stem cells from human dental tissues. Tissue Eng., 13: 767–773.

Jones DA, Newham DJ, Round JM, Tolfree SEJ (1986). Experimental human muscle damage: Morphological changes in relation to other indices of damage. J. Physiol., 375:435-448.

Jones BJ, Brooke G, Atkinson K, McTaggart SJ (2007). Immunosuppressant by placental indoleamine 2,3- dioxygenase: a role for mesenchymal stem cells. Placenta, 28: 1174–1181.

Kagami H, Agata H, Tojo A (2011). Bone marrow stromal cells (bone marrow-derived multipotent mesenchymal stromal cells) for bone tissue engineering: Basic science to clinical translation. Int. J. Biochem. Cell Biology, 43: 286-289.

Kang XQ, Zang WJ, Bao LJ, Li DL, Xu XL, Yu XJ (2006). Differentiating characterization of human umbilical cord blood derived mesenchymal stem cells in vitro. Cell Bio. Int., 30: 569-575.

Kern S, Eichler H, Stove J, Kluter H, Bieback K (2006). Comparative analysis of mesenchymal stem cells from bone marrow, Umbilical cord blood, or Adipose tissue. Stem Cells, 24: 1294-1301.

Khademhosseini A, Vacanti J, Langer R (2009). Next generation tissue constructs and challenges to clinical practice. Sc. Ame., 300: 64-71.

Labarge MA, Blau HM (2002). Biological progression from adult bone marrow to mononucleated muscle stem cell to multinucleated muscle fibre in response to injury. Cell, 111: 589-601.

Lanfer B, Seib FP, Freudenberg U, Freudenberg D, Stamov T (2009). The growth and differentiation of mesenchymal stem and progenitor cells cultured on aligned collagen matrices. Biomat, 30: 5950–5958.

Lanza RP, Langer RS, Vacanti J (2007). Principles of tissue engineering. Amsterdam, the Netherlands: Elsevier Academic Press.

Lazarus HM, Haynesworth SE, Gerson SL, Rosenthal NS, Caplan AI (1996). Ex vitro expansion and subsequent infusion of human bone marrow-derived stromal progenitor cells (mesenchymal progenitor cells): implications for therapeutic use. Bone Marrow Transpl. 16:557-564.

Lee MW, Choi J, Yang MS, Moon J, Park JS, Kim HC, Kim YJ (2004a). Mesenchymal stem cell from cryopreserved human umbilical cord blood. Biochem. Biophys. Res. Commun. 320: 273-278.

Lee OK, Kuo TK, Chen WM. Lee KD, Hsich SL, Chen TH (2004b). Isolation of multipotent MSCs from UCB. Blood, 103(5):966-975.

Lee K, Silva K A, Moonen DJ (2011). Growth factor delivery based tissue engineering: general approaches and a review of recent developments. J. R. Soc. Interface, 8: 153-170.

Leask A, Abraham DJ (2004). TGF-beta signalling and the fibrotic response. FASEB J., 18: 816–827.

Leong KF, Cheah CM, Chua CK (2003). Solid freeform fabrication of three-dimensional scaffolds for engineering replacements tissues and organs. Biomat, 24: 2363-2378.

Limbert C, Couri CE, Foss MC, Voltarelli JC (2006). Secondary prevention of type 1 diabetes mellitus: stopping immune destruction and promoting beta cell regeneration. Braz. J. Med Biol Res., 39: 1271-1280.

Lo H, Ponticiciello MS, Leong KW (1995). Fabrication of controlled release biodegrable foams by phase separation. Tissue Eng., 1: 15-28.

Magro E, Regidor C, Cabrera R, Sanjuan I, Fores R, Garcia JA, Ruiz E, Gil S, Bautista G, Millan I, Madrigal A, Fernandez MN (2006). Early hematopoietic recovery after unit unrelated cord blood transplantation in adults supported by co-infusion of mobilized stem cells from a third party donor. Haematol., 91:640-648.

Mantovani A, Sica A, Sozzani S, Allavena P, Vecchi A, Locati M (2004). The chemokine system in diverse forms of macrophage activation and polarization. Trends Immunol., 25: 677-686.

Martinez C, Hofmann TJ, Marino R (2011) Human bone marrow mesenchymal stromal cells express the neural ganglioside GD2: A novel surface marker for the identification of MSCs. Blood, 109: 4245–4248.

Mauro A (1961). Satellite cells of skeletal muscle fibers. J. Biophys. Biochem. Cytol. 9: 493-495.

Miao Z, Jin J, Chen L, Zhu J, Huang W, Zhao J, Qian H, Zhang X (2006). Isolation of mesenchymal stem cells from human placenta: Comparison with human bone marrow stem cells. Cell Bio. Int., 30: 681-687

Mikos AG, Lyman MD, Freed LE, Langer R (1994). Wetting of poly [l-Lactic acid) and poly (DL-lactic acid co-glycolic acid) foams for tissue culture, Biomat, 15: 55-58.

Miot S, Woodfield T, Daniels AU, Suetterlin R, Peterschmitt I, Heberer, M (2005). Effects of scaffold composition and architecture on human nasal chondrocyte redifferentiation and cartilaginous matrix deposition. Biomat, 26: 2479-2489.

Mondrinos MJ, Dembzynaki R, Lu L, Venkata KC, Wootton DM, Lelkes PI, Zhou J (2006). Progen based solid freeform fabrication of polycaprolactone-calcium phosphate scaffolds for tissue engineering. Biomat, 27: 4399-4408.

Mooney DJ, Kaufmann PM, Sano K, McNamara KM, Vacanti JP, Langer R (1997). Transplantation of hepatocytes using porous, biodegradable sponges. Transpl. Proc., 26: 3425-3426.

Moroni L, de Wijn JR, van Blitterswijk CA (2006). 3D fiber-deposited scaffolds for tissue engineering: Influence of pores geometry and architecture on dynamic mechanical properties. Biomat. 27: 974-985.

Muir AR, Kanji AHM, Allbrook D (1965). The structure of the satellite cells in skeletal muscle. J. Anat., 99:435–444.

Mulder de ELW, Buma P, Hannink G (2009). Anisotropic porous biodegradable scaffold for skeletal muscle regeneration. Materials, 2: 1674-1696

Musaro A, Giacinti C, Pelosi L, Pelosi B, Molinaro (2007). Cellular and molecular bases of muscle regeneration: The critical role of insulin-like growth factor-1. Int. Cong. Series, 1302: 89-100.

Nasef A, Mazurier C, Bouchet S, Fr Musaro A, Giacinti C, Pelosi L, Pelosi B, Molinaro M (2008). Leukemia inhibitory factor: role in human mesenchymal stem cells mediated immunosuppression. Cell Immunol., 253: 16–22.

Oh W, Dal SK, Yang YS, Lee, JK (2008). Immunological properties of umbilical cord blood-derived mesenchymal stromal cells. Cellular Immunol., 251: 116-123.

Oh SH, Kang SG, Lee JH (2006). Degradation behaviour of hydrophilized PLGA scaffold prepared by melt –modelling particulate-leaching. J. Matet. Sc. Med., 17: 131-137.

Opitz CA, Litzenburger UM, Lutz C, Lanz TV, Tritschler I, Köppel A, Tolosa E, Hoberg M (2009). Toll-like receptor engagement enhances the immunosuppressive properties of human bone marrow-derived mesenchymal stem cells by inducing indoleamine-2,3-dioxygenase-1 via interferon-beta and protein kinase R. Stem Cells, 27: 909–919.,

Patel AN, Park E, Kuzman M, Benetti F, Silva J, Allickson JG (2008). Multipotent menstrual blood stromal stem cells: isolation, characterization, and differentiation. Cell Transplant, 17: 303–311.

Penn MS (2009). Importance of the SDF-1:CXCR4 axis in myocardial repair. Circul. Res., 104: 1133–1135.

Pittenger MF, Mackay AM, Beck SC, Jaiswal R K, Douglas R, Mosca J D, Moorman M A, Simonetti DW, Craig S, Marshak D R (1999). Multilineage potential of adult human mesenchymal stem cells. Science, 284:143-147.

Prockop DJ (1997). Marrow stromal cells as stem cells for Osteogenesis in transplants of bone marrow cells. J. Embryol. Exp. Morphol., 16: 381-90.

Puissant B, Barreau C, Bourin P,_Clavel, C, Corre J, Bousquet. L, Casteilla L, Blancher L (2005). Immunomodulatory effect of human adipose tissue-derived adult stem cells: comparison with bone marrow mesenchymal stem cells. British J. Haematol., 129: 118–129.

Qu Z, Balkir L, van Deutekom JC, Bobbins PD, Pruchnic R, Huard J (1998). Development of approaches to improve cell survival in myoblast transfer therapy. J. Cell Biol., 142: 1257-1267.

Renner P, Eggenhofer E, Rosenauer A, Popp FC, Steinmann JF Slowik P, geissler EK, Piso P, Schlitt HJ, Dahlke MH (2009). Mesenchymal stem cells require a sufficient, ongoing immune response to exert their immunosuppressive function. Transpl. Proc., 41: 2607–2611.

Ripoll CB, Flaat M, Eiermann J, Fisher-Perkins JM, Trygg CB, Scruggs BA (2011). Mesenchymal lineage stem cells have pronounced anti-inflammatory effects in the twitcher mouse model of Krabbe's disease. Stem Cell, 29: 67-77.

Riordan NH, Chan K, Marleau AM, Ichim TE (2007). Cord blood in regenerative medicine: do we need immune suppression? J. Translat. Med., 5: 8-14.

Rizzo R, Campioni D, Stignani M, Melchiorri L, Bagnara GP, Bonsi L, Alviano F, Lanzoni G, Moretti S, Cuneo A, Lanza F, Baricordi OR (2008). A functional role for soluble HLA-G antigens in immune modulation mediated by mesenchymal stromal cells. Cytotherapy. 10: 364–375.

Robertson TA, Maley MA, Grounds MD, Papadimitriou JM. (1993). The role of macrophages in skeletal muscle regeneration with particular reference to chemotaxis. Exp. Cell Res., 207: 321–31.

Robinson SN, Simons PJ, Yang H, Alousi AM, De Lima JM, Shpall JE (2011). Masenchcord blood exp vivo cord blood expansion . Best Prac. Clin. Haematol., 24:83-92.

Rubinstein P, Rosenfield RE, Adamson JW, Stevens CE (1993). Stored placental blood for unrelated bone marrow reconstruction. Blood, 8: 679-90.

Ryan JM, Barry F, Murphy JM, Mahon BP (2007). Interferon-gamma does not break, but promotes the immunosuppressive capacity of adult human mesenchymal stem cells. Clini. Exp. Immunol., 149: 353–363.

Sadat S, Gehmert S, Song YH, Yen Y, Bai X, Gaise S, Klein H, Alt E (2007). The cardioprotective effect of mesenchymal stem cells is mediated by IGF-I and VEGF. Biochem. & Biophys. Res. Comm., 363: 674–679.

Schulze PC, Spate U (2005). Insulin-like growth factor-1 and muscle wasting in chronic heart failure. Int. J. Biochem. Cell Biol., 37: 2023–2035.

Selmani Z, Naji A, Zidi I, Favier B, Gaiffe E, Obert L, Borg C, Saas P, Tiberghien P Rouas-Freiss N, Carosella ED, F Deschaseaux F (2008). Human leukocyte antigen-G5 secretion by human mesenchymal stem cells is required to suppress T lymphocyte and natural killer function and to induce CD4+CD25highFOXP3+ regulatory T cells. Stem Cells. 26: 212–222.

Semenov OV, Koestenbauer S, Riegel M, Zech N, Zimmermann R, Zisch AH, Malek A (2009). Multipotent mesenchymal stem cells from human placenta: critical parameters for isolation and maintenance of stemness after isolation. Am. J. Obstetri. Cynecol., 202: 193.e1-193.e13.

Shi X, Garry DJ (2006). Muscle stem cells in development, regeneration and disease, Genes Dev., 20: 1692–1708.

Sica A, Larghi P, Mancino A, Rubino L, Porta C (2008). Macrophage

polarization in tumour progression. Semin. Cancer Biol., 18: 349–355.

Spencer MJ, Walsh CM, Dorshkind KA, Rodriguez EM, Myonuclear JG (1997). Myonuclear apoptosis in dystrophic mdx muscle occurs by perforin-mediated cytotoxicity. J. Clin. Investi., 99: 2745-2751.

Spaggiari GM, Abdelrazik H, Becchetti F,_Moretta L (2009). MSCs inhibit monocyte-derived DC maturation and function by selectively interfering with the generation of immature DCs: central role of MSC-derived prostaglandin E2. Blood, 113: 6576- 6583.

Sun W, Lal P (2002). Recent development on computer aided tissue engineering - a review. Comp. Metho. Prog. Biomed., 67: 85–103.

Summan M, Warren GL, Mercer RR, Chapman R, Hulderman T, Van Rooijen N, Simeonova PP (2006). Macrophages and skeletal muscle regeneration: a clodronate-containing liposome depletion study. Ame. J. Physiol. Regul. Integ. Compa. Physiol. 290: R1488–95.

Tanaka T, Furutani S, Nakamura M, Nishida T (1999). Chpanges in extracellular matrix components after excimer laser photoablation in rat cornea. Jap. J. Ophthalmol., 43: 348-354.

Tan KH, Chua CK, Leong KF, Cheah CM, Cheang P, Abu Bakar MS, Cha SW (2003). Scaffold development using selective laser sintering of polyether etherketone hydroxy apatite bio composite blends. Biomat., 24: 3115-3123.

Tao XR, Li WL, Su J, Jin CX. (2009). Clonal mesenchymal stem cells derived from human bone marrow can differentiate into hepatocyte-like cells in injured livers of SCID mice. J. Cellul. Biochem., 108: 693–704.

Tedesco FS, Dellavalle A, Diaz-Manera J, Messina G, Cossu G (2010). Repairing skeletal muscle: regenerative potential of skeletal muscle stem cells. J. Clin. Invest., 120: 11-19.

Tervo K, van Setten GB, Beuerman RW, Virtanen I, Tarkkanen A. (1991). Expression of tenascin and cellular fibronectin in the rabbit nterior keratectomy. Immunohistochemical study of wound healing dynamics. Invest. Ophthalmol. Visual Sci., 32: 2912-2918.

Thomson RC, Yaszemski MJ, Powers J M, Mikos AG (1995). Fabrication of biodegradable polymer scaffold to engineer trabecular bone. J. Biomat. Sci. Polymer. Ed. 7: 23-38.

Tidball JG (1995). Inflammatory cell response to acute muscle injury. Medi. Sci. Sports Exercise, 27: 1022-1032.

Tidball JG (2002). Interactions between muscle and the immune system during modified musculoskeletal loading, Clini. Orthop. Relat. Res., 400: 100-109.

Tidball JG, Villalta SA (2010). Interactions between muscle and the immune system regulate muscle growth and regeneration. Am. J. Physiol., 298: R1173-1187.

Tidball JG, Berchenko E, Frenette J (1999). Macrophage invasion does not contribute to muscle membrane injury during inflammation. J. Leukoc. Biol., 65: 492–498.

Vaziri H, Dragowska , Allsopp RC, Thomas TE , Harley CB, Lansdorp PM (1994). Evidence for a mitotic clock in hematopoietic stem cells: Loss of telomeric DNA with afge. Proc. Nat. Acad. Sci., USA. 9857-9860.

van den Bos C, Mosca JD, Winkles J, Kerrigan L, Burgess WH, Marshak DR. (1997). Human mesenchymal stem cells respond to fibroblast growth factors. Human Cell, 10, 45-50.

Vasita R, Katti DS (2006). Growth factor delivery systems for tissue engineering: A materials perspective. Exp. Rev. Med. Devi. 3: 29-47.

Vetrone SA, Rodrigue ME, Kudryashova E, Kramerova I, Hoffman EP, Liu DS, Miceli MC, Pencer MJ (2009). Osteopontin promotes fibrosis in dystrophic mouse muscle by modulating immune cell subset and intramuscular TGE-beta. J. Clin. Invest., 119: 1583-1594.

Wagers AJ, Conboy IM (2005). Cellular and molecular signatures of muscle regeneration: Current concepts and controversies in adult myogenesis, Cell, 122, 659–667.

Wang Y, Wang YP, Zheng G , Lee VWS, Ouyang J, Chang DHH, Mahajan D, Coombs J, Wang Y M, Alexander SI, Harris DCH (2007). Ex vivo programmed macrophages ameliorate experimental chronic inflammatory renal disease. Kidney Internaltion., 72: 290–299.

Yamada Y, Yokoyama SC, Fukuda N, Kidoya H, Huang XY, Naitoh H, Satoh N, Takakura N (2007): A novel approach for myocardial regeneration with educated cord blood cells cocultured with cells from brown adipose tissue. Biochem. Biophys. Res. Commun., 353: 182-188

Yang YW, Bali H, Wang CB, Lin M, Wu LQ (2008). Experimental study on influence of bone marrow mesenchymal stem cells on activation and function of mouse peritoneal Macrophages, Zhonghua Xue Ye Xue Za Zhi, 29: 540–543.

Yan X, Gu P (1996). A review of rapid prototyping technologies and systems. Compputer Aided Design. 28: 307-318.

Yeong WY, Chua CK, Leong KF, Chandrasekharan M (2004). Rapid prototyping in tissue engineering: challenges and potential. Trends in Biotechnology, 22: 643-652.

Zammit PS, Heslop L, Hudon V, Rosenblatt JD, Tajbakhsh S (2002). Kinetics of myoblast proliferation shows that resident Satellite Cells are competent to fully regenerate skeletal muscle fibers. Exp. Cell Res., 281: 39-49.

Zannettino AC, Paton S, Arthur A, Khor F, Itescu S, Gimble JM, Gronthos S (2008). Multipotential human adipose-derived stromal stem cells exhibit a perivascular phenotype in vitro and in vivo. J. Cell, Physiol., 214: 413–421.

Zein I, Hutmacher DW, Tan KC, Teoh SH (2002). Fused deposition modeling of novel scaffold architectures for tissue engineering applications. Biomat, 23: 1169-1185.

Zhou DH, Huang SL, Wu Yf, Wei J, Li Y, Bao R (2003) Zhonghua Er. Ke. Za. Zhi. 41(8): 607-610.

Zhou K, Zhang H, Jin O, Feng X, Yao G, Hou Y, Sun L (2008). Transplantation of Human Bone Marrow Mesenchymal Stem Cell Ameliorates the Autoimmune Pathogenesis in MRL/lpr Mice Cell Molecular Immunology, 5: 417-424.

8

The role of biotechnology towards attainment of a sustainable and safe global agriculture and environment – A review

Soetan, K.O.

Department of Veterinary Physiology, Biochemistry and Pharmacology, University of Ibadan, Nigeria. E-mail: soetangboye@yahoo.com.

Biotechnology is producing great opportunities for the increase in global agricultural production and for protecting the environment through the reduced use of agro-chemicals like pesticides, fertilizers and rodenticides. Biotechnology has played an important role towards the attainment of environmental sustainability by using environment-friendly crops such as insect-resistant, herbicide-tolerant species and crops that can fix nitrogen leading to purification of the environment. Increasing global food production within existing land area and the use of modern plant breeding methods have enhanced increased production of crops like legumes to improve soil structure, organic matter and fertility. These lead to conservation of bioresources and prevent soil erosion. Some beneficial effects of livestock production on the environment are also discussed. However, fears and concerns about the environmental consequences of biotechnology are also discussed. The overall aim of this review is to emphasize the importance of biotechnology towards attaining a safe and sustainable environment for increased global agricultural production.

Key words: Biotechnology, environmental safety, agricultural production, heavy metal, pollution.

INTRODUCTION

Biotechnology can simply be defined as a technique that uses living organisms to make or modify and improve products (Olatunji, 2007). Biotechnology can also be defined as any technological application that uses biological systems, living organisms or derivatives thereof to make or modify products or processes for specific use (UNCBD, 1992). Traditionally, micro-organisms have been deliberately used to produce beverages and fermented foods (Olatunji, 2007).

Environmental biotechnology is the application of biotechnology to the study of natural environment. It can also imply trying to harness biological processes for commercial uses and exploitation (Wikipedia.org). The International Society for Environmental Biotechnology defines Environmental Biotechnology as the development, use and regulation of biological systems for remediation of contaminated environments (land, air, water) and for environment-friendly processes (green manufacturing technologies and sustainable development). It can also be described as "the optimal use of nature, in the form of plants, animals, bacteria, fungi and algae, to produce renewable energy, food and nutrients in a synergistic integrated cycle of profit-making processes where the waste of each process becomes the feedstock for another process" (Wikipedia.org).

Environmental biotechnology plays an important role in agroecology in the form of zero waste agriculture and most significantly through the operation of over 15 million biogas digesters worldwide (Wikipedia.org; Zylstraa and Kukor, 2005; Vidya, 2005). Agroecology is the application of ecological principles to the production of food, fuel, fibre and pharmaceuticals. The term encompasses a broad range of approaches and is considered "a science, a movement and a practice" (Wezel et al., 2009).

The roles of biotechnology in improving agricultural productivity and environmental conditions, removal of toxic chemicals and heavy metal pollution from the environment, desulphurization of fossil fuels, ecosystem modeling, control of oil spillage and saving of resources and energy will be discussed. Also, the fears and concerns about biotechnology approach to achieving a safe environment and agriculture will be mentioned in this review.

BIOTECHNOLOGY, AGRICULTURE AND ENVIRONMENTAL POLLUTION

Agriculture is the use of natural resources base for the improvement and increase in production of crops, livestock, fish and trees (Anderson, 1991; Ene-Obong, 2007a). In Agricultural biotechnology, improvement is accelerated and production is increased, using updated knowledge of living organisms including the genetic code. These include well-established conventional techniques as in biological pest control, fermentation, and production of vaccines and biofertilizers as well as modern techniques like tissue culture, genetic engineering (GE) also called genetic modification, recombinant DNA technology (rDNA), crop and animal transformation as a result of transgenesis (Ene-Obong, 2003). The importance of these new technologies like biotechnology in food security, environmental sustainability and economic development was captured at the United Nations General Assembly in 2005 (Ene-Obong, 2007b).

Global industrial explosion which is intended to cater for the needs of the world's increasing population is always associated with environmental pollution (Okpokwasili, 2007). Pollution occurs as a result of improper management of industrial by-products, their accumulation in the environment beyond acceptable limits therefore causes hazard and or nuisance to man. The industrial by-products that are pollutants may be either organic or inorganic compounds (Okpokwasili, 2007). Man's environment is composed of abiotic and biotic components (Okpokwasili, 2007).

Developing countries are faced with the challenge of rapidly increasing agricultural productivity to help feed their growing populations without depleting the natural resource base (Rege, 1996). In many African countries, agriculture is still subsistent and primitive and this raises concerns on food security, deforestation, rapid population growth, environmental protection, poor soils, stressed environments, unfavourable climatic conditions and improved crops and livestock (Ene-Obong, 2007a).

For instance, an environment in which pollution of a particular type is maximum. The effluents of a starch industry mixing up with a local water body like a lake or pond. These cause huge deposits of starch which are not easily degraded by micro-organisms except for a few exceptions. Through genetic engineering, a few micro-organisms were isolated from the polluted site and scanned for any significant changes in their genome like evolutions or mutations. The modified genes were then identified because the isolate would have adapted itself to utilize/degrade the starch better than other microbes of the same genus. As a result, the resultant genes are cloned onto industrially significant micro- organisms which are used for economically significant processes like fermentation and it can also be applied in pharmaceutical industries (Wikipedia.org).

Another case study is the incidence of oil spills in the oceans which require cleanup, microbes isolated from oil rich environments like oil transfer pipelines, oil wells etc have been discovered to have the potential to degrade or use it as an energy source and thus serve as a remedy to oil spills. Still another case study is the case of microbes isolated from pesticide rich soils. These microbes have the potential to utilize the pesticides as a source of energy and so when mixed along with bio-fertilizers, they would serve as a good insurance against increased pesticide-toxicity levels in agricultural processes (Wikipedia.org).

However, there are counter arguments that the newly introduced micro-organisms used for cleanup of oil spillage could create an imbalance in the natural environment concerned. There are also concerns that the mutual harmony in which the organisms in that particular environment existed may be altered and extreme caution should be taken so as not to disturb the mutual relationships that already existed in the environment to which these newly discovered and cloned micro-organisms are introduced. This leads to a suggestion that the positive and negative environmental consequences of environmental and agricultural biotechnology needs to be promptly addressed.

CHALLENGES IMPOSED ON THE ENVIRONMENT BY HUMAN ACTIVITIES

Human activities constitute one of the major means of introduction of heavy metals into the environment. One of the major development challenges facing this decade is how to achieve a cost effective and environmentally sound strategies to deal with the global waste crisis facing both the developed and developing countries (Parker and Corbilt, 1992; Jensen,1990; NEST, 1991; Oyediran,1994; Alloway and Aryes, 1997). The crisis has threatened the assimilative and carrying capacity of the earth, which is our life support system. Although the nutrient content of wastes makes them attractive as fertilizers, land application of many industrial wastes and sewage is constrained by the presence of heavy metals, hazardous organic chemicals, salts, and extreme pH (Cameron et al., 1997). Heavy metal pollution of the environment, even at low levels, and their resulting long-term cumulative health effects are among the leading health concerns all over the world. For example, the bioaccumulation of Pb in human body interferes with the functioning of mitochondria, thereby impairing respiration, and also causes constipation, swelling of the brain, paralysis and eventual death (Chang, 1992). This problem is even more pronounced in the developing countries where research efforts towards monitoring the environment have not been given the desired attention by the stake holders. Heavy metals concentration in the environment cannot be attributed to geological factors alone, but human activities do modify considerably the mineral composition of soils, crops and water. The recent population and industrial growth has led to increasing production of domestic, municipal and industrial wastes, which are indiscriminately dumped in landfill and water bodies without treatment. Ogunyemi et al. (2003) reported that the use of dumpsites as farm land is a common practice in urban and sub-urban centers in Nigeria because of the fact that decayed and composted wastes enhance soil fertility and these wastes often contain heavy metals in various forms and at different contamination levels. Some of these heavy metals like As, Cd, Hg and Pb are particularly hazardous to plants, animals and humans (Alloway and Ayres, 1997).

Municipal waste contains such heavy metals as As, Cd, Co, Cu, Fe, Hg, Mn, Pb, Ni, and Zn which end up in the soil as the sink when they are leached out from the dump sites. Soil is a vital resource for sustaining two human needs of quality food supply and quality environment (Wild, 1995). Plants grown on a land polluted with municipal, domestic or industrial wastes can absorb heavy metals in the form of mobile ions present in the soil solution through their roots or through foliar absorption. These absorbed metals get bioaccumulated in the roots, stems, fruits, grains and leaves of plants (Fatoki, 2000). Plants are known to take up and accumulate heavy metals from contaminated soils (Madejon et al., 2003). The consumption of such plants could particularly be hazardous because the accumulated metals in edible plants may end up in human food chain with the attendant adverse effects on human and animal health. A promising cost-effective plant-based technology for the cleanup of heavy metal pollution is phytoremediation. Lombi et al. (2001) stated that this technology has attracted attention in recent years because of the low cost of implementation and is particularly attractive in the tropics, where normal climatic conditions favour plant growth and microbial activity. Plants that sprout and grow in metal laden soils are tolerant to metal pollution in soil and are 'candidates' for remediation strategies and management for heavy metals contaminated soils. Fulekar (2004) reported that in recent decades, the mangrove forests have been affected mainly by human developmental activities, pollution discharge from industrial and domestic waste resulting into impact on the mangrove forests and coastal aquatic ecosystem. The potential environmental and public health impacts of biotechnology approach to livestock production have been reviewed by Soetan and Oluwayelu (2011).

CONTRIBUTIONS OF BIOTECHNOLOGY TO IMPROVED AGRICULTURAL PRODUCTIVITY

Biotechnology is regarded as a means to the rapid increase in agricultural production through addressing the production constraints of small-scale or resource-poor farmers who contribute more than 70% of the food produced in developing countries (Rege, 1996). Biotechnology is

applicable to all areas and fields of human endeavours. The dynamic and ubiquitous nature of biotechnology has been reviewed by Soetan (2008a). Agricultural biotechnology as the solution to the problem of global food insecurity has also been reviewed by Soetan (2008b). Agricultural biotechnology has the potential to address some of the problems of developing countries like food insecurity, unfavourable environmental and climatic conditions etc mentioned above and also improve agricultural productivity. Agricultural biotechnology has provided animal agriculture with safer, more efficacious vaccines against pseudo rabies, enteric collibacilosis and foot-and-mouth disease (FMD) (Stenholm and Waggoner, 1992). Disease detection in crops and animals are more efficiently and rapidly done using DNA probes. Biotechnology as a key tool to breakthrough in medical and veterinary research has been reviewed by Soetan and Abatan, (2008).

Crops are now routinely genetically modified for insect and pest resistance, delayed ripening, herbicide tolerance and maximal production under stressed environments. Molecular mapping of crops and farm animals has markedly cut down breeding time and enhanced man's understanding and manipulation of genes (Ene-Obong, 2007b). Application of modern technology to agriculture as a catalyst to sustainable food production and industrial growth in Nigeria has been reviewed (Soetan, 2008c).

Nutrition is one of the most serious limitations to livestock production in developing countries, especially in the tropics (Rege, 1996). Plants generally contain antinutrients acquired from fertilizers, pesticides and several naturally-occuring chemicals (Igile, 1996). Some of these chemicals are known as secondary metabolites and they have been shown to be highly biologically active (Zenk, 1991). Examples of these secondary plant metabolites are saponins, tannins, flavonoids, alkaloids, oxalates, phytates, trypsin (protease) inhibitors, cyanogenic glycosides etc. Some of these chemicals have been shown to be deleterious to health or evidently advantageous to human and animal health if consumed at appropriate amounts (Kersten et al., 1991; Sugano et al., 1993). These antinutritional factors affect the overall nutritional value of human foods and animal feeds (Osagie, 1998). Some of these plant components have the potential to precipitate adverse effects on the productivity of farm livestock (D'Mello, 2000). Conventional plant breeding methods has been used to reduce and in some cases, eliminate such antinutritive factors (ANF) (Rege, 1996). An example is the introduction of cultivars of oilseed rape which are low in or free from erucic acid and glucosinolates.

A combination of genetic engineering and conventional plant breeding methods could lead to substantial reduction or removal of the major antinutritive factors in plant species of importance in animal feeds (Rege, 1996). Transgenic rumen microbes could also play a role in the detoxification of plant poisons or inactivation of antinutritional factors (Rege, 1996). Successful introduction of a caprine rumen inoculum obtained in Hawaii into the bovine rumen in Australia to detoxify 3 hydroxy 4(IH)pyridine (3,4 DHP), a breakdown product of the non-protein amino acid mimosine found in Leucaena forage.

Jones and Megarrity (1986) demonstrate the possibilities. However, the pharmacological and other beneficial effects of these anti-nutritional factors in plants have been reviewed by Soetan (2008d).

In animal production, biotechnology techniques applied include gene cloning, embryo transfer, artificial insemination, milk modification etc. In animal health, biotechnology techniques are used for the fast and accurate diagnosis and treatment of diseases. Gene therapy, vaccine production, production of recombinant pharmaceuticals etc are examples (Soetan and Abatan, 2008; Soetan, 2009). According to Ene-Obong (2007b), biotechnology can help promote sustainable and safe agriculture and environment respectively globally in two ways:

1. By increasing food production within existing land area under plough, making it unnecessary to use marginal land or environmentally-sensitive methods and areas. This leads to conservation of bioresources thereby avoiding soil erosion.
2. Using environment-friendly crops such as insect-resistant, herbicide tolerant species, as well as crops that can fix nitrogen lead to purification of the environment. Consequently, less chemicals like pesticides, herbicides and synthetic nitrogen fertilizers are used.

Agricultural biotechnology has long been a source of innovation in the production and processing of agricultural products and has profound impact on the livestock sector (Jutzi et al., 2003). Improved agricultural technology as the catalyst to sustainable food production and industrial growth in Nigeria has been reviewed by Soetan (2008a). Globally, if hunger and malnutrition, is to be reduced drastically, agriculture must be tailored to meet the future demands of increased population. The increase in human population increase the demand for land, space and available resources and primitive agricultural practices cause desertification, environmental pollution and produces resultant effects on climate, ecosystems, biogeochemical cycles and human health (Ene-Obong, 2007b). Sustainable agricultural practices targeted towards improved agricultural productivity, under clean, safe and environment- friendly conditions must be introduced into the global agricultural system, in order to reduce the adverse effects of environmental pollution on human health and the climate (global warming) (Soetan, 2008c). The new techniques of biotechnology provide innovations that complement the weaknesses of conventional agricultural practices and should be adopted for increased food production (Ene-Obong, 2003; 2005; 2007a).

BIOTECHNOLOGY AND LIVESTOCK PRODUCTION IN THE IMPROVEMENT OF ENVIRONMENTAL CONDITIONS

Livestock recycle nutrients on the farm, produce valuable output from land that is not suitable for sustained crop production and provide energy and capital for successful farm operations (Delgado et al., 1999). Livestock can also help maintain soil fertility in soils lacking adequate organic content or nutrients (Ehui et al., 1998). Adding animal manure to the soil increases the nutrient retention capacity (or cation-exchange capacity), improves the soil's physical condition by increasing its water-holding capacity and improves soil structure (Delgado et al., 1999). Animal manure also helps maintain or create a better climate for micro- flora and fauna in soils. Grazing animals improve soil cover by dispersing seeds, controlling shrub growth, breaking up soil crusts and removing biomass that otherwise might be fuel for bush fires (Delgado et al., 1999). These activities stimulate grass tilling and improve seed germination and thus improve land quality and vegetation growth. Livestock production also enables farmers to allocate plant nutrients across time and space by way of grazing to pro-duce manure, land that cannot sustain crop production. This makes other land more productive (Delgado et al., 1999). Grazing livestock can also accelerate transforma-tion of nutrients in crop by-products to fertilizer, thus speeding up the process of land recovery between crops. As disease constraints are also removed, large breeds of livestock can be integrated into crop operations for providing farm power and manure (Delgado et al., 1999). Biotechnology has enhanced increased animal produc-tion through Artificial insemination (AI) and also improved animal health and disease control through the production of DNA recombinant vaccines (Soetan and Abatan, 2008).

BIOTECHNOLOGY AND THE REMOVAL OF TOXIC CHEMICALS FROM THE ENVIRONMENT

Micro-organisms have broadened the environments they live in by evolving enzymes that allow them to metabolize numerous man-made chemicals (that is, xenobiotics) (Okpokwasili, 2007). Bioremediation is the use of micro-organisms or microbial processes to detoxify and degrade environmental contaminants. Micro-organisms have been used for the routine treatment and transforma-tion of waste products for several decades (Okpokwasili, 2007). The fixed-film and activated sludge treatment systems depend on the metabolic activities of micro-orga-nisms which degrade the wastes entering the treatment facility. Specialized waste treatment plant containing selected and acclimated microbial populations are often used to treat industrial effluents (Okpokwasili, 2007). The innovation in bioremediation has been applied to the remediation of soils, groundwater and similar environmental media (Okpokwasili, 2006 a, b).

Bioremediation techniques depend on having the right micro-organisms in the right place with the right environmental condition for degradation to occur. The right micro organisms are those bacteria and fungi which possess the physiological and metabolic capability to degrade the contaminants (Okpokwasili, 2007). Already, bacteria with natural abilities to digest certain chemicals are being used to clean up industrial sites (Su, 1998). By means of genetic engineering, biotechnology has brought about the rapid production of bacteria.

BIOTECHNOLOGY AND THE REMOVAL OF HEAVY METAL POLLUTION FROM THE ENVIRONMENT

Basically, the heavy metals of environmental interest include mercury, vanadium, nickel, cobalt, lead, cadmium, chromium, tin etc (Okpokwasili, 2007). Some harmful compounds that cause serious environmental pollutions and disaster like Dichlorodiphenyltrichloroethane (DDT) and lead (Pb) could be safely removed by means of genetic engineering of bacteria manufactured for that purpose. The ability of micro-organisms to accumulate metals and their potential use in the decontamination of environments impacted by toxic metals has been reported by (Kelly et al. (1979) and Aiking et al. (1985).

Micro-organisms remove toxic metals by various mechanisms such as adsorption to cell surfaces, complexation of exopolysaccharides, intracellular accumulation, biosynthesis of metallothionins and other proteins that trap metals and transform them to volatile compounds (Bitton and Freichoffer, 1978; Highman et al., 1984; Meissner and Falkinham, 1984; Mullen et al., 1989). *Micrococcus luteus and Azotobacter sp.* have been shown to immobilize large quantities of lead from sites containing high concentrations of lead salts, without a detectable effect on viability (Tornabene and Edwards, 1972). Volatization of mercury by *Klebsiella aerogenes* has also been reported (Magos *et al.*, 1964). Uranium, copper and cobalt could be removed by polyacrylamide-immobilized cells of *Streptomyces albus*. Microbial processes can also mediate the precipitation of metals from aqueous solutions. Certain bacteria extracellular products may interact with free or absorbed metal cations forming insoluble metal precipitates (Okpokwasili, 2007). The major mechanism involved in such precipitation is through the formation of hydrogen sulphide and the immobilization of metal cations as metal sulphides. Certain fungi that produce oxalic acid (oxalates) facilitate the immobilization of metals such as metal oxalate crystals (Okpokwasili, 2007). Microbes can also catalyze a range of metal transformations which are useful for waste treatment. These transformations include oxidation, reduction and alkylation reactions. Bacteria, fungi, algae

or protozoa, in the oxidation reactions, can deposit ferrous and manganese ions. *Geobacter metallireducens* remove uranium, a radioactive waste, from drainage waters in mining operations and from contaminated ground waters (Okpokwasili, 2007).

BIOTECHNOLOGY AND DESULPHURIZATION OF FOSSIL FUELS

The removal of inorganic sulphur from coal is mediated by microbial oxidation of sulphur (Okpokwasili, 2007). The direct oxidation of inorganic sulphur by *Thiobacillus sp.* is a membrane-bound reaction and requires direct contact of the substrate with the bacterium. As a result of this, the attachment of the culture to coal particle is the absolute requirement. Mixed and pure cultures of a variety of micro-organisms (heterotrophic bacteria) can be used to remove organic sulphur from coal and oil (Okpokwasili, 2007). However, sulphur removal has also been reported under anaerobic microbial action (Fligwe, 1988).

BIOTECHNOLOGY AND ECOSYSTEM MODELLING

An ecosystem consists of producers, consumers, decomposers and detritivores and their physical environment, all interacting through energy flow and materials recycling (Starr and Taggart, 1995). A food web is a network of crossing, interlinked food chains involving primary producers, consumers and decomposers (Starr and Taggart, 1995).

Disturbances to one part of an ecosystem can have unexpected effects on other, seemingly unrelated parts (Starr and Taggart, 1995). Ecosystem modeling is an approach to predict unforeseen effects. By this method, researchers identify crucial bits of information about different ecosystem components. They use computer programs and models to combine the information and then use the resulting data to predict the outcome of the next disturbance. Biotechnology techniques like bioinformatics are useful in ecosystem modeling.

Bioinformatics deals in gene database management, gene mapping, coding, sequence alignment etc (Abd-Elsalam, 2003; Olukosi, 2006).

MICROBIAL BIOTECHNOLOGY IN THE MONITORING OF ENVIRONMENTAL POLLUTION

A number of microbial parameters are used for the detection and monitoring of pollutants, especially in water bodies (Okpokwasili, 2007). Some micro-organisms serve as indicators of organic pollution while others serve as indicators of inorganic pollution.

Some of the parameters used for the monitoring of

organic pollution are heterotrophic bacteria, total and faecal coliforms and faecal streptococci. Parameters used for monitoring inorganic pollution include nitrifying bacteria, sulphur oxidizing bacteria, sulphate reducing bacteria (SRB), iron bacteria etc (HTC, 1993; Odokuma and Okpokwasili, 1997). The presence of faecal coliforms in numbers above the World Health Organization (WHO) standard for portable water is indicative of faecal contamination of human origin (Okpokwasili, 2007).

Heterotrophic micro-organisms are organisms that derive their energy from the oxidation of organic molecules. Their presence in large numbers in aquatic systems is indicative of organic pollution. The presence of biodegradable carbon sources supports the proliferation of heterotrophs in aquatic systems. In the case of xenobiotics, the few species that can degrade them may produce by-products during metabolism that may support other microbial species (Okpokwasili, 2007). Thus, a high heterotrophic microbial count is suggestive of high level of organics in aquatic system while low count is suggestive of either a low level of organic pollution or the presence of persistent organic matter within the aquatic system (Okpokwasili, 2007).

MICROBIAL BIOTECHNOLOGY IN THE BIOASSAY OF ENVIRONMENTAL TOXICITY

Toxic industrial wastes are a threat to both the biological waste treatment systems and the environment of their ultimate disposal (Okpokwasili, 2007). As a result, bioassays are very necessary to generate data that could be used for the prediction of environmental effects of waste and regulation of discharges (Okpokwasili, 2007). Although fishes have been the most popular test organisms, standard organisms for aquatic bioassays also include phytoplankton, zooplankton, molluscs, insects and crustaceans (Wang and Reed, 1983; APHA, 1998). The use of microbes (especially bacteria) as bioassay organism is gaining wide acceptance and offers a number of advantages over the standard organisms (Williamson and Johnson, 1981; Wang and Reed, 1983). Bacteria are easily handled and require relatively small space for culturing and/or testing, compared with other bioassays.

Moreover the short life cycle means fast experimental results, thus enabling the laboratory to process more samples (Okpokwasili, 2007). The simple and rapid bacteria bioassay techniques include Nitrobacter assay, Microtox tests, the Toxi-chromotest and the Ames/Salmonella test (Ames et al, 1975; Williamson and Johnson, 1981; Bullic, 1984; Dutton et al., 1990; Okpokwasili and Odokuma, 1993, 1996a, 1996b). The Nitrobacter bioassay relies on the quantification of Nitrobacter activity determined by measuring the toxicant effect on the rate of nitrite utilization (Okpokwasili, 2007).

Photobacterium phosphoreum is the basis of the Microtox assay, toxichromotest is based on the inhibition of beta

galactosidase biosynthesis in E. coli or biosynthesis of enzymes, such as tryptophanse and alpha-glucosidase, under the control of operons other than the Lac operon by environmental pollutants. The Ames/Salmonella assay measures the mutagenic activities of pollutants. It involves the detection of histidine-negative, ampicillin-sensitive and ultra-violet (UV) resistant revertants in frame shift and base pair mutations of Salmonella TA 1537, Ta 1538 and TA 98 strains (Ames et al., 1973).

BIOTECHNOLOGY AND CONTROL OF OIL SPILLAGE

Micro-organisms can now be genetically engineered for use in oil recovery, pollution control, mineral leaching and recovery (Daini, 2000). In the petroleum industry, micro organisms can also be genetically engineered to produce chemicals useful for enhanced oil recovery (Daini, 2000). Cleaning up oil spills could in the future be left to genetically- engineered bacteria (Su, 1998). In the mining industries, micro-organisms with the property of enhanced leaching ability could be designed. Micro-organisms can bind metals to their surfaces and concentrate them internally. As a result of this, genetically improved strains can be used to recover valuable metals or remove polluting metals from dilute solution as in industrial waste (Daini, 2000). Research is already being carried out to improve the naturally-occurring bacteria that can 'eat oil', for use following an oil spill. By applying bacteria to oil-covered beaches, the complex oil molecules would be broken down into harmless sugars (Su, 1998).

Many micro-organisms can degrade various kinds of environmental pollutants into relatively harmless materials before the death of the micro-organisms. This property could also be used in overcoming the environmental hazards of DDT, lead and other environmental pollutants like toxic wastes globally (Soetan, 2008c). Strains of bacteria which can degrade fuel hydrocarbons have been designed and the use of genetically engineered micro-organisms to clean up oil spillages or treat sewages has been proposed and is undergoing production/manufacturing.

BIOTECHNOLOGY AND THE SAVING OF RESOURCES AND ENERGY

Breeding of insect and pest–resistant crop strains help promote a safe environment, saves money and conserve resources (Su, 1998). Industrial processes are very complex and chemists make use of inorganic catalysts which speed up the rate of reaction when making new chemicals (Su, 1998). These catalysts often need high temperatures, and acid or alkaline conditions, in order to work efficiently. In future, genetically engineered organisms may be able to work effectively at lower temperatures, and require less extreme conditions. This will save money and resources, and will also produce fewer

fewer hazardous by-products (Su, 1998). For example in paper-making, the wood pulp has to be treated with chemicals which break up the fibres and remove the lignin (the substance that makes up the wood). The pulp is bleached so that the finished paper is white. This process produces a large volume of chemical waste that has to be treated before it is ready for disposal (Su, 1998). Enzymes have been discovered in fungi which may be suitable for use as biological alternatives to some of these chemicals (Su, 1998). In the near future, using the modern plant breeding techniques, it may be possible to breed trees which have less lignin, and so require fewer chemicals and less energy to produce the pulp. Plastic is made from oil and its manufacture uses a lot of energy and produces a lot of polluting by-products (Su, 1998). There is now hope that some forms of plastic will be made by living organisms. One biodegradable plastic, called Biopol (trade name) is made by bacteria (Su, 1998). One way to make larger quantities of this plastic at lower cost might be to insert the gene into potatoes. This would save energy, and reduce both cost and pollution.

As the supply of fossil fuels (oil, gas and coal) dwindles as a result of the global financial crisis, genetically engineered organisms may be manufactured to produce far more materials like plastics at less energy, reduced cost and minimal environmental pollution.

FEARS AND CONCERNS ABOUT BIOTECHNOLOGY APPROACH TO ACHIEVING A SAFE ENVIRONMENT AND AGRICULTURE

There are some fears and concerns about biotechnology and safe environment and agriculture. For instance, plant breeding, an agricultural biotechnology approach has some concerns. Genetically engineered organisms are living things and so are much less predictable than artificial materials and chemicals (Su, 1998). They can reproduce, move and even mutate. Developments in genetic engineering take place in carefully controlled laboratory conditions. However, once a new or modified organism has been developed, it is likely to be grown outside and once released into the environment, it cannot be recalled. The organism could change or interbreed with others, creating new specie. Geneticists should therefore be cautious and assess the possible risks involved so that genetically engineered plants cause no more harm than the chemicals they are replacing (Su, 1998). It has been found that genetic alteration of plants to resist viruses can stimulate the virus to mutate into a more virulent form, one that might even attack other plant species (Su, 1998). If the genes for insect- and weed-killer resistance, introduced into crop plants, find their way into weeds, this could result in development of super-weeds which would be impossible to kill using traditional weed killers. There are also concerns that the genetically engineered soyabeans poses an unacceptable risk to human health and the environment (Su, 1998).

There are many questions about this soyabeans, some of which are:

1. Will the new bean force out other plants?
2. Will it enter other ecosystems?
3. Will it genetically contaminate wild relatives or traditional strains?
4. Will it change in the long term due to its resistance to toxic substances?

Some scientists think that genetically engineered plants and animals will threaten the survival of other species and reduce diversity (the number of different plant and animal species) (Su, 1998). The genetically engineered plants, which may be more resistant to disease and pests, when grown around the world, may cross to the wild species. The wild population would be contaminated which could affect local habitats and the species that grow there. For example, oilseed rape cross-breeds easily with wild relatives. This may mean that genetically engineered oilseed rape would breed with related plants, and the new gene for resistance to weed killer could spread into the wild population. Some scientists predict that, within just one year, a large percentage of weeds growing near the crop would have acquired this gene (Su, 1998). There is no way of stopping genetically engineered crops from breeding with wild plants.

FEARS AND CONCERNS ABOUT ENVIRONMENTAL IMPACT OF BIOREMEDIATION

Bioremediation is the use of micro-organisms or microbial processes to detoxify and degrade environmental contaminants (Okpokwasili, 2007). Micro-organisms have been used for the routine treatment and transformation of waste products for several decades. Although bioremediation represents a promising and largely untapped environmental biotechnology, it has some disadvantages. Additives added to enhance the functioning of one particular group of micro-organisms during *in situ* bioremediation, may be disruptive to other organisms inhabiting that same environment. Also, stimulated microbial population or genetically-modified micro-organisms introduced into the environment after a certain point of time may become difficult to remove. Bioremediation is generally very costly and labour- intensive, and can take several months for the remediation to reach acceptable levels (Okpokwasili and Oton, 2006).

CONCLUSION

The way these fears and concerns about application of biotechnology to achieving a safe environment and agriculture are addressed will have a remarkable impact on the future of biotechnology. A detailed analysis of both the advantages and the disadvantages would assist in directing the future of environmental and agricultural biotechnology, since the overall goal is to achieve a safe environment and improved agricultural productivity.

REFERENCES

Abd-Elsalam K.A (2003). Bioinformatic tools and guideline for PCR primer design. African J. Biotechnol., 2 (5): 91-95, Available online at www.academicjournals.org/AJB.

Aiking H, Covers H, Van Riet J (1985). Detoxification of medrcury, cadmium snd lead in Klebsiella aerogenes NCTC 48 growing in continous culture. Appl. Environ. Microbiol., 50: 1262-1267.

Alloway BJ, Ayres DC (1997). Chemical Principles of Environmental pollution, Blackie Academic and Professional. pp. 353-359.

Ames BN, Lee FO, Durston WE (1973). An improved bacterial test system for the detection and classification of mutagens and carcinogens. Proc. Natl. Acad. Sci. USA., 70: 782-786.

Ames BJ, McCann J, Yamasaki E (1975). Methods for detecting carcinogens and mutagens with the Salmonella/mammalian microsome mutagenicity test. Mutation Res., 31: 347-363.

Anderson WT (1991). The past and future of Agricultural Biotechnology. In: Agricultural Biotechnology at the Cross roads (ed. MacDonald June F.) NABC, National Agricultural Biotechnology Council, Ithaca, New York. Report, 3: 53-64

*APHA (1998). Standard Methods for the Examination of Water and Wastewater. 20th Edition. APHA, AWWA and WEF, Washington, DC, page number

Bitton G, Frechoffer V (1978). Influence of extracellular polysaccharides on the toxicity of copper and cadmium towards K. aerogenes. Microb. Ecol., 4: 119-125.

*Bulic AA (1984). Microtox- A bacterial toxicity test with several environmental applications. In: Toxicity Screeing Procedures using Bacterial Systems, Liu V and Dukta BJ (Eds). Marcell Dekker, New York.

Cameron KC, Di HJ, McLaren RG (1997). Is soil an appropriate dumping ground for our wastes? Aust. J. Soil Res., 35: 995-1035.

Chang LW (1992). The Concept of Toxic Metal / Essential Element interactions as a common biomechanism underlying metal toxicity. In: Toxins in Food. Plenum Press, New York, p.61.

Daini OA (2000). Fundamentals of Genetic Engineering. Samrol Ventures and Printing Co.,Ijebu-Igbo, Ogun State, Nigeria,

Delgado C, Rosegrant M, Steinfeld H, Ehui S, Courbois C(1999). Livestock to 2020: The next Food revolution. Food, Agriculture, and the Environment. Discussion Paper 28. IFPRI/FAO/ILRI,IFPRI, Washington, D.C.

D'Mello JPF (2000). Anti-nutritional factors and mycotoxins in J.P.F. D'Mello ed. Farm animal metabolism and nutrition, Wallingford, U.K.,CAB International, pp. 383-403.

Dutton RM, Bitton G, Koopman B, Aami O (1990). Effect of environmental toxicants on enzyme biosynthesis. A comparison of ß-galactosidase, α-glucosidase and tryptophanase. Arch. Environ. Contamin. Toxicol. 19: 395-398.

Ehui S, Li-Pun H, Mares V, Shapiro B (1998). The role of livestock in food security and environmental protection. Outlook on Agriculture 27(2):81-87.

Ene-Obong EE 2003). Current Issues in Agricultural Biotechnology. Quarterly Public Lecture of the Nigerian Academy of Science, p. 83.

*Ene-Obong EE (2005). Institutional and Public concerns in Crop Genetics, Transformation and Breeding. Keynote Address at the 30th Annual Conference of the Genetics Society of Nigeria, University of Nigeria, Nsukka.

Ene-Obong EE (2007a). Tailoring Tropical African Agriculture Towards the Millenium Development Goals: A Plant Breeder's Perspective. Third Inaugural Lecture, Michael Okpara University of Agriculture, Umudike, p. 40.

*Ene-Obong EE (2007b). Achieving the Millenium Development Goals (MDGS) in Nigeria: The Role of Agricultural Biotechnology. Proc. of the 20th Annl. Conf. Biotechnology Society of Nigeria (BSN), 14th – 17th Nov, 2007 at the Ebonyi State University, Abakaliki, Nigeria.

Fatoki OS (2000). Trace zinc and copper concentration in road side

vegetation and surface soils: A measurement of local atmospheric pollution in Alice, South Africa. Int. J. Env. Stud., 57: 501-513.

Fligwe CA (1988). Microbial desulfurization of coal. Fuel, 67: 451-458.

Fulekar MH (2004). Effects of human activities on mangroves ecosystem; International Conference on Biogeochemistry of Estuaries- Mangroves and the Coastal Management, Book of Abstracts, Envis Centre JNU, 9: 15.

Highman DP, Sadler PJ, Scawen MD (1984). Cadmium resistant P. putida synthesizes novel cadmium proteins. Science 225: 1043-1046.

HTC (1993). Environmental Management in the Oil and Petrochemical Industries. Hybrid Technologies Limited, Lagos.

Igile GO (1996). Phytochemical and Biological studies on some constituents of Vernonia amygdalina (compositae) leaves. Ph.D thesis, Department of Biochemistry, University of Ibadan, Nigeria.

Jensen P (1990). Sorting and solution to waste source 2(2) United Nations Development Programme (UNDP), New York.

Jutzi SC, Otte J, Wagner HG (2003). The potential impact of biotechnology on the Global livestock sector. Food and Agricultural Organisation (FAO) of the United, Nations (UN), Rome. Kelly DP, Norris PR, Brierley CL (1979). Microbial methods for the extraction and recovery of metals. In: Microbial Technology: Current State, Future Prospect. Bull AT, Ellowood DC, and Ratted C (Eds). Cambridge University Press. Cambride.

Kersten GF, Spiekstra A, Beuvery EC, Crommelin DJ (1991). On the structure of immune-stimulating saponin-lipid complexes (iscoms). Biochimica et. Biophysica. Acta., 1062 (2): 165-171.

Lombi E, Zhao FJ, Dunham SJ, McGrath SP (2001). Phytoremediation of heavy metal-contaminated soils: Natural hyperaccumulation versus chemically enhanced phytoextraction. J. Environ. Qual., 30: 1919-1926.

Madejon P, Murillo JM, Maranon T, Cabrera F, Soriano MA (2003). Trace Element and nutrient accumulation in sunflower plants two years after the Aznalcollar mine spill. Sci. Total Eviron., 307 (1-3) : 239-257.

Magos I, Tuffery AA, Clarkson TW (1964). Volatilization of mercury by bacteria. Br. J. Ind. Med., 21: 294-298.

Meissner PS, Falkinham IO (1984). Plasmid-encoded mercuric reductase in Mycobacterium scrofulaceum. J. Bacteriol., 157: 669-672.

Mullen MD, Wolf DC, Ferris FG, Beveridge TJ (1989). Fleming CA, and Barley GW: Bacterial sorption of heavy metal. Appl. Environ. Microbiol., 55: 3143-3149.

NEST (1991). Nigeria's Threatened Environmental, National Profile of Nigerian Environmental Study / Action TEAM (NEST), Ibadan, pp. 58-69.

Ogunyemi S, Awodoyin RO, Opadeji T (2003). Urban agricultural production: heavy metal contamination of Amaranthus cruenties L. grown on domestic refuse landfill soil in Ibadan, Nigeria. Emir. J. Agric. Sci., 15(2): 87-94.

Okpokwasili GC (2006a). Bioremediation, Potential Entrepreneurship opportunities and the problem of oil spills in the Niger Delta. A paper presented at the UNUINRA/NABDA Training programme on Plant Taxonomy, Systematics and Indegenous Bioresources Management, 7th – 25th August, 2006, University of Nigeria, Nsukka.

Okpokwasili GC (2006b). Microbes and the Environmental Challenge. Inaugural Lecture Series No. 53, University of Port Harcourt, Uniport Press, Port Harcourt.

Okpokwasili GC (2007). Biotechnology and Clean Environment. Proc. of the 20th Annl. Conf. of the Biotechnology Society of Nigeria (BSN), 14th – 17th, November, 2007 at the Ebonyi State University, Abakaliki, Nigeria.

Okpokwasili GC, Odokuma LO (1993). Tolerance of Nitrobacter specie to toxicity of some Nigerian Crude oils. Environ. Contamn. Toxicol., 52: 388-395.

Okpokwasili GC, Odokuma LO (1996a). Response of Nitrobacter specie to toxicity of drilling chemicals. J. Petrol. Sci. Eng., 18: 81-87.

Okpokwasili GC, Odokuma LO (1996b). Tolerance of Nitrobacter specie to toxicity of hydrocarbon fuels. J. Petrol. Sci. Eng., 18: 89-93.

Okpokwasili GC, Oton NS (2006). Comparative applications of bioreactor and shake-flask systems in the laboratory treatment of oily sludge. Int. J. Environ. Waste Mgt., 1(1): 49-60.

Olatunji O (2007). Biotechnology and Industries in Nigeria. Proc. 20th Annual Conf., Biotech. Soc. of Nig. (BSN), 14th – 17th Nov, 2007 at Ebonyi State University, Abakaliki, Nigeria, pp. 36-38.

Olukosi A (2006). Introductory Bioinformatics. A lecture paper at theDanifol Biotechnology Training on Bioinformatics, November, pp. 28-30.

Osagie AU (1998). Anti-nutritional factors. In: Nutritional Quality of plant foods. Pp. 221-244.

Oyediran ABO (1994). Waste generation and disposal in Nigeria a keynote address presented at a workshop on waste generation and disposal in Nigeria NEST Annual conference 1994, NEST Ibadan, pp. 95-100.

Parker SP, Corbilt RA (1992). McGram-Hill Encyclopedia of environmental Science and Engineering 3rd Edition, McGraw-Hill, Inc. pp. 210-211, 541-595, 675-678.

Rege JEO (1996). Biotechnology options for improving livestock production in developing countries, with special reference to Sub-Saharan Africa. In: Lebbie S.H.B.and Kagwini E. 1996. Small Ruminant Research and Development in Africa. Proc. Of the Third Biennial Conf. of the African Small Ruminant Research Network, UICC, Kampala, Uganda, 5th – 9th Dec, 1994. ILRI, Nairobi, Kenya,p. 322. http://www.fao.org/wairdocs/ilri/x5373b/x5473b05.htm.

Soetan KO (2008a). The Dynamic and Ubiquitous nature of Biotechnology-A Review. African J. Biotechnol., **7**(16): 2768-2772. Available online at http://www.academicjournals.org/AJB.

Soetan KO (2008b) Agricultural Biotechnology: The Solution to the Problem of Global Food Insecurity. Proceedings of the 1st International Society BioTechnology Conference (ISBT), 28th – 30th December, 2008, at Gangtok, Sikkim, India.

Soetan KO (2008c). The catalyst to sustainable Food Production and Industrial Growth in Nigeria. Proc. of the Intn'l. Conf. of the Nigerian Society for Experimental Biology (NISEB), held at the Lead City University, Ibadan from May 5-8,2008.

Soetan KO (2008d). Pharmacological and other Beneficial effects of Antinutritional factors in Plants. Afr. J. Biotechnol., 7(25): 4713- 4721. Available online at http://www.academicjournals.org/AJB.

Soetan KO (2009). Biotechnology in the Improvements of Animal Production and Health- A Review. In: Advances in Biotechnol. Bentham Science Publishers Limited, USA. Eds. Pankaj K. Bhowmik, Saikat K Basu and Aakash Goyal, 1: 1-34.

Soetan KO, Abatan MO (2008). Biotechnology: A key tool to Breakthrough in Medical and Veterinary Research- A Review. Biotechnol. Mol. Biol. Rev., 3(4): 088-094. Available online at http://www.academicjournals.org/BMBR

Soetan KO, Oluwayelu DO (2011). The Potential Environmental and Public Health impacts of Biotechnology Approach to Agriculture – A Review. Int. J. Appl. Agric. Res., (In Press).

Stenholm CW, Waggoner DB (1992). Public Policy and Animal Biotechnology in the 1990s: Challenges and Opportunities. In: Animal Biotechnology – Opportunities and Challenges (Ed MacDonald, June F.) NABC National Agricultural Biotechnology Council, Ithaca, New York, Report 4: 25-35.

Su S (1998). Genetic Engineering. Evans Brothers Limited, 2a Portman Mansions, Chiltern Street, London W1M 1LE.

Sugano M, Goto S, Yaoshida K, Hashimoto Y, Matsno T, Kimoto M (1993). Cholesterol-lowering activity of various undigested fractions of soybean protein in rats. J. Nutr., 20(9): 977-985.

Tornabene TG, Edwards HH (1972). Microbial uptake of lead. Science 176: 1334-1335.

UNCBD (1992). United Nations Conference on Biological Diversity (Earth Summit) held in Rio de Janeiro, Brazil.

Vidya SK (2005). National Conference on Environmental Biotechnology, Bangalore.

Wang W, Reed P (1983). Nitrobacter bioassay for aquatic toxicity. Illinois State Water Survey, Peoria, Illinois, USA.

Wezel A, Bellon S, Dore T, Francis C, Vallod D, David C (2009). Agroecology as a science, a movement or a practice. A Review. Agronomy for Sustainable Development (published online). Wikipedia.org. Environ. Biotechnol. http://en.wikipedia.org/wiki/Environmental_biotechnology.

Wild A (1995). Soil and the Environment: An Introduction. Cambridge University press, pp. 109-165.

Williamson KJ, Johnson OG (1981). A bacterial bioassay for assessment of wastewater toxicity. Water Res., 15: 383-390.

Zenk HM (1991). Chasing the enzymes of secondary metabolism: Plant cell cultures as a pot of goal. Phytochemistry, 30(12): 3861-3863. Zess Nauk, UMK Tornu,13: 253-256.

Zylstraa GJ, Kukor JJ (2005). What is Environmental biotechnology? Curr. Opin. Biotechnol., 16(3): 243-245.

9

Biotechnology a key tool to breakthrough in medical and veterinary research

Soetan K. O.* and Abatan M. O.

Department of Veterinary Physiology, Biochemistry and Pharmacology, University of Ibadan, Ibadan, Nigeria.

The elucidation of the structure, function and metabolism of Deoxyribonucleic acid (DNA) has led to the current global revolution in the recombinant DNA technology, with the possibility to modify these molecules in many ways for the benefit of man and animals. In this review, we considered the basic principles of genetic engineering (gene cloning), bioinformatics, and its applications in medical and veterinary sciences. The issue of ethical questions and fears about biotechnology is also discussed. The ultimate goal of this paper is to re-invigorate the interest of medical and veterinary personnels, biochemists and related scientists to this technology, which is well able to have an effect on the way all the biosciences will be practiced in this new millennium.

Key words: Biotechnology, Medical, Veterinary research

TABLE OF CONTENTS

INTRODUCTION

The discovery and elucidation of the structure and function of DNA by Watson and Crick in the early 1950s led to the present day recombinant DNA technology (Kwaga and Kabir, 1999).

Biotechnology is the technical applications of biological systems for the production of natural substances (biogas, antibiotics, enzymes organic acids etc) it involves the use of living organisms deliberately to carry out defined che-mical processes and exploitation of biological pro-cesses for man's use (Olasupo, 2005). Modern biotechnology is

hinged on the tools of recombinant technology also called gene manipulation or genetic engineering.

Importance and applications of molecular biology to veterinary medicine and medical sciences

Recombinant DNA technology offers a rational approach to the understanding of the molecular basis of a number of diseases e.g. sickle cell disease, cystic fibrosis etc.

Human proteins can be produced in abundance for therapeutic purposes e.g. insulin, growth hormone, recombinant factor VIII etc. Proteins for vaccines (e.g. hepatitis B) and for diagnostic tests (e.g. AIDS Test) can be obtained. It's used to diagnose existing diseases

*Corresponding author. E-mail: soetangboye@yahoo.com.

Table 1. Human Proteins Synthesized By Recombinant DNA Technology

Protein	Therapeutic Use
Insulin	diabetes
Somatostatin	growth disorders
Somatotropin	growth disorders
Factor VIII	haemophilia
Factor IX	Christmas disease
Interferon ∂	leukaemia and other cancers
Interferon β	cancers, AIDS
Interferon y	cancers, rheumatoid arthritis
Interleukins	cancers, immune disorders
Granulocyte colony Stimulating factor	cancers
Tumour necrosis factor	cancers
Epidermal growth factor	ulcers
Fibroblats growth factor	ulcers
Erythropoietin	anaemia
Tissue plasminogen activator	Heart attack
Superoxide dismutase	Free radical damage in kidney transplants
Lung surfactant protein	respiratory distress
∂1-antiytypsin	emphysema
Serum, albumin	used as a plasma supplement
Relaxin	used to aid child birth

and predict the risk of developing a given disease.

Special techniques have led to remarkable advances in forensic medicine.

Gene therapy for sickle cell disease, the thalassaemias, adenosine deaminase deficiency and other diseases may be devised (Murray et al., 2000).

Gene cloning

Gene cloning is the technique whereby multiple copies of a plasmid or other cloning vehicles are produced by inserting the plasmid into a suitable host capable of producing multiple copies and growing in a bulk culture. The bacterium *Escherichia coli* is often used as the host organism for this purpose (Coombs,1992).The word gene cloning is often used in place of genetic engineering (Soetan, 2007).The basic steps in a gene cloning experiment has been reported by Brown (1998).

The list of human proteins synthesized by recombinant DNA technology is fast growing rapidly. (Table 1)

Production of recombinant vaccines

Development of vaccines for protozoan and helminths parasites of livestock has not been successful. This is because of difficulties encountered in identifying antigens which induce protective immune responses and in obtaining sufficient quantities of vaccine trials (Gamble and Zarlenga, 1986). The use of monoclonal antibodies and genetic engineering technologies could provide the essential tools to help overcome these difficulties (Soetan, 2007).

African swine fever, a disease for which there is no effective prophylaxis nor vaccine, could be effectively conquered with the application of biotechnology techniques used in the production of vaccines through the use of monoclonal antibodies and genetic engineering (Soetan, 2007).

Molecular biology has brought more light into the principle and causes of ageing in mammals and this offers the prospect for both understanding and control of ageing (Brash et al., 1979). Reverse transcriptase polymerase chain reactions (RT-PCR) has brought about new ideas on the detection of RNA Viruses in tissues and body fluids. RNA viruses can now be detected at a high level of sensitivity in infected materials. (Wambura, 2006).

The detection of Newcastle disease virus (NDV) in infective allantoic fluids using PCR were first discovered by Jestin and Jestin, 1991. Later, several RT-PCR methods were developed and used in molecular studies of NDV (Belak and Ballagi-Pordany, 1993; Kant et al., 1997; Cavanagh 2001; Wang et al., 2001; Aldous and Alexander, 2001). Successful extraction and purification of RNA is very important and is the initial step in RT-PCR for successful detection of infectious agents by this technique (Wambura, 2006). Rinderpest (RP), a highly contagious disease of cattle, buffaloes and some wild animals is caused by the rinderpest virus (RPV) and is characterized by a very high mortality (Scott, 1964; Plowright, 1968) and is an economically important disease in Africa, Asia and the middle-East (Couacy-Hyman et al., 2006). Peste-des-petits ruminant (PPR), a highly contagious disease of sheep and goats, similar to rinderpest, is also characterized by a very high mortality and it is caused by PPR virus (PPRV).

There is a close relationship between the RPV and PPRV and they both belong to the morbilli virus genus of the paramyxoviridae family (Gibbs et al., 1979). In infected animals, both diseases are clinically very similar, thus making it very difficult to differentiate them on the field. Molecular biology techniques have been developed to differentiate RPV from PPRV. Diallo et al. (1989), Pandey et al. (1992) and Libeau et al. (1994) have all developed different molecular techniques to differentiate RPV from PPRV. Couacy-Hyman et al., 2006, describe the use of a reverse transcription –PCR (RT-PCR) technique for the specific diagnosis of RPV and also report the use of NP gene of RPV to distinguish all RPV strains from those of PPRV and the use of the technique in epidemiological surveys for rinderpest.

DNA fingerprinting, also called DNA profiling is a DNA identification technique that is based on similarity investtigation of two nucleotide sequences. This is a molecular biology method that has application in Agriculture and the medical sciences (Iwalokun, 2005).The use of DNA fingerprinting technique is now regarded as a milestone in diagnosis and surgical pathology (Persing, 1993).

Gene therapy in medicine and veterinary medicine

This is another application of gene cloning in medicine and veterinary medicine. Gene therapy is the name given to methods that aim to cure an inherited disease by providing the patient with a correct copy of the defective gene. Gene therapy has been successful with experimental animals and clinical trials with humans have been approved by the relevant regulatory agencies (Brown, 1998).

Gene therapy is a therapeutic technique in which a functioning gene is inserted into a cell to correct a metabolic abnormality or to introduce a new function is one of the outcomes of breakthroughs in molecular biology. Gene therapy is a promising approach to the treatment of cancer and other genetic diseases in human and animals.

Polymerase chain reaction (PCR)

The PCR is a very sensitive and specific in-vitro technique for the amplification of a DNA or RNA seg-

ment, through a succession of incubation steps at different temperature, making use of a thermostable DNA polymerase, two primers unique to and which flank the particular segment and the four deoxynucleoside triphosphate. The PCR is one of the most versatile techniques in molecular genetics with a wide range of applications in the medical and veterinary sciences (Saiki et al., 1988; Erlich et al., 1991).

Applications of PCR

PCR has multiple applications in gene cloning, biodiversity studies, diagnostics, forensics etc. Biotechnology has great potentials for harnessing the genetic potential of animals and for enhancing their genetic performance. Animal breeding techniques like embryo transfer have been used in safely propagating animals of similar genotype and phenotype. Recent DNA based techniques have facilitated the identification of specific gene sequences, called genetic markers, that labels desirable or undesirable traits in animals (Nyira, 1995).

A gene marker differentiating Bos taurus from Bos indicus cattle was recently identified using PCR amplification of pooled DNA and RFLP (Restriction fragment length polymorphism). This marker is potentially useful for the breeding of high yielding temperate animals for resistance to trypanosomosis (Kemp and Teale, 1994).

Markers are also employed to determine the genetic relatedness in animals (Grunder et al., 1994; Medjugorac et al., 1994) using RFLP and other DNA fingerprinting methods, for purposes of trade, litigation or phylogenetic studies.

DNA finger printing can also be used accurately to trace offsprings to parents or genetic source.

Bioinformatics and its application in medicine and veterinary medicine

Bioinformatics is the use of Information Technology (IT) in biotechnology for data storage, data warehousing and DNA sequence analysis. It is the comprehensive application of mathematics (e.g. probability and statistics), science (e.g. biochemistry) and a core-set of problem-solving methods (e.g. computer algorithms) to the understanding of living systems (Iwalokun, 2006). The knowledge gained from the study of bioinformatics is crucial for the understanding of the code and evolution of life as well as applications in other areas of life like Agriculture, Health, Environment, Energy etc. (Iwalokun, 2006).

Bioinformatics is one of the latest additions to the scientific world. The word suggests a bridge between biology and information technology. Biochemistry was the first to serve as a bridge between physical and biological sciences. Bioinformatics is also known as computational molecular biology. The origin of bioinformatics can be traced to the development by Sanger and Coulson (1973). Bioinformatics is essentially a theoretical discipline which attempts to make predictions about biological functions from sequence data. It is a powerful tool in experimental design. Kaikabo and Kalshingi (2007) reported that bioinformatics has advanced the course of research and future veterinary vaccines development because it has provided new tools for identification of vaccine targets from sequenced biological data of organisms. Bewaji (2003) defined bioinformatics as application of information technology to the domain of biology. The aims of bioinformatics are:

1. To organise data in a way that allows researchers to access existing information and submits new entries as they are produced (Bernstein et al., 1997; Beran et al., 2000).
2. To develop tools are sources that aid in the analysis of data (Pearson and Lipman, 1988; Altschul et al., 1990; Thompson et al., 1994).
3. To use tool(s) to analyze the data and interpret the results in a biologically meaningful manner.

According to Kaikabo and Kalshingi (2007), in veterinary research, bioinformatics tools could be used to generate biological data for research (Pongor and Landsman, 1999), retrieve and analyze biological data, predict and identify protein(s) in a sequence and also has laboratory application

In veterinary research, bioinformatics tools were used in the detection of new castle diseases (NDV). Alfonso et al. (2006) used bioinformatics to examine the genome of NDV to determine which sequence mismatches have the potential to produce false negative results. Similarly, Kumar (2003) used bioinformatics approach to identify antigenic epitopes from rabies virus glycoprotein G, which could be used to develop antirabies sub-unit vaccine. All the above examples reveal how bioinformatics may be used to identify diagnostic problems and to generate novel solutions for the continued improvement and development of molecular diagnostics (Kiakabo and Kalshingi, 2007).

Berge et al. (2004) reported the use of antibiotic susceptibility Patterns and Pulsed –field Gel Electro-phoresis (PFGE) to compare historic and contemporary isolates of muilti-drug resistant *Salmonella enterica* sub-spenterica serovar newport.

Chen et al. (2004) reported the characterization of multiple antimicrobial resistant salmonella serovars isolated from retail meats purchased in the United States and the peoples Republic of China. A better understanding of the molecular mechanisms by which antimicrobial resistance emerges and spreads should enable us in the future to design intervention strategies to reduce its progression (Chen et al., 2004). Because antimicrobial resistant bacteria may be transferred to humans through the food chain (Threlfall and Ward, 2001; Witten, 1998), selection of novel antimicrobial resistance mechanisms in Salmonella in animals (Threlfall and Ward, 2001), which

specify resistance to antibiotics used in humans is troubling (Chen et al., 2004).

Aminov et al. (2001) reported the molecular ecology of tetracycline resistance: Development and validation of primers for detection of tetracycline resistance genes encoding ribosomal protection proteins. The report by Aminov et al. (2001) was the first demonstration of the applicability of molecular ecology techniques to estimation of the gene pool and the flux of antibiotic resistance genes in production.

Adah (2007) reported that in Africa rotavirus gastro-entesitis remains a major cause of high mortality among human infants and young animals. Rotavirus infection in young animals could also constitute a meaningful threat to the control of the disease in humans. Therefore, the application of a safe and effective rotavirus vaccine which incorporates the diversity of circulating strains in the target population will expectedly reduce the disease prevalence. Adah (2007) also reported that the application of advanced molecular techniques of nucleic acid probe hybridization, restriction endonuclease analysis of PCR, generated cDNA copies, PCR and sequencing analysis to the characterization of African rotavirus strains has changed the epidemiology of rotavirus diversity in the African continent.

Ducatez et al. (2007) reported the molecular and antigenic evolution and geographical spread of H5NI highly pathogenic avian influenza viruses in Western Africa (Ducatez et al., 2006b; Igbokwe et al., 1996; Owoade et al., 2006, 2004 a, b). Owoade et al., 2007, reported the cross reactivities of rabbit anti-chicken horse radish peroxidase conjugate with sera of chic-kens, ducks, geese, guinea fowl, hawks, pigeons and turkeys in indirect enzyme linked immunosorbent assay (ELISA) system. Owoade et al. (2006) reported the seroprevalence of avian influenza virus (AIV), infectious bronchitis virus (IBV), Reovirus, Avian Pneumovirus (APV), infectious laryngotracheitis virus (ILTV) and avian leukosis virus in Nigerian poultry. Their report was the first report of serological evidence of the above viruses in West Africa. In the study, they made use of ELISA, a molecular biology technique.

In recent years, various approaches such as mutational analyses and biochemical and pharmacological characterization have yielded significant information about the relationship of structure and function of P-glycoprotein (Ambudkar et al., 1999). Molecular biology has helped to advance the management of cancer in humans and animals with the discovery and knowledge of P-glycoprotein.

Cadmus et al. (2006) reported the molecular analysis of Human and Bovine Tubercle bacilli from a local setting in Nigeria. Their study was the first molecular analysis of M. tuberculosis complex strains circulating among humans and cattle in Nigeria and the results have significant implications for disease control. Oluwayelu et al. (2005) reported the isolation and characterization of

Table 2a. Examples of international movement of embryos

Purpose	Product
Improved dairy breeds	North American Holsteins
Improve beef breeds	European cattle
High milk production	North American Holsteins
Rapid growth rates in cattle	Large European breed
Rapid growth rates in swine	North American or Danish swine

Table 2b.

Heat tolerance	Bos indicus breeds from Latin America and India
Disease resistance	N'dama cattle from Africa
More hair production	Angora goats from Australia
High ovulation rate in sheep	Booroola Merinos form Australia
High ovulation rate in swine	Prolific Chinese breeds.
Animals for game farming	Elk from North America

of chicken anaemia virus (CAV) from chickens in Nigeria. This was the first time CAV was isolated from the Nigerian chicken population. In a similar study, Oluwayelu (2006) reported a molecular analysis study of chicken anaemia virus (CAV) in backyard chickens in Nigeria using molecular cloning and sequence analysis to characterize chicken anaemia virus strains obtained from commercial chickens and Nigerian backyard chickens. Luther et al., 2007, reported the use of PCR to detect the genome of African swine fever virus (ASFV) from natural infection in a Nigerian baby warthog (Phacochoereus aethiopicus). They stated that application of PCR for the detection of organs of ASFV genomic DNA presents a sensitive and specific method of identifying the virus (Saiki et al., 1985, 1988). Their communication is reported to be the first documented report of the detection of ASFV from a Nigerian warthog reported hitherto only in eastern southern African countries (Scott, 1965; Plowright et al., 1969; Lither et al., 2001).

Embryo transfer

Embryo transfer is a technique by which embryos are collected from a donor female and transferred to a recipient female which serves as a surrogate mother for the remainder of pregnancy. Such techniques have been applied to nearly every specie of domestic animals, of wild life and exotic animals as well as to humans and other primates. Embryo transfer is used in buffalo and dromedary camel (Camelus dromedaries) (Musa, 1992). (Table 2a and 2b)

Embryo transfer is useful for rapidly increasing numbers of an imported breed or line animals. There are many applications of embryo transfer. They are training and research, testing for deleterious recessive genes, management of disease, conservation of native breeds,

exotic and wild animals (Kuzan and Seidel, 1986).

Ethical questions raised by gene therapy and genetic engineering in general.

Morals are the norms or values considered or judged as being good or bad, right or wrong. Morals are societal values and may differ between societies or cultures. Ethics deals with morals and moral rules. Ethical conduct is that perceived to be morally right by the society Osuntoki, (2005).

Every area of human endeavor generates unique ethical challenges. In biotechnology, ethical conducts are based on four key factors:

1. Beneficence
2. Risk prevention
3. Fundamental principles of respect for persons
4. Justice (Osuntoki, 2005).

Advances in medical biotechnology raise some ethical questions. For example;
1. Should gene therapy be used to cure human disease!
2. Is it right to manipulate genes that generations unborn may inherit!
3. Does the embryo have the moral status to prescribe its genetic alteration or exploitation regardless of the potential benefit!
4. Is it right to generate genetic information on humans which could have negative socio-psychological impacts!
5. Is it right to screen children and teens for adult onset genetic pathologies!
6. Can genetic information about individuals be used for unethical things!
7. Do we have a right to "play God"!
8. Who defines normality!
9. Why alter what is not fully known! (Osuntoki, 2005).

It is very important to exploit ways of resolving these ethical and moral issues. However, ethical dilemmas are minimized when mechanisms that sustain a solid foundation of trust and safety are incorporated into research design and implementation and also in the biotechnological production processes (Osuntoki, 2005).

Dr. Norman Borlaug, who was awarded the Nobel Peace Prize for his development of study, high-yielding cereal grains for use in developing countries had this to say about biotechnology:

"----- Too many opponents of biotechnology too easily dismiss the many safety and regulatory checks that govern whether a new agricultural product brought to the market-place is worthless. Unfortunately, they willfully choose to emphasize highly potential risks rather than recognize the years of experienced research and regulatory oversight that govern the safe use of these new technologies".

The immediate past Vatican was quoted as making this statement regarding biotechnology:

"---- We are increasingly encouraged that the advantages of genetic engineering of plants and animals are greater than the risks. The risks should be carefully followed through openness, analysis and controls, but without a sense of alarm".

Finally, former President Jimmy Carter of the United States of America had this to say:

"------Responsible biotechnology is not the enemy; starvetion is. Without adequate food supplies at affordable prices, we can't expect world health and peace".

Conclusion

In conclusion, Recombinant DNA technology has provided highly sensitive and specific tools for the diagnosis, prognosis and disease surveillance in humans and animals. In the future, preventive human and veterinary medicine will focus on genetically engineered animals resistant to specific endemic diseases. Gene's confering resistance to specific diseases may be copied, sequenced and propagated using PCR, and inserted into the genes of recipient humans and animals (Kwaga and Kabir, 1999).

RECOMMENDATIONS

1. An urgent, serious and strong Government commitment is essential for the success of biotechnology project.
2. The introduction of Biotechnology to the Nigerian Environment has a unique strategic significance, in that it can contribute a lot considerably to improve the quality of life of Nigerians by providing solutions to survival problems such as diseases, foods, fertilizers, fuels etc.
3. There is need for bioinformatics training in Nigeria. This should be introduced into the curriculum of the Nigerian universities, as this will aid rapid research and development in the country.
4. Establishment of small software groups, biotechnology

groups and companies should be encouraged.

REFERENCES

Adah IM (2007). Strategies for rotavirus strain characterization in Africa. Problems and prospects, Tropical Veterinarian.25 (1): 1-14.

Aldous EW, Alexander DJ (2001). Detection and differentiation of Newcastle disease virus (avian paramyxovirus type 1). Avian.pathol. 30: 117-128.

Altschul SF, Guish W, Miller W, Myers EW, Lipman DJ (1990). Basic alignment search tool. Mol. Biol. 215: 403-410.

Ambudkar SV, Dey S, Hrycyna CA, Ramachandra M, Rastan I, Grottersman MM (1999). Biochemical Cellular and Pharmacological aspects of the multidurg transporter. Ann. Rev. Pharmacol. Toxicol. 39: 361-398.

Aminov RI, Garrigues SN, Mackie RI (2001). Molecular ecology of tetracycline resistance. Development and validation of primers for Detection of Tetracycline resistance genes encoding ribosomal protection proteins I. Applied and environ. Microbial. 67(1): 22-32.

Belak S, Ballagi-Pordany A (1993). Application of the Polymerase chain reaction (PCR) in Veterinary diagnostic virology. Vet. Res. Comm. 17:55-72.

Berge ACB, Adaska SM, Sischo WM (2004). Use of antibiotic susceptibility patterns and Pulsed Field Gel Electrophoresis to compare historic and contemporary isolate of multi-drug resistant *Salmonella enterica* subsenterica serovar Newport J. Appl. Environ. microbial. pp. 318-323.

Beran HM, Westbrook J, Feng Z, Gilliland G, Bhat TN and Weissig H (2000). The Protein Data Bank. Nucleic. Acids. Res. 28(1): 235-242.

Bernstein FC, Koetzle TF, Williams GJ, Meyer EF, Brice JR, Rodger JR (1997). The Protein Data bank: A Computer-based Archival file Macromolecular Structures. Env. J. Biochem. 80(2):319-324.

Bewaji CO (2003). Bioinformatics and Computational Molecular Biology Techniques Manual 2003 UNAAB, Summer Course in Practical Biotechnology. University of Agriculture Abeokuta, Nigeria. 28th July-1st August; 1-5.

Brown TA (1992). Gene Chaning. An Introduction 3rd Edition Stanley Thornes Publisher Ltd. U.K.

Cadmus S, Palmer S, Okker M, Dale J, Grover K, Smith N, Jahans K, Hewinson RG and Gordon, SV (2006). Molecular analysis of Human & Bovine Tubercle Bacilli from a local setting in Nigeria. J. Chin. Microbiol. 44(7): 29-34.

Cavanagh D (2001). Innovation and Discovery: The application of nucleic acid-based technology to avian virus detection and characterization. Avian. pathol. 30: 581-598.

Chen S, Zhao S, White DG, Schroeder CM, Lv RE, Tang H, McDermott PF, Ayers S, Merng S (2004). Characterization of multiple antimicrobial resistant Salmonella serovars isolated from retail meats I Appl. Env. Microbiol. 70(1):1-7.

Coombs J (1992). Dictionary of biotechnology, 2nd Edition. The Macmillan Press Ltd, London and Basingstoke.

Couacy-Hymann E, Bodjo SC, Danho T, Koffi MY, Akoua-Koffi C (2006). Diagnosis and surveillance of rinderpest using reverse transcription-PCR. Afr. J. Biotechnol. 5(19):1717-1721.

Diallo A, Barrett T, Barbron M, Shaila MS, Taylor WP (1989). Differentiation of rinderpest and peste des petits ruminants viruses using cDNA clones. J. Virol. Methods. 23:127-136.

Ducate MF, Owoade AA, Abiola SO, Miller CP (2006b). Molecular epidemiology of chicken anaemia virus in Nigeria. Arch. Virol 151: 97-111.

Ducatez MF, Olinger CM, Owoade AA, Tarnagda Z, Tahita MC, Sow A, Delandtsheer S, Ammerlaan W, Ovedraogo SB, Osterhavus ADME, Fouchier RAM, Miller CP (2007). Molecular and antigenic evolution and geographical spread of H5N1 highly pathogenic avian influenza virus in Western Africa J. General Virolog. 88:2297-2306.

Erlich HA, Gelfard D, Sminsky JJ (1991). Recent advances in polymerase chain reaction. Science 252: 1643-1651.

Gamble HR and Zarlenga DS (1986). Biotechnology in the development of vaccines for animal parasites. Veterinary Parasitol. 20: 237-250.

Gibbs EPJ, Taylor WP, Lawman MPJ, Bryant J (1979). Classification of peste des petits ruminants virus as a fourth member of the Genus morbillivirus. Intervirology. 11: 268-274.

Grunder AA, Sabour MP, Govova JS (1994): Estimate of relatedness and inbreeding in goose strains from DNA fingerprints. Animal genetics 25: 81-88.

Igbokwe SO, Salako MA, Rabo SS, Hassan SV (1996). Outbreak of infections bursal disease associated with acite septicaemic colibacillosis in adult prelayers hens. Rev. Elev. Med Vet pays Trop. 49: 110-113.

Iwalokun BA (2005). DNA fingerprinting: A tool in Agriculture, Crime monitoring, health care delivery and industries. Proceedings of the workshop on DNA fingerprinting and blotting techniques. Organized by Danifol Biotechnology Consult. August 9-11.

Iwalokun BA (2006). Bioinformatics: Biological databases and tools. A lecture paper delivered at DANIFOL Biotechnology Training workshop. Lagos, Nigeria, Nov. 28-30.

Jestin V, Jestin A (1991). Detection of Newcastle disease virus RNA in infected allantoic fluid by in vitro enzymatic amplification (PCR). Arch-virol 118:151-161.

Kaikabo AA, Kalshingi HA (2007). Concepts of bioinformatics and its applications in veterinary research and vaccines development. Nigerian Vet. J.28(2): 39-46.

Kant A, Roch DJ, Roozelaar V, Balk F, Huurne AT (1997). Differentiation of Virulent and non-virulent strains of Newcastle disease virus within 24 h by polymerase chain reaction. Avian. Pathol. 26: 837-849.

Kemp SJ, Teale AJ (1994). Randomly primed PCR amplification of pooled DNA. reveals polymorphism in a ruminant receptive DNA sequence which differentiates Bos indicus and Bos taurus. Animal Genetics. 25: 83-88.

Kumar D (2003). Identification of promiscuous MHC Class-1 and MHC Class-11 binding epitopes of rabies virus glycoprotein. Unpublished Ph.D. Thesis submitted to Deemed University, Indian Veterinary Research Institute, Lzatnagar (UP)-243122; 1-49.

Kuzan FB, Seidel GE Jr (1989). Embryo transfer in animals. In: Guvatkin RBL (ed) Manipulation of Mammalian Dvpt. Plenum, New York, 1986, pp. 249-279.

Kwaga JKP, Kabir J (1999). Basic principles of Genetic Engineering and Applications in Veterinary Medicine. Nigerian Vet. J. Vol. 20(2):17-33.

Land RB (1986). Genetic resource requirements under favourable production, marketing systems, priorities and organization.3rd World congress on genetics applied to livestock production (Lincoln) XII. pp. 486–491.

Libeau G, Colas F, Guerre L (1994). Rapid differential diagnosis of rinderpest and pests des petits ruminants using an immunocapture ELISA. Vet Rec. 19:300-304.

Luther NJ, Udeama PG, Majiyagbe KA, Shamaki D, Antiabong JF, Bitrus Y, Nwosh CI, Owolodun OA

Medjugorac L, Krustermann W, Lazar P, Russ L, Pirchner F (1994). Marker-derived phylogeny of European Cattle supports demic expressions of Agriculture. Animal Genetics 25: 19-27.

Meyn K (1992). Legal and social aspects of biotechnology application in developing countries. Consideration into livestock patenting. Paper presented at a symposium on potentials and limitations of biotechnology in developing countries, Mariensee, Germany, May 14th –16th.

Murray RK, Granner DK, Mayer PA, Rodwell VW (2000). Harperly Biochemistry, 25th Edition, McGraw-Hill. Health Profession Division U.S.A.

Murray RK, Granner DK, Mayes PA, Rodwell VW (2000). Harper's Biochemistry,25th Edition.

Musa BE (1992). Embryo transfer in the Dromedary camel (Camelus dromedarius). In potentials and limitations of biotechnology in livestock production in developing countries Part Animal reproduction and breeding, Mariensee, Germany, May 14th-16th, 1992.

Nyira ZM (1995). Challenges and objectives for Biotechnology and Agriculture in Africa. In: Koman J, Cohen JI, Ofir Z (eds). Turning priorities into feasible programs. Agricultural Biotechnology for East Africa. Proceedings of a policy seminar. No 2 held in South Africa. 23-24.

Olasupo NA (2005). Food Biotechnology and fortification. Proceedings of the workshop on Molecular Biology techniques (Theory and Practicals) organized by Danifol Biotechnol. Consult. March 23-25.

Oluwayelu DO (2006). Isolation and characterization of chicken infectious anaemia virus in southwestern Nigeria Ph.D. Thesis, University of Ibadan, Nigeria.

Oluwayelu DO, Todd D, Ball MN, Scott ANS, Oladele OA, Emikpe BO, Fagbohun OA, Owoade AA, Olaleye OD (2005). Isolation and Preliminary characterization of chicken anaemia virus from chickens in Nigeria. J. Avian. Disease. 49:446-450.

Osuntoki AA (2005). A review of Molecular Biology Techniques. Proceedings of the workshop on DNA Fingerprinting and Blotting techniques. Organized by Danifol Biotechnology Consult. August 9-11.

Owoade AA, Ducatez MF, Muller CP (2006). Seroprevalence of avian influenza irus, infectious bronchitis virus, reovirus, avian pneumovirus, infections laryngotracheitis virus and avian leucosis virus in Nigerian poultry. Avian. Dis. 50:222-227.

Owoade AA, Fagbohun OA, Oluwayelu DO (2007). Cross reactivities of rabbit antichicken horse radish peroxidase conjugate with sera of some other avian species in ELISA system. Africa J. Biomed. Res. 10:193-196.

Owoade AA, Mulders MN, Kohnen J, Ammeriaan W, Muller CP (2004a). High sequence diversity in infectious bursal disease virus serotype 1 in poultry and turkey suggest West-African origin of very virulent strains. Arch. Virol. 149: 653-672.

Owoade AA, Oluwayelu DO, Fagbohun OA, Ammerlaan W, Mulders MN, Muller CP (2004b). Serologic evidence of chicken infectious anemia in commercial chicken flocks in Southwest Nigeria. Avian Dis 48:202-205.

Pandey KD, Baron MD, Barrett T (1992). Differential diagnosis of rinderpest and PPR using biotinylated cDNA probes. Vet Rec. 131:199-200.

Pearson WR, Lipman DJ (1988). Improved tools for biological sequence Comparsion. Proc. Natl. Acad. Sci. (USA). 85(8):2444-2448.

Persing E (1993). Diagnostic molecular microbiology: Principles and Applications. American Society of Microbiology. California, USA. pp. 1-105.

Plowright W (1968). Rinderpest Virus: Spring-verlag, Wien, New-York. Virol.Monographs 3:25-110.

Plowright W, Parker I, Peirce MA (1969). The epiozotiology of African swine fever in Africa. Vet. Rec. 83: 668-674.

Pongor S, Landsman D (1999). Bioinformatics and the developing world Biotechnology and Development Monitor. 40:10-13.

Saiki RK, Gelfard DH, Stoffel S, Scharf SJ, Higuchi R, Horn GT, Mullis KB, Erich HA (1988). Primer-directed enzymatic amplification of DNA with a thermostable DNA Polymerase. Sci. 239: 487-491.

Saiki RK, Scharf S, Faloona F, Millis KB, Horn GT, Erlich HA, Arnheim N (1985). Enzymatic amplification of A. Globin Genomic sequence and restriction site analysis for diagnosis of sickle cell anaemia Sci. 230: 1350-1354.

Sanger F, Coilson AR (1975). A rapid method for determining sequences in DNA by primed synthesis with DNA polymerase J. Mol. Biol. 94: 444-448

Scott GR (1964).Rinderpest Adv. Vet. Sci. 9: 113-224.

Scott, GR (1965). The smallest stowaways I. African swine fever. Vet. Rec. 77: 1421-1422.

Soetan KO (2007). Personal communication.

Threlfall EJ, Ward LR (2001). Decreased susceptibility to ciprofloxacin in Salmonella enterica serotype typhi, United kingdom. Emerg. Infect. Dis 72. pp.448-450.

Wambura PN (2006). Use of virus suspensions without RNA extraction as RT-PCR templates for detection of Newcastle disease virus. African journal of Biotechnology. Vol. 5 (19): 1722-1724.

Wang Z, Vreede FT, Mitchell JO, Viljoen GJ (2001). Rapid detection of Newcastle disease virus isolates by a triple one-step. RT-PCR. Onderstepoort J. Vet.Res. 68:131-134.

Witten W (1998). Medical consequences of antibiotic use in agriculture. Sci. 279: 996-997.

10

Recent advances in salt stress biology – a review

Town Mohammad Hussain[1]*, Thummala Chandrasekhar[2], Mahamed Hazara[3], Zafar Sultan[4], Brhan Khiar Saleh[1] and Ghanta Rama Gopal[2]

[1]Department of Horticulture, Hamelmalo Agricultural College, Keren, P.O. Box 397, Eritrea, North East Africa.
[2]Department of Botany, Sri Venkateswara University, Tirupati-517 502, Andhra Pradesh, India.
[3]School of Life Sciences, Department of Molecular Biology, University of Skövde, Skövde, Sweden.
[4]Department of Plant Protection, Hamelmalo Agricultural College, Keren, P.O. Box 397, Eritrea, North East Africa.

Soil salinity is one of the major abiotic stresses that adversely affects crop productivity and quality. Hence developing salt tolerant crops is essential for sustaining food production. Understanding of the molecular basis of salt stress signaling and tolerance mechanisms are essential for breeding and genetic engineering of salt tolerance in crop plants. The modern approaches being used to impart salt tolerance involves exploitation of natural genetic variations and/or the generation of transgenic plants. This review discusses the challenges and opportunities provided by recently developed molecular tools in the development of salt tolerant crops.

Key words: Soil salinity, plasma membrane antiporter (*AtSOS1*), AtHKT1, vacuolar antiporter (*AtNHX1*), compatible solutes, reactive oxygen species (ROS), antioxidants.

TABLE OF CONTENTS

INTRODUCTION

Agricultural productivity is severely affected by soil salinity. Environmental stress due to salinity is one of the most serious factors limiting the productivity of agricultural crops, most of which are sensitive to the presence of high concentrations of salts in the soil. About 20% of irrigated agricultural land is adversely affected by salinity (Flowers and Yeo, 1995).

The problem of soil salinity is further aggravated through the use of poor quality water for irrigation and inadequate drainage. Soil type and environmental factors, such as vapor pressure deficit, radiation and temperature may also alter salt tolerance. In clay soils, improper ma-

nagement of salinity and subsequent accumulation of sodium salts may lead to soil sodicity. In sodic soils sodium binds to negatively charged clay particles, causing swelling and dispersal, thus making the soil less fit for crop growth (Chinnusamy et al., 2005). The loss of farmable land to salinization is in direct conflict with the needs of the world population, projected to increase by 1.5 billion in the next 20 years (Blumwald and Grover, 2006).

Adverse effects of salinity on plant growth may be due to ion cytotoxicity (mainly due to Na^+, Cl^- SO_4^-) and osmotic stress (reviewed by Zhu, 2002). Metabolic imbalances caused by ion toxicity, osmotic stress and nutritional deficiency under saline conditions may also lead to oxidative stress (Zhu, 2002). Hence, engineering crops that are resistant to salinity stress is critical for sustaining food prod-

*Corresponding author. E-mail: md_hussain_2000@yahoo.com.

uction and achieving future food security. However, progress in breeding for salt tolerant crops has been hampered by the lack of complete understanding of the molecular basis of salt tolerance and lack of availability of genes that confer salt tolerance.

Two major approaches being used to improve stress tolerance are: (1) Exploitation of natural genetic variations and (2) generation of transgenic plants with novel genes or altered expression levels of the existing genes. Zhang et al. (2004) and Zhu (2001, 2002) recently reviewed signaling and transcriptional control in plants under salt stress. In this review, we discuss the approaches that have led to increased salinity tolerance, with particular emphasis on ion homeostasis, synthesis of compatible solutes and oxidative stress management.

Ion homeostasis

Plants respond to salinity using two different types of responses. Salt-sensitive plants restrict the uptake of salt and adjust their osmotic pressure through the synthesis of compatible solutes (e.g. proline, glycinebetaine, soluble sugars; Greenway and Munns, 1980). Salt-tolerant plants sequester and accumulate salt into the cell vacuoles, controlling the salt concentrations in the cytosol and maintaining a high cytosolic K^+/Na^+ ratio in their cells (Glenn et al., 1999).

A high NaCl concentration in tissues is toxic for growth of glycophytes (Glenn et al., 1999). The alteration of ion ratios in plants could result from the influx of Na^+ through pathways that also function in the uptake of K^+ (Blumwald et al., 2000). The maintenance of a high cytosolic K^+/Na^+ ratio and precise regulation of ion transport is critical for salt tolerance (Glenn et al., 1999).This can be achieved by extrusion of Na^+ ions from the cell or vacuolar compartmentation of Na^+ ions. Three classes of low-affinity K^+ channels have been identified (Sanders, 1995), these are K^+ Inward rectifying channels (KIRC); K^+ outward rectifying channels (KORCs) and Voltage-independent cation channels (VIC).

K^+ outward rectifying channels (KORCs) could play a role in mediating the influx of Na^+ into plant cells. These channels, which open during the depolarization of the plasma membrane, could mediate the efflux of K^+ and the influx of Na^+ ions. Na^+ competes with K^+ uptake through Na^+ - K^+ co-transporters and may also block the K^+ specific transporters of root cells under salinity (Zhu, 2003). This could result in toxic levels of sodium as well as insufficient K^+ concentration for enzymatic reactions and osmotic adjustment. The influx of Na^+ is controlled by AtHKT1, a low affinity Na^+ transporter (Rus et al., 2001; Uozumi et al., 2000). The knockout mutant (hkt1) from Arabidopsis suppressed Na^+ accumulation and sodium hypersensitivity (Rus et al., 2001), suggesting that AtHKT1 is a salt tolerance determinant, while the efflux is

controlled by Salt Overly Sensitive1 (SOS1), a plasma membrane Na^+/H^+ antiporter (Shi et al., 2000). This antiporter is powered by the operation of H^+ -ATPase (Blumwald et al., 2000). In addition to its role as an antiporter, the plasma membrane Na^+/K^+ SOS1 may act as a Na^+ sensor (Zhu, 2003). The overexpression of SOS1 improved salt tolerance in Arabidopsis (Shi et al., 2003) (Table.1).

The compartmentation of Na^+ ions in vacuoles provides an efficient and cost effective mechanism to prevent the toxic effects of Na^+ in the cytosol. The transport of Na^+ into the vacuoles is mediated by a Na^+/H^+ antiporter (AtNHX1) that is driven by the electrochemical gradient of protons generated by vacuolar H^+-translocating enzymes, the H^+-ATPase and the H^+-PPiase (Blumwald, 1987). The overexpression of AtNHX1, resulted in the generation of transgenic arabidopsis (Apse et al., 1999), tomato (Zhang and Blumwald, 2001), Brassica napus (Canola) (Zhang et al., 2001), rice (Ohta et al., 2002), tobacco (Wu et al., 2004), maize (Yin et al., 2004), tall fescue plants (Luming et al., 2006) that were not only able to grow in significantly higher salt concentration (200 mM NaCl) but could also flower and set fruit. Also transgenic plants over expressing AVP1, coding for the vacuolar H^+-pyrophosphatase, showed enhanced salt tolerance (Gaxiola et al., 2001).

Synthesis/overexpression of compatible solutes

The cellular response of salt-tolerant organisms to both long- and short-term salinity stresses includes the synthesis and accumulation of a class of osmoprotective compounds known as compatible solutes. These relatively small organic molecules are not toxic to metabolism and include proline, glycinebetaine, polyols, sugar alcohols, and soluble sugars. These osmolytes stabilize proteins and cellular structures and can increase the osmotic pressure of the cell (Yancey et al., 1982). This response is homeostatic for cell water status, which is perturbed in the face of soil solutions containing higher amounts of NaCl and the consequent loss of water from the cell. Glycinebetaine and trehalose act as stabilizers of quaternary structure of proteins and highly ordered states of membranes. Mannitol serves as a free radical scavenger. It also stabilizes sub cellular structures (membranes and proteins), and buffers cellular redox potential under stress. Hence these organic osmolytes are also known as osmoprotectants (Bohnert and Jensen, 1996; Chen and Murata, 2000).

Genes involved in osmoprotectant biosynthesis are up-regulated under salt stress and concentrations of accumulated osmoprotectants correlate with osmotic stress tolerance (Zhu, 2002). Although enhanced synthesis and accumulation of compatible solutes under osmotic stress is well known, little is known about the signaling cascades that regulate compatible solute biosynthesis in hig

Table 1. Salt tolerant transgenic plants expressing genes involved in synthesis/over expression of ion transporters and compatible solutes.

Gene	Gene product	Source	Cellular role(s)	Target plant	Reference
Ion transporters					
AtNHX1	Vacuolar Na$^+$/H$^+$ antiporter	Arabidopsis thaliana	Na$^+$ vacuolar sequestration	Arabidopsis	Apse et al., 1999.
AVP1	Vacuolar H$^+$ pyrophosphatase	A. thaliana	Vacuolar acidification	Arabidopsis	Gaxiola et al., 200:
AgNHX1	Vacuolar Na$^+$/H$^+$ antiporter	Atriplex gmelini	Na$^+$ vacuolar sequestration	Rice	Ohta et al., 2002
AtSOS1	Plasma membrane Na$^+$/H$^+$ antiporter	A. thaliana	Na$^+$ extrusion	Arabidopsis	Shi et al., 2003.
GhNHX1	Vacuolar Na$^+$/H$^+$ antiporter	Gossypium hirsutum	Na$^+$ vacuolar sequestration	Tobacco	Wu et al., 2004.
AtNHX1	Vacuolar Na$^+$/H$^+$ antiporter	A. thaliana	Na$^+$ vacuolar sequestration	Festuca arundinacea	Luming et al., 2006
Compatible solutes					
COX	Choline oxidase	Arthrobacter globiformis	Glycinebetaine	Arabidopsis	Hayashi et al., 1997
ProDH	Proline dehydrogenase	Arabidopsis thaliana	Proline	Arabidopsis	Nanjo et al., 1999.
EctA	L-2,4-diaminobutyric acid acetyltransferase	Halomonas elongata	Ectoyne	Tobacco	Nakayama et al., 2000.
betA	Choline Dehydrogenase	E. coli	Glycinebetaine	Tobacco	Holmstrom et al., 2000.
Stpd1	Sorbitol-6-phosphate dehydrogenase	apple	Sorbitol	Japanese persimmon	Gao et al., 2001.
BADH	Betaine dehydrogenase	Atriplex hortensis	Glycinebetaine	Tomato	Jia et al., 2002.
TPSP	Trehalose-6-phosphate phosphatase	E. coli	Trehalose	Rice	Garg et al., 2002.
PmSDH1	Sorbitol dehydrogenase	Plantago major	Sorbitol	Plantago major	Pommerrenig et al. 2007.

plants. Salt tolerance of transgenic tobacco engineered to over accumulate mannitol was first demonstrated by Tarczynski et al. (1993). The other examples of compatible solute genetic engineering includes the transformation of genes for Ectoine synthesis with enzymes from the halophilic bacterium Halomonas elongata (Nakayama et al., 2000; Ono et al., 1999) and trehalose synthesis in potato (Yeo et al., 2000), rice (Garg et al., 2002), and sorbitol synthesis in plantago (Pommerrenig et al., 2007) (Table 1) . Initial strategies aimed at engineering higher concentrations of proline began with the overexpression of genes encoding the enzymes pyrroline-5-carboxylate (P5C) synthetase (P5CS) and P5C reductase (P5CR), which catalyze the two steps between the substrate (glutamic acid) and the product (proline). P5CS overexpression in transgenic tobacco dramatically elevated free proline (Kishor et al., 1995). However there is strong evidence that free proline inhibits P5CS (Roosens et al., 1999). Hong et al. (2000) achieved a two-fold increase in free proline in tobacco plants by using a P5CS modified by site directed mutagenesis.The procedure alleviated the feedback inhibition of P5CS activity by proline and resulted in improved germination and growth of seedlings under salt stress. Also, Nanjo et al. (1999) used antisense cDNA transformation to decrease ProDH (Proline dehydrogenase) expression in order to increase free proline levels.

The enhancement of glycinebetaine syntheses in target plants has received much attention (Rontein et al., 2002). In spinach and sugar beet which naturally accumulate glycinebetaine, the synthesis of this compound occurs in the chloroplast. The first oxidation to betaine aldehyde is catalyzed by choline mono-oxygenase (CMO). Betaine aldehyde oxidation to glycinebetaine is catalyzed by betaine aldehyde dehydrogenase (BADH) (Rathinasabapathi, 2000). In Arthrobacter globiformis, the two oxidation steps are catalyzed by one enzyme, choline oxidase (COD), which is encoded by the codA locus (Sakamoto and Murata, 2000). Hayashi et al. (1997) used choline oxidase of A. globiformis to engineer glycinebetaine synthesis in Arabidopsis and subsequently tolerance to salinity during germination and seedling establishment was improved markedly in the transgenic lines. Huang et al (2000) used COX from *A. panescens*, which is homologous to the *A. globiformis* COD, to transform *arabidopsis*, *B. napus* and tobacco. In this set of experiments COX protein was directed to the cytoplasm and not to the chloroplast. Improvements in tolerance to salinity, drought and freezing were observed in some transgenics from all three species, but the tolerance was variable. The results offered the possibility that the protection offered by glycinebetaine is not only osmotic but also function as scavengers of oxygen radicals. This point was also raised by Bohnert and Shen (1999) and further supported by the results of Alia et al. (1999). The level of glycinebetaine production in transgenics could be limited by choline. A dramatic increase in glycinebetaine levels (to 580 mmol/g dry weight in arabidopsis) was achieved when the growth medium was supplemented with choline (Huang et al., 2000).

Antioxidant protection

Stress induces production of reactive oxygen species (ROS) including superoxide radicals, hydrogen peroxide (H_2O_2) and hydroxyl radicals (OH^-) and these ROS cause oxidative damage to different cellular components including membrane lipids, protein and nucleic acids (Halliwell and Gutteridge, 1986). Reduction of oxidative damage could provide enhanced plant resistance to salt stress. Plants use antioxidants such reduced glutathione (GSH) and different enzymes such as superoxide dismutases (SOD), CAT, APX, glutathione-S-transferases (GST) and glutathione peroxidases (GPX) to scavenge ROS.

Transgenic tobacco plants overexpressing both GST and GPX showed improved seed germination and seedling growth under stress (Roxas et al., 1997). A major function of glutathione in protection against oxidative stress is the reduction of H_2O_2 (Foyer and Halliwell, 1976). Ruiz and Blumwald (2002) investigated the enzymatic pathways leading to glutathione synthesis during the response to salt stress of wild-type and salt-tolerant *B. napus* L. (Canola) plants overexpressing a vacuolar Na^+/H^+ antiporter (Zhang et al., 2001).

Wild-type plants showed a marked increase in the activity of enzymes associated with cysteine synthesis (the crucial step for assimilation of reduced sulfur into organic compounds such as glutathione) resulting in a significant increase in GSH content. On the other hand, these activities were unchanged in the transgenic salt-tolerant plants and their GSH content did not change with salt stress. These results clearly showed that salt stress induced an increase in the assimilation of sulfur and the biosynthesis of cysteine and GSH in order to mitigate salt-induced oxidative stress.

Conclusions

The evaluation of salt tolerance in transgenic experiments has mostly been carried out using a limited number of seedlings or mature plants under laboratory and/or greenhouse conditions different from what the plants would naturally be exposed to (e.g. high-salinity soils, high diurnal temperatures, presence of other sodic salts etc). Thus, the evaluation of field performance under salt stress is difficult because of the variability of salt levels under field conditions (Daniells et al., 2001; Richards, 1983) and the potential for interactions with other environmental factors (including soil fertility, temperature, light intensity and water loss through transpiration).

Conventional breeding programs for generation of salt-tolerant genotypes have met with limited success. This lack of success is due in part to breeders preferring to evaluate their genetic materials under ideal conditions and also, the use of constitutive promoters like the CaMV 35S promoter, ubiquitin and actin promoters (Grover et al., 2003). However, stress-induced or tissue-specific promoters result in a better phenotype than overexpressing the same genes under a constitutive promoter (Kasuga et al., 1999; Zhu et al., 1998). There is a clear and urgent need to begin to introduce these tolerance genes into crop plants, in addition to establishing gene stacking or gene pyramiding.

Although progress in improving stress tolerance has been slow, there are a number of reasons for optimism. These include recent developments in the area of plant molecular biology, specifically 1) the development of molecular markers and gene tagging methodologies (2) the complete sequencing of model plant genomes (3) the production of T-DNA insertional lines of arabidopsis for gene tagging (4) the availability of forward genetics tools (Tilling) (Colbert et al., 2001) and (5) the availability of microarray analysis tools offers advantages and solutions to the complex intriguing questions of salt resistance.

REFERENCES

Alia KY, Sakamoto A, Nonaka H, Hayashi H, Saradhi PP, Chen THH, Murata N (1999). Enhanced tolerance to light stress of transgenic arabidopsis plants that express the codA gene for a bacterial choline oxidase. Plant Mol. Biol. 40: 279–288.

Apse MP, Aharon GS, Snedden WA, Blumwald E (1999). Salt tolerance conferred by overexpression of a vacuolar Na^+/H^+ antiporter in Arabidopsis. Sci. 285: 1256–1258.

Blumwald E (1987). Tonoplast vesicles for the study of ion transport in plant vacuoles. Physiologia Plantarum. 69: 731–734.

Blumwald E, Aharon GS, Apse MP (2000).Sodium transport in plant cells. Biochimica et Biophysica Acta. 1465: 140–151.

Blumwald E, Grover A (2006). Salt Tolerance. In: Nigel Halford (ed) Plant Biotechnology, Current and Future Applications of Genetically Modified Crops, Agritech Publications, New York. pp 206-224.

Bohnert HJ, Jensen RG (1996). Strategies for engineering water stress tolerance in plants. Trends. Biotechnol. 14: 89–97.

Bohnert HJ, Shen B (1999).Transformation and compatible solutes. Scientia Horticulturae. 78: 237–260.

Chen THH, Murata N (2000).Enhancement of tolerance of abiotic stress by metabolic engineering of betaines and other compatible solutes. Curr. Opin. Plant Biol. 5: 250–257.

Chinnusamy V, Jagendorf A, Jhu J-K (2005).Understanding and improving salt tolerance in plants. Crop Sci. 45: 437-448.

Colbert T, Till BJ, Tompa R, Reynolds S, Steine MN, Yeung AT, McCallum CM, Comai, L, Henikoff S (2001). High-throughput screening for induced point mutations Plant. Physiol. 126: 480–484.

Daniells IG, Holland JF, Young R., Alston CL, Bernardi AL (2001). Relationship between yield of grain sorghum (Sorghum bicolor) and soil salinity under field conditions. Aust. J. Exp. Agric. 41:211–217.

Flowers TJ, Yeo AR (1995). Breeding for salinity resistance in crop plants—where next? Aust. J. Plant. Physiol. 22: 875–884.

Foyer C, Halliwell B (1976).The presence of glutathione and glutathione reductase in chloroplasts: a proposed role in ascorbic acid metabolism. Planta. 133: 21–25.

Gao M, Tao R, Miura K, Dandekar AM, Sugiura A (2001). Transformation of Japanese persimmon (Diospyroskaki Thunb.) with apple cDNA encoding NADP-dependent sorbitol-6-phosphate dehydrogenase. Plant Sci. 160: 837–845.

Garg AK, Kim JK, Owens, TG, Ranwala AP, Do Choi Y, Kochian LV, Wu, RJ (2002). Trehalose accumulation in rice plants confers high tolerance levels to different abiotic stresses. PNAS 99:15898–15903.

Gaxiola RA, Li J, Unurraga S, Dang LM, Allen GJ, Alper, SL, Fink GR (2001). Drought- and salt-tolerant plants result from overexpression of the AVP1 H^+-pump. PNAS 98: 11444–11449.

Glenn EP, Brown JJ, Blumwald E (1999).Salt tolerance and crop potential of halophytes. Crit. Rev. Plant. Sci. 18: 227–255.

Greenway H, Munns R (1980). Mechanisms of salt tolerance in nonhalophytes. Annu. Rev. Plant. Physiol.. 31: 149–190.

Grover A, Aggarwal PK, Kapoor A, Katiyar-Agarwal S, Agarwal M, Chandramouli A (2003). Addressing abiotic stresses in agriculture through transgenic technology. Curr. Sci. 84: 355–367.

Halliwell B, Gutteridge JMC (1986). Oxygen free radicals and iron in relation to biology and medicine: some problems and concepts. Arch. Biochem. Biophys. 246: 501–514.

Hayashi H, Alia ML, Deshnium PG, Id M, Murata N (1997). Transformation of Arabidopsis thaliana with the codA gene for choline oxidase; accumulation of glycinebetaine and enhanced tolerance to salt and cold stress. Plant J. 12: 133–142.

Holmstrom KO, Somersalo S, Mandal A, Palva TE, Welin B (2000). Improved tolerance to salinity and low temperature in transgenic tobacco producing glycinebetaine. J. Exp. Bot. 51: 177–185.

Hong ZL, Lakkineni K, Zhang ZM, Verma DPS (2000). Removal of feedback inhibition of delta (1)-pyrroline-5-carboxylate synthetase results in increased proline accumulation and protection of plants from osmotic stress. Plant Physiol. 122: 1129–1136.

Huang J, Hirji R, Adam L, Rozwadowski KL, Hammerlindl JK, Keller WA, Selvaraj G (2000). Genetic engineering of glycinebetaine production toward enhancing stress tolerance in plants: metabolic limitations. Plant Physiol. 122: 747–756.

Jia W, Wang Y, Zhang S, Zhang J (2002). Salt-stress-induced ABA accumulation is more sensitively triggered in roots than in shoots. J. Exp. Bot. 53: 2201–2206.

Kasuga M, Liu Q, Miura S, Yamaguchi-Shinozaki K, Shinozaki K (1999). Improving plant drought, salt, and freezing tolerance by gene transfer of a single stress-inducible transcription factor. Nature Biotechnology 17: 287–291.

Kishor PBK, Hong Z, Miao GH, Hu CAA, Verma DPS (1995). Overexpression of 1-pyrroline-5-carboxylate synthetase increases proline production and confers osmotolerance in transgenic plants. Plant Physiol.108: 1387–1394.

Luming T,, Conglin H, Rong Y, Ruifang L, Zhiliang L, Lusheng Z, Yongqin W, Xiuhai Z Zhongyi W (2006). Overexpression AtNHX1 confers salt-tolerance of transgenic tall fescue. Afr. J. Biotech. 5 (11): 1041-1044.

Maathuis FJM, Sanders D (1995). Contrasting roles in ion transport of two K^+ channel types in root cells of Arabidopsis thaliana. Planta. 197: 456–464.

Nakayama H, Yoshida K, Ono H, Murooka Y, Shinmyo A (2000).

Ectoine, the compatible solute of Halomonas elongata, confers hyperosmotic tolerance in cultured tobacco cells. Plant Physiol. 122: 1239–1247.

Nanjo T, Kobayashi M, Yoshiba Y, Kakubari Y, Yamaguchi-Shinozaki K., Shinozaki, K (1999). Antisense suppression of proline degradation improves tolerance to freezing and salinity in Arabidopsis thaliana. FEBS Letters. 461: 205–210.

Ohta M, Hayashi Y, Nakashima A, Hamada A, Tanaka A, Nakamura T, Hayakawa T (2002). Introduction of a Na^+/H^+ antiporter gene from Atriplex gmelini confers salt tolerance in rice. FEBS Letters 532, 279–282.

Ono H, Sawads K, Khunajakr N, Tao T, Yamamoto M, Hiramoto M, Shinmyo A, Takano M, Murooka Y (1999). Characterization of biosynthetic enzymes for ectoine as a compatible solute in a moderately halophilic eubacterium, Halomonas elongata. J. Bacteriol. 181:91–99.

Pommerrenig B, Papini-Terzi FS, Sauer N (2007). Differential regulation of sorbitol and sucrose loading into the phloem of Plantago major in response to salt stress. Plant. Physiol. pp. 1029 – 1038.

Rathinasabapathi B (2000). Metabolic engineering for stress tolerance: installing osmoprotectant synthesis pathways. Ann. Bot. 86: 709–716.

Richards R. (1983). Should selection for yield in saline regions be made on saline or non-saline soils? Euphytica 32: 431–438.

Rontein D, Basset G, Hanson AD (2002).Metabolic engineering of osmoprotectant accumulation in plants. Metabolic Engineering. 4:49–56.

Roosens NH, Willem R, Li Y, Verbruggen II, Biesemans M, Jacobs M (1999). Proline metabolism in the wild-type and in a salt-tolerant mutant of Nicotiana plumbaginifolia studied by $(13)^C$ -nuclear magnetic resonance imaging. Plant Physiol. 121: 1281–1290.

Roxas VP, Smith RK, Allen ER, Allen RD (1997). Overexpression of glutathione S-transferase/glutathione peroxidase enhances the growth of transgenic tobacco seedlings during stress. Nat. Biotechnol. 15: 988–991.

Ruiz JM, Blumwald, E (2002). Salinity-induced glutathione synthesis in Brassica napus. Planta. 214: 965–969.

Rus A, Yokoi S, Sharkhuu A, Reddy M, Lee B, Matsumoto TK, Koiwa H, Zhu JK, Bressan RA, Hasegawa PM (2001). AtHKT1 is a salt tolerance determinant that controls Na^+ entry into plant roots. PNAS. 98: 14150–14155.

Sakamoto A, Murata N (2000). Genetic engineering of glycinebetaine synthesis in plants: current status and implications for enhancement of stress tolerance. J. Exp. Bot. 51: 81–88.

Shi H, Lee BH, Wu SJ, Zhu JK (2003). Overexpression of a plasma membrane Na^+/H^+ antiporter gene improves salt tolerance in Arabidopsis thaliana. Nat. Biotechnol. 21: 81–85.

Tarczynski M, Jensen R, Bohnert H (1993). Stress protection of transgenic tobacco by production of the osmolyte mannitol. Sci. 259: 508–510.

Uozumi N, Kim EJ, Rubio F, Yamaguchi T, Muto S, Tsuboi A, Bakker EP, Nakamura T, Screoder JI (2000). The Arabidopsis HKT1 gene homolog mediates inward Na^+ currents in Xenopus laevis oocytes and Na^+ uptake in Saccaharomyces cerevisiae. Plant Physiol. 122: 1249–1259.

Wu CA, Yang GD, Meng QW, Zheng CC (2004).The cotton GhNHX1 gene encoding a novel putative tonoplast Na^+/K^+ antiporter plays an important role in salt stress. Plant.Cell.Physiol. 45: 600–607.

Yancey PH, Clark ME, Hand SC, Bowlus RD, Somero GN (1982). Living with water stress: evolution of osmolyte systems. Science 217: 1214–1222.

Yeo ET, Kwon HB, Han SE, Lee JT, Ryu JC, Byu MO (2000). Genetic engineering of drought resistant potato plants by introduction of the trehalose-6-phosphate synthase (TPS1) gene from Saccharomyces cerevisiae. Mol. Cells. 10: 263–268.

Yin XY, Yang AF, Zhang KW, Zhang JR (2004).Production and analysis of transgenic maize with improved salt tolerance by the introduction of AtNHX1 gene. Acta Botanica Sinica. 46: 854–861.

Zhang HX, Blumwald E (2001). Transgenic salt-tolerant tomato plants accumulate salt in foliage but not in fruit. Nat. Biotechnol. 19: 765–768.

Zhang HX, Hodson JN, Williams JP, Blumwald E (2001). Engineering salt-tolerant Brassica plants: characterization of yield and seed oil quality in transgenic plants with increased vacuolar sodium accumulation. PNAS 98: 6896–6901.

Zhang JZ, Creelman RA, Zhu JK (2004). From laboratory to field. Using information from Arabidopsis to engineer salt, cold and drought tolerance in crops. Plant. Physiol. 135: 615–621.

Zhu B, Su J, Chang M, Verma DPS, Fan YL, Wu R (1998). Over expression of a delta1-pyrroline-5-carboxylate synthetase gene and

analysis of tolerance to water- and salt-stress in transgenic rice. Plant Sci. 139: 41–48.

Zhu JK (2001).Cell signaling under salt, water and cold stress.Curr. Opin. Plant Biol. 4: 401–406.

Zhu JK (2002). Salt and drought stress signal transduction in plants. Annu. Rev. Plant Biol. 53: 247–273.

Zhu JK (2003). Regulation of ion homeostasis under salt stress. Curr. Opin. Plant Biol. 6: 441–445.

11

High fructose corn syrup: Production, uses and public health concerns

Kay Parker, Michelle Salas and Veronica C. Nwosu*

Department of Biology, College of Science and Technology, North Carolina Central University, Durham, NC 27707, USA.

High fructose corn syrup (HFCS) is a liquid alternative sweetener to sucrose that is made from corn, the "king of crops" using chemicals (caustic soda, hydrochloric acid) and enzymes (α-amylase and glucoamylase) to hydrolyze corn starch to corn syrup containing mostly glucose and a third enzyme (glucose isomerase) to isomerize glucose in corn syrup to fructose to yield HFCS products classified according to their fructose content: HFCS-90, HFCS-42, and HFCS-55. HFCS-90 is the major product of these chemical reactions and is blended with glucose syrup to obtain HFCS-42 and HFCS-55. HFCS has become a major sweetener and additive used extensively in a wide variety of processed foods and beverages ranging from soft and fruit drinks to yogurts and breads. HFCS has many advantages compared to sucrose that make it attractive to food manufacturers. These include its sweetness, solubility, acidity and its relative cheapness in the United States (US). The use of HFCS in the food and beverage industry has increased over the years in the US. The increase in its consumption in the US has coincided with the increase in incidence of obesity, diabetes, and other cardiovascular diseases and metabolic syndromes. This study examines literature on the production and properties of HFCS and the possible health concerns of HFCS consequent to its consumption in a wide variety of foods and beverages in the typical US diet.

Key words: High fructose corn syrup, sweeteners, soft and juice drinks, baked goods, obesity, diabetes, metabolic syndrome, mercury; honey bees, colony collapse disorder.

INTRODUCTION

The bulk of the United States (US) diet comes from four crops: corn, wheat, soybean, and rice. Of the four crops, corn is arguably the most dominant and most profitable to farm with its cultivation being highly subsidized by the US government elevating corn to the "king of crops". Corn has been subjected to various genetic modifications that have resulted in a crop that is resistant to pesticides, a feature that has increased the productivity of corn for farmers. Corn is not only food for humans, but is also feed for farm animals as diverse as cattle, pigs, and poultry that are the major sources of meats for the US diet. Corn is the primary source of high fructose corn syrup (HFCS) in the US. Marshall and Kooi (1957) developed the process for making HFCS. HFCS is made by the chemical and enzymatic hydrolysis of corn starch containing amylose and amylopectin to corn syrup containing mostly glucose followed by the isomerization of the glucose in corn syrup to fructose to yield HFCS (Figure 1). Three categories of HFCS are in common use: HFCS-90 (90% fructose and 10% glucose) which is used in specialty applications but more importantly is blended with glucose syrup to yield HFCS-42 (42% fructose and 58% glucose) and HFCS-55 (55% fructose and 45% glucose). HFCS is called isoglucose in England and glucose-fructose in Canada, and was first introduced to the food and beverage industry in the late 1960s (HFCS-42 in 1967) and 1970s (HFCS-55 in 1977) to improve stability and functionality of various foods and beverages. Carbohydrate sweeteners are craved for their sweetness because they enhance the taste and enjoyment of various foods. They are mostly monosaccharides such as glucose, fructose, and galactose; and disaccharides such as sucrose, lactose, and maltose. They come in various forms such as cane and beet sugar, cane juice, molasses, honey, fruit juice concentrates, corn syrups,

*Corresponding author. E-mail: vcnwosu@nccu.edu.

and HFCS. Non carbohydrate sweeteners that have been in use include saccharin (discovered in 1879); cyclamate (discovered in 1937); aspartame (discovered in 1965); acesulfame (discovered in 1967) and sucralose (discovered in 1976). Sweeteners are measured on a sweetness index using sucrose as the baseline sugar with a sweetness index of 1.0. The impetus for the search for alternative and non-caloric sweeteners to sucrose has historically been for better health for diabetics and also for weight control. Cost is another major factor behind this search. The development of the relatively inexpensive HFCS has made it possible for it to become a viable alternative to sucrose and other natural sugars in a very short time. HFCS represents approximately 40% of all added caloric sweeteners in the US diet (Putnam and Allshouse, 1999). Sucrose contains fructose and glucose in equal amounts linked by glucosidic bond. This bond has to be broken to release both monosaccharides for metabolism. HFCS-55 contains more fructose than glucose and this fructose is more immediately available because it is not bound up in sucrose. There are differences in the metabolism of glucose and fructose with that of glucose being better understood than that of fructose. The use of HFCS in the food and beverage industry has increased over the years in the US. This study examines literature on the production and properties of HFCS and the possible health concerns of HFCS consequent to its consumption in a wide variety of foods and beverages in the typical US diet. The health implycation of HFCS consumption is subject to intense debate. The increase in its consumption in the US has coincided with the increase in incidence of obesity, diabetes, and other cardiovascular diseases and metabolic syndromes. Thus published literature was surveyed to collate data on the impact of HCFS on human health. Of concern is the possible contamination of HFCS with mercury during processing. Also of concern is the possible toxicity of HFCS and its by-products to honey bees.

PRODUCTION AND USES OF HFCS

The schematic of HFCS production is shown in Figure 1. HFCS is produced from corn. The corn grain undergoes several unit processes starting with steeping to soften the hard corn kernel followed by wet milling and physical separation into corn starch (from the endosperm); corn hull (bran) and protein and oil (from the germ). Corn starch composed of glucose molecules of infinite length, consists of amylose and amylopectin and requires heat, caustic soda and/or hydrochloric acid plus the activity of three different enzymes to break it down into the simple sugars glucose and fructose present in HFCS. An industrial enzyme, α-amylase produced from *Bacillus* spp., hydrolyzes corn starch to short chain dextrins and oligosaccharides. A second enzyme, glucoamylase (also called amyloglucosidase), produced from fungi such as *Apergillus*, breaks dextrins and oligosaccharides to the simple sugar glucose. The product of these two enzymes is corn syrup also called glucose syrup. The third and relatively expensive enzyme used in the process is glucose isomerase (also called D-glucose ketoisomerase or D-xylose ketolisomerase), that converts glucose to fructose.

While α-amylase and glucoamylase are added directly to the processing slurry, pricey glucose isomerase is immobilized by package into columns where the glucose syrup is passed over in a liquid chromatography step that isomerizes glucose to a mixture of 90% fructose and 10% glucose (HFCS-90). Whereas inexpensive α-amylase and glucoamylase are used only once, glucose isomerase is reused until it loses most of its enzymatic activity. The α-amylase and glucoamylase used in HFCS processing have been genetically modified to improve their heat stability for the production of HFCS. In the US, four companies control 85% of the $2.6 billion HFCS business—Archer Daniels Midland, Cargill, Staley Manufacturing Co, and CPC International.

With clarification and removal of impurities, HFCS-90 is blended with glucose syrup to produce HFCS-55 (55% fructose) and HFCS-42 (42% fructose). Both HFCS-55 and HFCS-42 have several functional advantages in common, but each has unique properties that make them attractive to specific food manufacturers. Because of its higher fructose content, HFCS-55 is sweeter than sucrose and is thus used extensively as sweetener in soft, juice, and carbonated drinks. HFCS-42 has a mild sweetness and does not mask the natural flavors of food. Thus it is used extensively in canned fruits, sauces, soups, condiments, baked goods, and many other processed foods. It is also used heavily by the dairy industry in yogurt, eggnog, flavored milks, ice cream, and other frozen desserts. The use of HFCS has increased since its introduction as a sweetener (Figure 2). Although, its use peaked in 1999, it rivals sucrose as the major sweetener in processed foods. The US is the major user of HFCS in the world, but HFCS is manufactured and used in many countries around the world (Vuilleumier, 1993). HFCS has functional advantages relative to sucrose.

These include HFCS's relative cheapness (at 32 cents/lb versus 52 cents/lb for sucrose); greater sweetness with HFCS being sweeter than sucrose (Table 1), better solubility than sucrose (Table 2) and ability to remain in solution and not crystallize as can sucrose under certain conditions. Moreover, HFCS is liquid and thus is easier to transport and use in soft drink formulations (Hanover and White, 1993). It is also acidic and thus has preservative ability that reduces the use of other preservatives. HFCS has little to no nutritional value other than calories from sugar (Table 3). Analysis of food consumption patterns using USDA (2008) food consumption tables for the US from 1967 to 2000 (Bray et al., 2004) showed that HFCS consumption increased

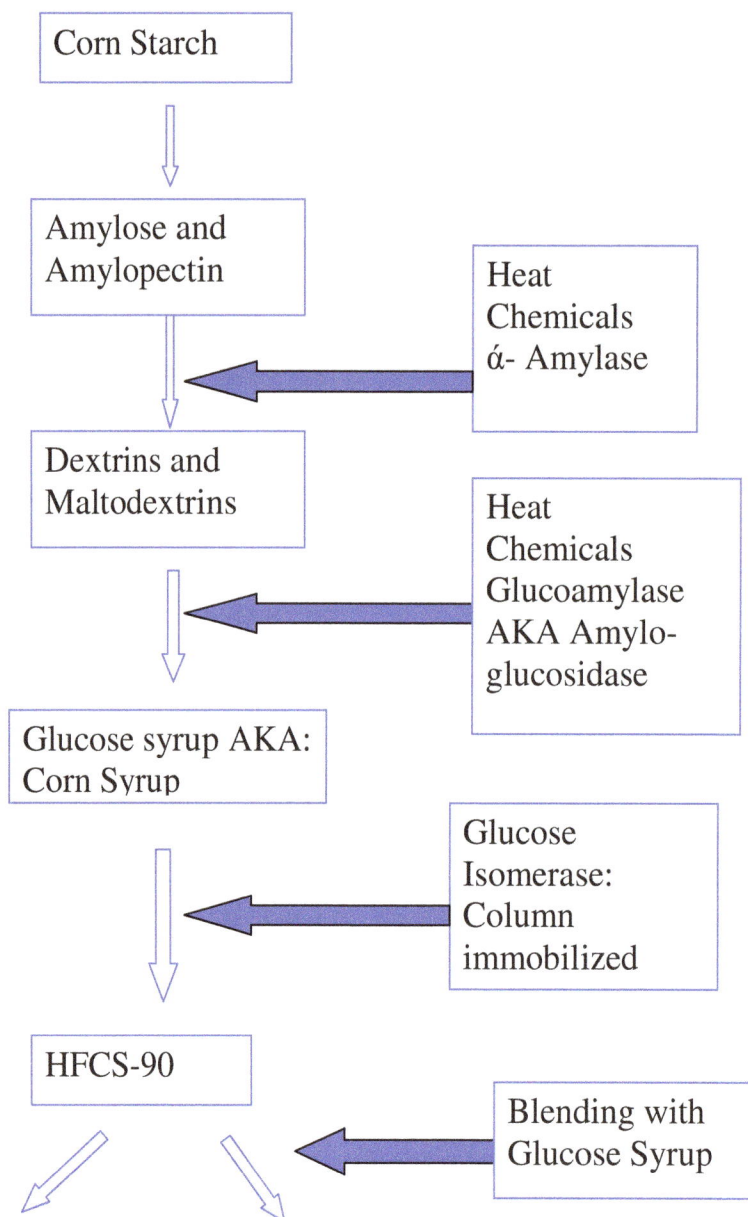

Figure 1. Schematic of HFCS production from corn starch. Amylose and Amylopectin are the two components of starch. The production of glucose syrup from corn starch is dependent on the activity of various amylases and glucoamylase (also known as amyloglucosidase), heat and chemicals such as caustic soda and/or hydrochloric acid. Glucose syrup produced is then passed through an immobilized column of glucose isomerase where glucose is isomerized to fructose to yield HFCS, primarily HFCS-90 which is then blended with glucose syrup to produce HFCS-55 and HFCS-42. (Authors' original schematic).

1000% between 1970 and 1999 with HFCS representing greater than 40% of all sweeteners added to foods and beverages and the sole sweetener in soft drinks. The average daily consumption of HFCS for all Americans 2 years or older is about 50 g/person or about 132 kcal/person with the top 20% of HCFS consumers ingesting as much as 316 kcal/day. Thus HFCS is a major source of dietary fructose.

PUBLIC HEALTH CONCERNS

There are three major concerns about the use of HFCS related to public health. The first is its possible role in

Dry weight, pounds per capita per year

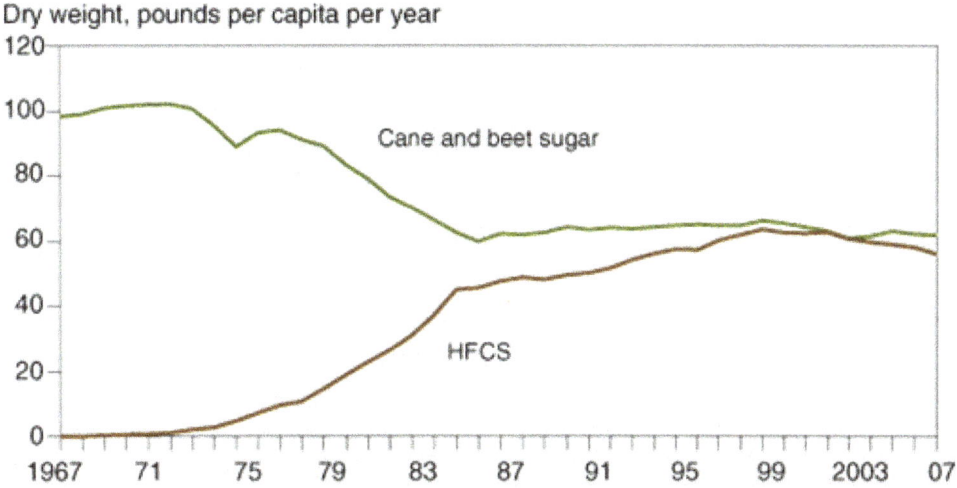

Figure 2. The use of HFCS, sucrose and other sweeteners in the US. Note the growth of use of HFCS between 1985 and 2005 with its use peak in 1999 as consumers began to question its extensive use: Data from USDA Economic Research Service (Putnam and Allshouse, 1999; and USDA: Economic Research Services Amber Waves, Feb, 2008).

Table 1. Relative sweetness of selected sugar solutions (5%) and other sweeteners. Sweetness is measured against sucrose as the reference sugar with a sweetness index of 1.0. Figures compiled from multiple sources including Godshall (1997).

Sugar or sweetener	Relative sweetness
Sucrose	1.0
Invert sugar	0.85 - 1.0
Fructose	1.3
Glucose	0.56
Galactose	0.4-0.6
Maltose	0.3-0.5
Lactose	0.2-0.3
Xylitol	1.01
Cyclamates	30-80
Acesulfame K (Sunnette®)	200
Aspartame (Equal®, Nutrasweet®)	100-200
Saccharin (The Pink Stuff)	200-300
Stevioside	300
Sucralose (Splenda®)	600
Thaumatin (Talin®)	2000-3000

Table 2. Solubility of selected sugars at 50°C. Solubility measured as grams of sugar dissolved in 100 ml water. Data from McWilliams (2008).

Sugar	Grams of sugar dissolved in 100 ml of water
Fructose	86.9
Sucrose	72.2
Glucose	65.0
Maltose	58.3
Lactose	29.8

Table 3. Nutritional values calculated per 100 g of HFCS. Percentages are relative to US recommendations for adults. Data from USDA nutrient database (USDA.gov).

Nutritional Items	Value
Energy	1,176 kJ (281 kcal)
Carbohydrates	76 g
Dietary fiber	0 g
Fat	0 g
Protein	0 g
Water	24 g
Riboflavin (Vitamin B 2)	0.019 mg (1%)
Niacin (Vitamin B3)	0 mg (0%)
Pantothenic acid (Vitamin B5)	0.011 mg (0%)
Vitamin B6	0.024 mg (2%)
Folic acid (Vitamin B9)	0 µg (0%)
Vitamin C	0 mg (0%)
Calcium	6 mg (1%)
Iron	0.42 mg (3%)
Magnesium	2 mg (1%)
Phosphorus	4 mg (1%)
Potassium	0 mg (0%)
Sodium	2 mg (0%)
Zinc	0.22 mg (2%)

obesity, cardiovascular disease, and other metabolic syndromes. The second is mercury contamination of HFCS samples during production and the third its toxicity to honey bees with possible contribution to colony collapse disorder (CCD) of honey bees.

Role in metabolic syndromes: obesity, diabetes, and other cardiovascular diseases

Several studies published in the last 10 years present data that suggest a correlation between increased consumption of HFCS in the past three decades with increased incidence of obesity and cardiovascular diseases in the US. Others studies have been published in defense of HFCS and emphasizing the absence of strong evidence that HFCS and sucrose have differing metabolic effects; and suggesting no causal role for HFCS in obesity. Proponents point to the problem of obesity to be due primarily to high caloric intake coupled with inactivity in the general population. White and Foreyt (2006) published ten myths associated with HFCS in an effort to underscore that "claims that HFCS bears a unique responsibility for the current obesity epidemic in the US are based on misunderstanding". There has been a reassessment of the overall intake of high caloric sweeteners by several scientific organizations such as the American medical association (AMA), the American dietetic association (ADA) and the International life sciences institute (ILSI). The consensus is that HFCS

should not be singled out from other sweeteners as the cause of increasing obesity in the US, and that the broader focus should be on combating the increase in consumption of high caloric diets coupled with increased inactivity in the general population. However, HFCS is a relatively recent addition to the US diet and studies to understand its functionality and possible adverse effects are warranted. Although, HFCS contains the same monosacharides as sucrose, the glycosidic linkage between fructose and glucose in sucrose is cleaved to initiate digestion, whereas both monosacharides are free and unlinked in HFCS. The digestion, absorption and metabolism of fructose are different from those of glucose. Whereas glucose is absorbed in the upper gastro-intestinal tract by a sodium-glucose cotransporter system, fructose is absorbed lower in the intestinal tract by a non-sodium-dependent process (Bray et al., 2004). Following absorption both glucose and fructose enter the hepatic portal system to the liver where fructose can be converted to glucose or passed into the general circulatory system. Petersen et al. (2001) presented evidence that fructose can modulate carbohydrate metabolism in the liver. They reported that the addition of small catalytic amounts of fructose to orally ingested glucose increased glycogen synthesis in the liver in human subjects and reduced glycemic responses in subjects with type 2 diabetes mellitus. The problem arises when large amounts of fructose are ingested such as from HFCS sources. The excess fructose thus provides a ready source of carbon for lipogenesis in the liver which can have negative health consequences. Glucose entry into cells is through insulin dependent Glut-4 transport system whereas fructose enters cells through a Glut-5 insulin independent pathway (Elliott et al., 2004). Once inside cells, glucose enters the glycolytic pathway through phosphorylation to glucose-6-phosphate by glucokinase, an enzyme that tightly controls the production of glucose-6-phosphate that is ultimately converted to two pyruvate molecules. Fructose on the other hand, inside cells, is phosphorylated to fructose-1-phosphate, a molecule that is readily cleaved by aldolase to trioses that form the backbone structure for the synthesis of triglycerides and phospholipids (Mayes, 1993). Glucose contributes to the feeling of satiety because its ingestion influences insulin release which increases leptin release (Saad et al., 1998). Fructose does not influence insulin release, thus its ingestion may lead to a low insulin concentration that results in low leptin levels. Leptin is a satiety hormone that curbs appetite, hence low levels of leptin would be expected to increase food intake. Low level of leptin in humans is associated with increased weight gain and obesity (Farooqi et al., 2001; Rosenbaum et al., 2002). Numerous other studies have been published on the role of HFCS in obesity, diabetes, and other metabolic syndromes. The major findings of these studies implicate metabolic syndromes that include the following: caloric over consumption (Bray et al., 2004); weight gain and obesity (Bray et al., 2004; Forshee et al., 2007; Jurgens et al., 2005; Monsivais et al., 2007; Shapiro et al., 2008); insulin resistance (Elliot et al., 2004 ; Faeh et al., 2005) stimulation of the liver (Faeh et al., 2005; Stanhope and Havel, 2008); lipogenesis and enhanced production of triglycerides (Petersen et al., 2001; Bray et al., 2004); leptin resistance and decreased ability to regulate fullness (Shapiro et al., 2008); increased glycosylation of proteins and possible onset of type 2 diabetes (Gross et al., 2004). There is a great deal of variation in study designs in published reports. Animal-based metabolic studies that used pure fructose showed very adverse metabolic effects. There is need to separate effect of fructose alone from effect of HFCS in the diet. Monsivais et al. (2007) studied hunger, appetite and food intake of participants in five groups and given beverage sweetened with the non-caloric sweetener aspartame; or soft drink sweetened with sucrose, HFCS-55 or HFCS-45; or 1% milk; or no-beverage control. They found no difference in how the four caloric beverages affected appetite and food intake and concluded that a calorie from HFCS is no different than a calorie from sucrose or from milk. Melanson et al. (2007) studied thirty lean women on randomized 2-day visits during which participants were given beverages sweetened either with sucrose or with HFCS as 30% of energy on an isocaloric diet. They found no significant differences between the two sweeteners on fasting plasma glucose, insulin, leptin, and ghrelin and concluded that when fructose is consumed in the form of

HFCS, the measured metabolic responses do not differ from sucrose in lean women. They, however, called for further research to see if the findings hold true for obese individuals, males, and for long-term studies. Stanhope et al. (2008) called for carefully controlled and long-term studies to fully understand the role of HFCS in metabolic disorders associated with ill-health. In both their long and short-term studies using pure fructose, they showed that consumption of fructose-sweetened beverages substantially increased postprandial triglyceride levels compared with glucose-sweetened beverages. They also reported increases in apolipoprotein B levels in their long-term studies. In a subsequent study with thirty-four men and women given sucrose and HFCS-sweetened beverages, they reported gender differences in post-prandial triglyceride profiles. There is no doubt of the need for ongoing studies in this area not just on HFCS, but other sugars and their contributions to high caloric intake that lead to weight gain, obesity and associated metabolic syndromes.

Mercury contamination

A second concern related to HFCS consumption is the presence of trace amounts of mercury in HFCS manufactured in the US. Caustic soda used in HFCS production is typically made at chlor-alkali plants that use mercury cells. Mercury is a potent neurological toxin (Dufault et al., 2009) that has been shown to be toxic to humans. Dufault et al. (2009) collected and analyzed twenty HFCS samples from three different manufacturers and found that 11 of 20 samples contain levels of mercury that were below detectable limits of 0.005 µg of mercury/g of HFCS while 9 of 20 had levels that ranged between 0.065 to 0.570 µg of mercury/g of HFCS. Since the average daily consumption of HFCS is approximately 50 g/person, Dufuault et al. (2009) stated that there was need to account for mercury from this source in the diet of sensitive populations such as children and others when examining total exposure to mercury. Of interest in this study is that 9 of the 11 below detection level samples came from 1 of the 3 manufactures indicating manufacturing process using caustic soda produced by a membrane chlor-alkali plant which does not use mercury. Eight of the 9 samples that had measurable mercury levels came from the other 2 manufacturers indicating the use of mercury grade caustic soda or hydrochloric acid in the manufacturing process for HFCS. Thus manufacturers need to use processing methods that mitigate the presence of mercury in the finished HFCS product.

Toxicity to honey bees

The discovery of the attraction of bees to HCFS was accidental when workers at HCFS plants noticed that honey bees clustered and feed on HCFS spills during loading of the product into shipping thanks (Barker and Lehner, 1978). Since then HFCS has become a sucrose alternative for honey bees. It is used by commercial beekeepers as food for honey bees to promote brood production in the spring for commercial pollination. It is also used to feed honey bees when sources of pollen and nectar are scarce. Hydroxymethylfurfural (HMF) is formed at high temperatures from dehydration of fructose. HMF in honey is an indication of its aging. Codex Alimentarius Commission prohibits the sale of honey meant for human consumption with HMF levels greater than 40 ppm. Leblanc et al. (2009) found that at temperatures above 45°C, HFCS begins to form HMF, a byproduct that is very toxic to bees. In addition, levulinic and formic acids which are byproducts of HMF are also toxic to bees. Toxicity is seen as dysentery-like symptoms in bees. Could the feeding of HFCS to honey bees be a contributory factor in the colony collapse disorder (CCD) of honey bees? The carbohydrate composition of HFCS and honey are more similar than that of honey and sucrose solution. The history of CCD began in 1971 with observations of a dramatic, but steady reduction in the number of wild honeybees in the United States and a somewhat gradual decline in the number of colonies maintained by beekeepers. CCD is a little-understood phenomenon in which worker bees from a Western honey bee colony abruptly disappear. CCD was originally found in Western honey bee colonies in North America in late 2006. The exact causes(s) of CCD are unknown, but factors suspected to be involved include: poor nutrition, immunodeficiency (or immunosuppression), overuse or misuse of pesticides, diseases caused by pathogens, mites, or fungi, and poor beekeeping practices. Characteristics of collapsed colony include: the complete absence of adult bees, with very little dead bees present, presence of still capped brood cells (indicating the bee colony collapsed leaving developing bee larva behind: this is hallmark), the queen is either gone (or dead), and minimal effort to defend the hive against predators or competitors such as wax moths. The history of CCD goes back to the 1970's but in terms of the severity of CCD; the incidence has been highest since 2006 with reports of loss of fifty to ninety percent of colonies by beekeepers around the US. Research is needed to exclude HFCS as a contributory factor in CCD.

Food items that contain HFCS

Grocery foods items found to contain HFCS are numerous. These include baked goods such as pastries; biscuits, breads, cookies, and shortcakes; soft drinks; juice drinks; carbonated drinks; jams and jellies; dairy products including ice creams, flavored milks, eggnog, yogurts and frozen desserts; canned ready to eat foods including sauces and condiments; cereals and cereal bars; and many other processed foods. Majority of processed foods in the US contain HFCS to meet some

functionality in the foods.

CONCLUSION

Fructose and glucose are monosaccharides found in equal proportion in sucrose but in slightly unequal amounts in HFCS. The metabolism of glucose is well understood while that of fructose requires further research especially in light of its over consumption through HFCS in the US diet. Makers of HFCS under the banner of the corn refiners association have mounted very strong advertising blitz to assure the public that HFCS is safe especially since the use of HFCS peeked and started to decline in 1999. The public largely remains skeptical and there has been push back from health conscious individuals in the US against the ubiquitous presence of HFCS in the US diet. Several companies are responding to the push back and some are starting to offer foods and beverages without added HCFS giving individuals choices in selecting sweeteners in their diets.

REFERENCES

Barker RJ, Lehner Y (1978). Laboratory comparison of high fructose corn syrup, grape syrup, honey, and sucrose syrup as maintenance food for caged honeybees. Apidolo., 9 (2): 111-116.

Bray GA, Nielson SJ, Popkin BM (2004). Consumption of high-fructose corn syrup in beverages may play a role in the epidemic of obesity. Am. J. Clin. Nutr., 79 (4): 537-543.

Dufault R, LeBlanc B, Schnoll R (2009). Mercury from Chlo-alkali plants: measured concentrations in food product sugar. Environ. Health 8: 2 doi10.1186/1476-069X-8-2.

Elliot SS, Keim NL, Stern JS, Teff K, Havel PJ (2004). Fructose, weight gain, and the insulin resistance syndrome. Am. J. Clin. Nutr., 79 (4): 537-543.

Faeh D, Minehira K, Schwarz JM, Periasamy R, Parks S, Tappy L (2005). Effect of fructose overfeeding and fish oil administration on hepatic de novo lipogenesis and insulin sensitivity in healthy men. Diabetes, 54 (7): 1907-1913.

Farooqi IS, Keogh JM, Kamath S, et al. (2001). Partial leptin deficiency and human adiposity. Nature, 414: 34-35.

Forshee RA, Story ML, Allison DB (2007). A critical examination of the evidence relating high fructose corn syrup and weight gain. Crit. Rev. Food Sci. Nutr., 47(6): 561-582.

Godshall MA (1997). How carbohydrates influence food flavors. Food Technol., 51 (1): 63.

Gross LS, Li L, Ford ES, Simin L (2004). Increased consumption of refined carbohydrates and the epidemic of type 2 diabetes in the United States: an ecologic assessment. Am. J. Clin. Nutr., 79 (5): 774-779.

Hanover LM, White JS (1993). Manufacturing, composition, and applications of fructose. Am. J. Clin. Nutr. 58 (suppl 5): 724S-732S.

Jurgens H, Haass W, Castaneda TR (2005). Consuming fructose-sweetened beverages increases body adiposity in mice. Obesity Res., 13: 1146-1156.

Leblanc W, Eggleston G, Sammataro D, Cornett C, Dufault R, Deeby T, St Cyr E (2009). Formation of hydroxymethylfurfural in domestic high-fructose-corn syrup and its toxicity to the honey bee (Apis mellifera). J. Agric. Food Chem., 57: 7369-7376.

Marshall RO, Kooi ER (1957). The enzymatic conversion of d-glucose to d-fructose. Sci., 125 (3249): 648-649.

Mayes PA (1993). Intermediary metabolism of fructose. Am. J. Clin. Nutr., 58: 754S-765S.

McWilliams M (2008). Foods: Experimental Perspectives, sixth edition, Prentice Hall, Upper Saddle River, New Jersey and Columbus, Ohio,

p. 144.

Melanson KJ, Zukley L, Lowndes J, Nguyen V, Angelopoulos TJ, Rippe JM (2007). Effects of high-fructose corn syrup and sucrose consumption on circulating glucose, insulin, leptin, and ghrelin and on appetite in normal-weight women. Nutr., 23(2): 103-112.

Monsivais P, Perrigue MM, Drewnowski A (2007). Sugars and satiety: does the type of sweetener make a difference? Am. J. Clin. Nutr., 86(1): 116-123.

Petersen KF, Laurent D, Yu c, Cline GW, Shelman GI (2001). Stimulating effects of low-dose fructose on insulin-stimulated hepatic glycogen synthesis in humans. Diabetes, 50: 1263-1268.

Putnam JJ, Allshouse JE (1999). Food consumption, prices and expenditures, 1970-1997. US Department of Agriculture Economic Research Service statistical bulletin no. 965, April. Washington, DC: US Government Printing Office.

Rosenbaum M, Murphy EM, Heymsfield SB, Matthews DE, Leibel RL (2002). Low dose leptin administration reverses effects of sustained weight-reduction on energy expenditure and circulating concentrations of thyroid hormones. J. Clin. Endocrinol. Metab., 87: 2391-2394.

Saad MF, Khan A, Sharma A, et al. (1998). Physiological insulinemia acutely modulated plasma leptin. Diabetes, 47: 544-549.

Shapiro A, Mu W, Roncal CA, Cheng KY, Johnson RJ, Scarpace PJ (2008). Fructose-induced leptin resistance exacerbates weight gain in response to subsequent high fat feeding. Am. J. Physiol. Regul. Integr. Comp. Physiol., 295 (5): R1370-R1375.

Stanhope KL, Griffen SC, Bair BR, Swarbrick MM, Keim NL, Havel PJ (2008). Twenty-four hour endocrine and metabolic profiles following consumption of high fructose corn syrup-, sucrose-, fructose-, and glucose-sweetened beverages with meals. Am. J. Clin. Nutr., 87(5): 1194-1203.

Stanhope KL, Havel PJ (2008). Endocrine and metabolic effects of consuming beverages sweetened with fructose, glucose, sucrose, or high-fructose corn syrup. Am J. Clin. Nutr., 88(6): 1733S-1737S.

USDA (2008). Economic Research Services. Amber Waves: The Economics of Food, Farming, Natural Resources, and Rural America, February: http:www.ers.usda.gov/Amber Wave/February 08/Findings/High Fructose.htm.

Vuilleumier S (1993). Worldwide production of high-fructose syrup and crystalline fructose. Am. J. Clin. Nutr., 58: 733S-736S.

White JS, Foreyt JP (2006). Ten myths about high-fructose corn syrup. Food Technol., 60(10) 96-96.

12

Comparative analysis of different immunological techniques for diagnosing fasciolosis in sheep: A review

Irfan-ur-Rauf Tak[1], Jehangir Shafi Dar[1], B. A. Ganai[1], M. Z. Chishti[1], R. A. Shahardar[2], Towsief Ahmad Tantry[1], Masarat Nizam[1] and Shoaib Ali Dar[3]

[1]Centre of Research for Development, University of Kashmir, Srinagar-190 006, India.
[2]Department of Veterinary Parasitology, SKUAST, Kashmir, India.
[3]Department of Zoology, Punjabi University Patiala.

Fasciolosis is a worldwide zoonotic infection caused by liver flukes of the genus Fasciola, of which *Fasciola hepatica* and a larger species, *Fasciola* gigantica are the most common representatives. These two food-borne trematodes usually infect domestic ruminants and cause important economic losses to sheep, goats and cattle. In commercial herds, fasciolosis is of great economic significance worldwide with losses estimated to exceed 2000 million dollars yearly, affecting more than 600 million animals, in articles reported a decade ago. In addition, *F. hepatica* causes an estimated loss of $3 billion worldwide per annum through livestock mortality, especially in sheep, and by decreased productivity via reduction of milk and meat yields in cattle. The parasitological diagnosis of fasciolosis is often unreliable because the parasite's eggs are not found during the prepatent period. Even when the worms have matured, the diagnosis may still be difficult since eggs are only intermittently released. Repeated examinations of stools are usually required to increase the accuracy of the diagnosis. Early diagnosis of fasciolosis is necessary for institution of prompt treatment before irreparable damage of the liver occurs. For these reasons, serology is the most dependable method for diagnosing fasciolosis. Attempts have been made to diagnose fasciolosis by detecting antibodies in the serum of sheep suspected of being infected with the flukes. Advances in immunodiagnosis have focused on detection of *Fasciola* antigens in host body fluid; these tests have an advantage over antibody detection because antigenemia implies recent and active infection. Similarly, somatic and excretory secretory (E/S) antigens of *Fasciola* sp. or their partially purified component are the commonest source of antigens used in protection trials and serodiagnosis. Thus, the aim of the present review is to encourage more young researchers to initiate work on this aspect of these economically cosmopolitan parasites.

Key words: Fasciolosis, Antigenemia, E/S antigens, serodiagnosis, immunoassay, *Fasciola* spp., zoonotic disease.

INTRODUCTION

Livestock infection by the liver flukes like *Fasciola hepatica* and *Fasciola gigantica* causes major economic losses worldwide. Mostly, the infection by members of the genus *Fasciola*, commonly known as liver flukes, may be responsible for morbidity and mortality in most mammal species, but have particular importance in sheep and cattle to livestock producers. Infection with *F. hepatica* and *F. gigantica* is regarded as one of the most common single helminth infection of ruminants in Asia and Africa (Hammond and Sewell, 1990). *Fasciola* spp. parasitizes a wide spectrum of domestic and wild animals (e.g., sheep, cattle, buffaloes and deer) and it causes a huge economic loss of $3 billion annually to the agriculture sector worldwide through losses of milk and meat yields (Mas-Coma et al., 2005; Robinson and Dalton, 2009). The immature flukes after penetrating the liver capsule migrate into the liver hepatic tissue. This migration usually cause trauma with hemorrhages, necrosis and subsequent granulation end by liver cirrhosis (Ozer et al., 2003). It has been reported that sheep and cattle do not develop strong immunity to infection by *Fasciola* species, or to re-infections, and this lack of resistance in ruminants is believed to be associated with the inability of their macrophages to produce nitric oxide. The parasitological diagnosis of fasciolosis is often unreliable because the parasite eggs are not found during the prepatent period (Noureldin et al., 2004). Even when the worms have matured, the diagnosis may still be difficult since eggs are only intermittently released. Repeated examinations of stools are usually required to increase the accuracy of the diagnosis. Early diagnosis of liver fluke infection is necessary for institution of prompt treatment before irreparable damage of the liver occurs (Rokni et al., 2004). For these reasons, serology is the most dependable diagnostic method. Attempts have been made to diagnose fasciolosis by detecting antibodies in the serum patient suspected of being infected with the flukes (Maleewong et al., 1999). Advances in immuno-diagnosis have focused on detection of parasite antigens in host body fluid; these tests have an advantage over antibody detection because antigenemia implies recent and active infection (Cornelissen et al., 1999). The somatic and E/S antigens of *Fasciola* spp. or their partially purified component are the commonest source of antigens used in protection trials and serodiagnosis (G'nen et al., 2004). Immunodiagnosis of parasitic disease is mainly based on antibody detection (Fagbemi et al, 1999) and revealed both recent and current infections with early diagnosis. To obtain reliable diagnostic method or to identify crude antigens, many authors prepared antigens from whole worm (Hillyer et al., 1987) or from tegument (Charmy et al., 1997) also coproantigen (Allan et al., 1996), egg antigen (Khalil et al., 1989 and Abdel-Rahman and Abdel Mageed, 2000), and excretory secretory products (Espino et al., 1994). Currently, haemaglutination (HA), indirect fluorescence antibody test (IFAT), immunoperoxydase (IP), counter-electrophoresis (CEP) and enzyme linked

immunosorbent assay (ELISA) are used in the early diagnosis of this disease, but they have some disadvantages such as cross reactions with other trematodes leading to false positive results. Therefore, the reliability of these tests is not high. In recent years, sodium dodecyl sulphate polyacrylamide gel electrophoresis (SDS-PAGE) and Western blotting procedures have initiated a new era in immunodiagnosis which greatly reduced cross-reactions. Enzyme-linked immunetrotransfer blot analysis (EITB) or Western blotting is evaluated in some research centers and encompasses remarkable sensitivity and specificity in diagnoses of the fasciolosis. Evaluation of sandwich ELISA and Dot ELISA as an immunological assay is used for detecting *Fasciola* copro-antigen and serum antigens in infected sheep, thus presenting an experimental trial that could be of value in providing a tool that may help in immunodiagnosis of fasciolosis.

OBJECTIVES OF IMMUNODIAGNOSIS

The objective of research in immunodiagnosis of fasciolosis is to develop rapid, cheap and technically easy tests that can be used in epidemiological surveys to evaluate the effects of various national or international schemes of control in areas where these infections are endemic. It should provide tests that have a high degree of sensitivity and that are specific for each infection, thus enabling their employment in immunodiagnosis even when few parasites are available for direct parasitological examinations. This is an important consideration in epidemiological surveys since it is recognized that in endemic areas, only a portion of the people carrying an infection may present clinical symptoms. Research in immuneassays also needs to provide tools that assess the effectiveness of chemotherapy or other curative measures, and thereby permit monitoring of treatment. Finally, it should provide tests that identify those individuals or animals that develop immunity to the infection. Such tests will be valuable in assessing the efficacy of vaccine programmes that can be expected in the future when anti-parasite vaccination becomes available.

NEW DEVELOPMENTS ENCOURAGED

Progress in the development of RIA, ELISA and related procedures has not yet, however, been so extensive as to replace conventional techniques. The main pitfalls have been the lack of commercial pressure to develop test kits, and the lack of good reagents. The specificity and sensitivity of the immunoassay depends on the

Table 1. Immunodiagnosis: Antigen-antibody detection.

S/N	Antigen catalogue	Techniques	Agents	Sensitivity (%)	Specificity (%)	References
1.	Somatic antigen	ELISA	Sheep serum		95	Cornelissen et al. (1992)
1a.	Somatic antigen	IHA	Sheep serum		86	Cornelissen et al. (1992)
2.	Crude excretory-secretory products (ES)	ELISA	Sheep serum		95	Cornelissen et al. (1992)
2a.	Crude excretory-secretory products (ES)	IHA	Sheep serum		86	Cornelissen et al. (1992)
3	28 kDa antigen (purified from ES)	ELISA		100		Dixit et al. (2002)
3a.	28 kDa antigen (purified from ES)	Western blot		100		Dixit et al. (2002)
3b.	28 kDa antigen (purified from ES)	Dipstick-ELISA		100		Dixit et al. (2002)
4.	mAb MM3	Copro ELISA	Sheep stool sample	0.3 ng/ml of F. hepatica ES antigen (100% with 1 fluke)	100	Mezo et al. (2004, 2007)

technology and on the reagents used. Limitations are now set by the lack of well-defined reagents and the false positive and false negative reactions that are seen as a consequence of this. The importance of the need for improved immunodiagnostic tests to use in individual or epidemiological studies is widely recognized and reflected in the priorities established by the Special Programme on Research and Training in Tropical Diseases of the UNDP, World Bank, and WHO, as well as by the IAEA's subprogramme component on parasitic diseases. Recent advances in separation techniques, and the production of antigens and antibodies using genetic engineering and biotechnology, augur well with development of improved serodiagnostic immunoassays using radionuclide and other tracers. This development is being encouraged by both WHO and IAEA.

TECHNIQUES USED FOR SERODIAGNOSIS OF FASCIOLOSIS IN SHEEP

Immunodiagnosis: Antibody detection

Infection with *Fasciola* spp. results in a specific antibody response. These antibodies can be detected in either serum or milk (Charlier et al., 2007). Several techniques have been described for the detection of antibodies against *Fasciola* spp. infection in sheep, such as the indirect hemagglutination test (IHA) (Levieux et al., 1992),

indirect immunofluorescence assay (IFA) (Hanna and Jura, 1977), ELISA and the Western immunoblot (Hillyer and Soler de Galanes, 1988), Dot- ELISA (Shaheen et al., 1989) and Micro-ELISA (Carnevale et al., 2001).

Antibodies to *Fasciola* spp. in infected hosts can be detected by ELISA (Ab ELISA) as early as one to two weeks post-infection (Hillyer et al., 1992), while eggs of flukes are found in faeces only after 12 - 14 weeks of infection (Burger, 1992). The Ab ELISAs have sensitivities and specificities of 87-100 and 86-100%, respectively. However, cross-reactions were seen with serum samples obtained from patients with hydatidosis and toxocariasis (Rokni et al., 2004) when using crude excretory secretory products (ES) of adult worms as the antigen. To improve the sensitivity and specificity of Ab ELISA, antigens purified from crude ES of flukes, recombinant antigens, or synthetic protein antigens should be used (Cornelissen et al., 2001; Silva et al., 2004; Yokanath et al., 2005) (Table 1).

Immunodiagnosis: Antigen detection

Active infection by *Fasciola* spp. can be demonstrated by the detection of metabolic products of flukes in the circulation. Such a test can also be used to confirm the efficacy of chemotherapy. Several assays have been developed to detect *Fasciola* spp. antigen in serum and faeces using monoclonal antibodies (Espino et al., 1994; Fagbemi et al., 1997; Mezo et al., 2004).

The antigen in blood can be detected by ELISA from one week post-infection onwards. However, Ag-ELISA has not been further developed because antigenaemia only develops when immature flukes are actively migrating through the liver parenchyma during 1-3 weeks post-infection and circulating antigens cannot be detected anymore once the flukes are established and mature to adult worms (Langley and Hillyer, 1989).

Copro Ag-ELISA has been applied to detect ES productions of *Fasciola* in stool samples by using mono-clonal antibodies. The antigen can be detected as early as 3-4 weeks post-infection when the flukes reside in the host liver (Fagbenmi et al., 1997). In addition, a strong correlation between copro-antigen levels and the numbers of flukes was seen (Abdel-Rahman et al., 1998). Circulating antigens were detected in 100% of sheep with 1 fluke and in 100% of cattle with 2 flukes, from five weeks post-infection (wpi) onwards (Mezo et al., 2004). The copro-antigen became undetectable from 1 - 3 weeks after treatment with a flukicide in sheep and cattle (Mezo et al., 2004) and from 2 months post-treatment in 78.6% of patients (Espino et al., 1994). The copro Ag-ELISA was demonstrated to have a sensitivity and specificity close to 100% (Table 1).

CONCLUSIONS

Although, coprological techniques based on the demonstration of eggs in faeces of the definitive host can be seen as a "gold-standard", these technique are not always adequate, especially for diagnosis of human fasciolosis, because during the long prepatent period immature flukes do not lay eggs, and in the case of ectopic migration of flukes, "false" positive results were seen in some cases of humans following eating from bovine livers harboring fluke eggs (Hillyer, 1999). Immuno-logical techniques provide the advantage of being applicable during all stages of the liver fluke lifecycle. These are reliable detection approaches, especially during the invasive or acute phases.

In spite of that, parasitological and immunological techniques are useful tools in epidemiological studies to estimate the prevalence and to map the presence of human and animal fasciolosis (Hillyer, 1999). Several serological techniques have also proved to be excellent methods for monitoring post-treatment evolution (Mas-Coma et al., 2005). The need to find and establish a new sensitive and specific method and to decrease as much as possible the cases of cross-reactions made us to evaluate the EITB test in this regard.

Despite the numerous above-mentioned assays, the serodiagnosis of naturally acquired fasciolosis in ruminants-in contrast to experimental infections is not yet entirely satisfactory and often rather limited. Recent research efforts have concentrated on the isolation of *F. hepatica* antigens by elution from polyacrylamide gels and on the isolation and translation of messenger RNA

from adult *F. hepatica*.

Future investigations will show whether continued deve-lopment of *F. hepatica* antigens by molecular biology techniques can lead to an improved, widely applicable and economical assay for the serodiagnosis of naturally acquired fasciolosis.

Conflict of Interests

The author(s) have not declared any conflict of interests.

REFERENCES

Abdel-Rahman EH, Abdel Mageed KN (2000). Molecular identity of major cross-reactive adult antigens in *Fasciola gigantica*, toxocara vitulorum and Moniezia expansa. J. Egypt. Soc. Parasitol. 30:561-571.

Abdel-Rahman SM, O'Reilly KL, Malone JB (1998). Evaluation of diagnostic monoclonal antibody based capture enzyme-linked immunosorbent assay for detection of a 26 to 28 kd *Fasciola hepatica* coproantigen in cattle. Am. J. Vet. Res. 59:533-537.

Allan JC, Valasquez-Tohom, Torres- Alvarez R, Yarrita P, and Garcia-Noval J (1996). Filed trial of the Coproantigen-based diagnosis of *Tanea solium* taeniaisis by enzyme-linked immunosorbent assay. Am. J. Trop. Med. Hyg. 54:352-356.

Burger HJ (1992). Helminthn, In: Veterinarmedizinische Parasitologie, Korting, W. (Ed.) 4th (Ed.: Parey V.P.), Berlin: 174.

Carnevale S, Rodriguez MI, Santillan G, Labbe JH, Cabrera MG, Bellegarde EJ, Velasquez JN, Trgovcic JE, Guarnera EA (2001). Immunodiagnosis of human fascioliasis by an enzyme-linked immunosorbent assay (ELISA) and a Micro-ELISA. Clin. Diagn. Lab. Immunol. 8:174-177.

Charlier J, Duchateau L, Claerebout E, Williams D, Vercruysse J (2007). Association between anti-*Fasciola hepatica* antibody levels in bulk-tank milk samples and production parameters in dairy herds. Prev. Vet. Med. 78:57-66.

Charmy RA, El-Kashef HS, ElGhorab NM, Gad HSM (1997). Identification of surface tegumental antigens of normal and irradiated schistosomula. J. Egypt. Soc. Parasitol. 27:479-491.

Cornelissen JB, De Leeuw WA, Vander Heijden PJ (1992). Comparison of and indirect haemagglutination assay and an ELISA for diagnosing *Fasciola hepatica* in experimentally and naturally infected sheep. Vet. Q. 14:152-156.

Cornelissen JB, Gaasenbeek CP, Borgsteede FH, Holland WG, Harmsen MM, Boersema, WJ (2001). Early immunodiagnosis of fascioliasis in ruminants usingrecombinant *Fasciola hepatica* cathepsin L-like protease. Int. J. Parasitol. 31:728-737.

Cornelissen JBWJ, Gaasenbeek CPH, Boersma W, Borgsteede FHM Van Milligen FJ (1999). Use of pre-selected epitope cathepsin –L1 in a highly specific peptide-based immunoassay for the diagnosis of *Fasciola hepatica* infection in cattle. Int. J. Parasitol. 29:685-696.

Dixit AK, Yadav SC, Sharma RL (2002). 28 kDa *Fasciola gigantica* cysteine proteinase in the diagnosis of prepatent ovine fasciolosis. Vet. Parasitol. 109:233-247.

Espino AM, Marcet R, Finlay CM (1994). Detection of circulating excretory secretory antigens in human fascioliasis by sandwich enzyme-linked immunosorbent assay. J. Clin. Microbiol. 28:2637-2640.

Fagbemi BO, Aderibigbe OA, Guobadia EE (1997). The use of monoclonal antibody for the immunodiagnosis of *Fasciola gigantica* infection in cattle. Vet. Parasitol. 69:231-240.

Fagbemi BO, Aderibigbe OA, Guobadia, EE (1999). The use of monoclonal antibody for the immunodiagnosis of *Fasciola gigantica* infection in cattle. Vet. Parasitol. 69:231-240.

G'nen B, Sarimehmetolu HO, Koro M, Kiracali, F (2004). Comparison of crude and excretory/secretory antigens for the diagnosis of *Fasciola hepatica* in sheep by western blotting. Turk. J. Vet. Anim. Sci. 28: 943-949.

Hammond JA, and Sewell MMH (1990). Deseases caused by Helminths. In: M. M. H. Sewell and D. W. Brocklesdy (Eds.), Handbook of Animal Diseases in the Tropics, 4th edn, (CTVM, Edinburgh University). pp. 119-123.

Hanna RE, Jura W (1977). Antibody response of calves to a single infection of Fasciola gigantia determined by an indirect fluorescent antibody technique. Res. Vet. Sci. 22:339-342.

Hillyer GM, Soler De Galanes M, Rodriguez-Perez J, Bjorland J, De Lagrava MS, Guzman SR, Bryan RT (1992). Use of the falcon assay screening test-enzyme-linked immunosorbent assay (FAST-ELISA) and the enzyme-linked immunoelectrotransfer blot (EITB) to determine the prevalence of human fascioliasis in the Bolivian Altiplano. Am. J. Trop. Med. Hyg. 46:603-609.

Hillyer GV (1999). Immunodiagnosis of human and animal. In: Fasciolosis (Dalton J.P.ed).CABI Publishing, Wallingford, UK: 345-449.

Hillyer GV, De Weil NS (1987) Partial purification of Fasciolia hepatica antigen for the immunodiagnosis of fascioliasis in rats. J. Parasitol. 63:430-433.

Hillyer GV, Soler de Galanes M (1988). Idenification of a 17-kilodalton Fasciola hepatica immunodiagnostic antigen by enzyme-linked immunoelectrotransfer blot technique. J. Clin. Microbiol. 26:2048-2053.

Khalil HM, Makled MKH, El-Missiry AG, Khalil NM, Sonobol SE (1989). The application of S. mansoni adult and soluble egg antigens for serodiagnosis of schistomiasis by CIEB, IHA and ELISA. J. Egypt. Soc. Parasitol. 19:872-843.

Langley RJ, Hillyer GV (1989). Detection of circulating parasitic antigen in murine fascioliasis by two-site enzyme linked immunosorbent assay. Am. J. Trop. Med. Hyg. 41:472-478.

Levieux D, Levieux A, Mage C, Venien A (1992). Early immunodiagnosis of bovine fascioliasis using the specific antigen f2 in a passive hemagglutination test. Vet. Parasitol. 42:77-86.

Maleewong W, Wongkhan C, Intapan PM, Pipitgool V (1999). Fasciola gigantic specific antigens: purification by a continuous-elusion method and its evaluation for the diagnosis of human Fascioliasis. Am. J. Trop. Med. Hyg. 61:648-651.

Mas-Coma S, Bargues MD, Valero MA (2005). Fasciolosis and other plant-borne trematode zoonoses. Int. J. Parasitol. 35:1255-1278.

Mezo M, González-Warleta M, Carro C, Ubeira, FM (2004). An ultrasensitivity capture ELISA for detection of Fasciola hepatica coproantigens in sheep and cattle using a new monoclonal antibody (MM3). J. Parasitol. 90:845-852.

Mezo M, González-Warleta M, Ubeira FM (2007). The use of MM3 monoclonal antibodies for the early immunodiagnosis of ovine fascioliasis. J. Parasitol. 93:65-72.

Noureldin MS, EL-Ganaini GA, Abou EL-Enin AM, Hussein EM, Sultan DM (2004). Evaluation of seven assays detectingserum immunoglobulin classes andsubclasses and salivary and faecal secretory IgG against Fasciola excretory/secretory(ES) antigen in diagnosing Fascioliasis. J. Egypt. Soc. Parasitol. 34:691-704.

Ozer BL, Ender SG, Yuksel G, Gurden YU, Sedat B. (2003). Endoscopic extraction of living Fasciola hepatica: Case report and literature review. Turk. J. Gastroenterol. 14(1):74-77

Robinson MW, Dalton JP (2009). Zoonotic helminth infections with particular emphasis on fasciolosis and other trematodiases. Philos. Trans. R. Soc. Lond. B Biol. Sci. 364:2763-2776

Rokni MB, Baghernejad A, Mohebali M, Kia EB (2004). Enzyme linked immunotransfer blot analysis of somatic and excretory-secretory antigens of Fasciola hepatica in diagnosis of human fascioliasis. Iran. J. Public Health 33:8-13.

Shaheen HI, Kamal KA, Farid Z, Mansour N, Boctor FN, Woody JN (1989). Dotenzyme linked immunosorbent assay (dot-ELISA) for rapid diagnosis of human fasciolosis. J. Parasitol. 75:549-552.

Silva E, Castro A, Lopes A, Rodrigues A, Dias C, Conceição A, Alonso J, Correia da Costa JM, Bastos M, Parra F, Moradas P, Moradas-Ferreira P and Silva M (2004). A recombinant antigen recognized by Fasciola hepatica-infected hosts. J. Parasitol. 90:746-751.

13

Recent molecular advances to combat abiotic stress tolerance in crop plants

J. Amudha* and G. Balasubramani

Central Institute for Cotton Research, Indian Council of Agricultural Research, Post Box 2, Shankar Nagar, Nagpur, Maharashtra-440 010, India.

Abiotic stress negatively influences survival, biomass production and crop yield. Being multigenic as well as a quantitative trait, it is a challenge to understand the molecular basis of abiotic stress tolerance and to manipulate it as compared to biotic stresses. Abiotic stresses including drought are serious threats to the sustainability of crop yields accounting for more crop productivity losses than any other factor in rainfed agriculture. Success in breeding for better adapted varieties to abiotic stresses depend upon the concerted efforts by various research domains including plant and cell physiology, molecular biology, genetics, and breeding. Use of modern molecular biology tools for elucidating the control mechanisms of abiotic stress tolerance, and for engineering stress tolerant crops is based on the expression of specific stress-related genes. Plant responses to water deficit can be analysed by systematically identifying genes that relate to drought tolerance followed by analysis to the cellular, biochemical and molecular basis of the gene (traits). Mechanism of drought tolerance and expression of these drought resistance genes in high yielding varieties will help to improve the drought condition. The genes conferring drought resistance provide foundation for scientific improvement of the crop's productivity under arid conditions and contribute to improvement and stabilization of cotton yield and farmers' income. Stress-induced gene expressions are of genes encoding proteins with known enzymatic or structural functions, proteins with as yet unknown functions, and regulatory proteins.

Key words: Abiotic stress, stress induced genes, regulatory proteins, transgenics.

INTRODUCTION

Drought plays a major role in destabilizing the productivity in crop plants. Regardless of whether it is irrigated or not, plants are often exposed to drought, which adversely affects both yield and lint quality. In this regard, conscious efforts are required to improve production in areas commonly exposed to abiotic stress especially drought. Yield improvement in crop plants may be possible by incorporating stable and ideal plant traits pertaining to drought tolerance in the plant system. The development of drought-tolerant crops through a direct selection has been hampered by the low heritability of traits such as

yield, particularly under drought, and by its large 'genotype × environment' interaction (Blum, 1988; Ceccarelli and Grando, 1996). The rainfed ecosystem has characteristic abiotic stress influences, particularly during reproductive ontogeny leading to considerable yield realization. Levit (1980) opined that resistance to water stress might be related to capacity to escape or tolerate adverse environments. In this context, the ability of crop to overcome drought impact is affected by its indeterminate growth habit, longer duration, and osmotic adjustments (Oosterhuis and Wullschilegar, 1987). Maintenance of relatively higher leaf water potential may be a desirable trait in lowering desiccation (Turner, 1986). Drought tolerance in plants is mostly characterized by avoidance and tolerance mechanisms. Morphological adaptations under stress environment generally on

---

*Corresponding author. E-mail: jamudhacicr@gmail.com.

avoidance in nature, whereas dehydration tolerance under low water potential may be a tolerance features. In this context osmo regulation is sustainable drought tolerant mechanism of dehydration tolerance when the plant water potential is extremely low.

Now it is realized that high throughput expression analysis of stress-specific genes is important for understanding gene function (Cushman and Bohnert, 2000). The recent years an enormous number drought stress response genes have been isolated and characterized using molecular techniques. Despite the fact that a large number of genes have to be contributing to the overall phenotypes, investigations on plant responses to environmental stresses have revealed relatively small number of major quantitative trait loci (Yano and Saski, 1997). The prospects of changing the phenotype through manipulation techniques of genetic engineering become much greater if one or few defined regions of chromosomes are of crucial importance. The identifications of QTLs have therefore practical importance in attempts to enhance stress tolerance (Koyama et al., 2001).

PLANT RESPONSES TO WATER DEFICIT

Plant resistance to water deficit may arise from escape, avoidance or tolerance strategies (Levitt, 1972; Turner, 1986). In most cases, plants may combine a range of response types (Chaves et al., 2003).

Drought escape

Drought escape relies on successful reproduction before the onset of severe stress. The plants combine short life cycles with high rates of growth and gas exchange, using maximum available resources while moisture in the soil lasts (Mooney et al., 1987; Maroco et al., 2000).

Drought avoidance

Drought avoidance involves minimizing water loss (closing stomata, reducing light absorbance through rolled leaves, and decreasing canopy leaf area) and maximizing water uptake (increasing investment in the root, reallocation of nutrients stored in older leaves, and higher rates of photosynthesis) (Chaves et al., 2003).

Drought tolerance

Drought tolerance appears to be the result of co-ordination of physiological and biochemical alternation at the cellular and molecular levels. These alterations may involve osmotic adjustment (Morgan, 1984), more rigid cell walls, or smaller cells (Wilson et al., 1980). Changes

occurring rapidly at the mRNA and protein levels lead to tolerant state (Ingram and Bartels 1996).

PHYSIOLOGICAL RESPONSES

Plants subjected to water stress, respond by number of physiological responses at the molecular, cellular, and whole-plant levels (Bray, 1993; Bartels et al., 1996; Chaves et al., 2003). Two physiological mechanisms, most relevant will be discussed subsequently.

Water use efficiency (WUE)

Water use efficiency (WUE) is a key factor determining plant productivity under limited water supply. In agronomic terms, it is defined as the ratio between total dry matter (DM) produced (or yield harvested) and water used (or applied) (Jones, 1993). In physiological terms, however, WUE is defined as the ratio between the rate of carbon fixed and the rate of water transpired. Index representing the number of moles of CO_2 assimilated by photosynthesis per mole of water transpired by the plant. C_4 plants and succulent plants with CAM metabolism show higher WUE than do C_3 plants. Carbon isotope ratio ($^{13}C/^{12}C$, $\delta^{13}C$) is commonly used as an indirect indicator of WUE (Araus et al., 2003). Water use efficiency (WUE), measured as the biomass produced per unit transpiration, describes the relationship between water use and crop production.

In water-limiting conditions, it would be important to produce a high amount of biomass, which contributes to crop yield, using a low or limited amount of water. Water scarcity can impose abiotic stresses like drought and salinity, which are among the most important factors limiting plant performance and yield worldwide. Plant resistance to drought stress can be improved through drought avoidance or drought tolerance, among which drought avoidance mechanisms tend to conserve water by promoting WUE.

Osmotic adjustment

Osmotic adjustment (OA) is the net increase in intercellular solutes in response to water stress (Morgan, 1984), which allows turgor maintenance at lower water potential. OA has been considered one of the crucial processes in plant adaptation to drought, because it sustains tissue metabolic activity and enables regrowth upon rewetting but varies greatly among genotypes. Plant productivity under arid conditions has been associated with OA in a number of species such as sorghum (Tangpremsri et al., 1995), wheat (Morgan, 1984; El Hafid et al., 1998) and oilseed brassicas (Kumar and Singh, 1998).

MOLECULAR RESPONSES

Genes induced during water-stress conditions are thought to function in protecting cells from water deficit by production of important metabolic proteins and regulation of genes for signal transduction in water-tress response. Recently, a number of droughts - responsive genes were cloned and characterized from different plant species (Nepomuceno et al., 2000). Transcription of many of these genes is unregulated by drought stress. Initial attempts to develop transgenics (mainly tobacco) for abiotic stress tolerance involved "single action genes" that is, genes responsible for modification of a single metabolite that would confer increased tolerance to salt or drought stress Stress-induced proteins with known functions such as water channel proteins, key enzymes for osmolyte (proline, betaine, sugars such as trehalose, and polyamines) biosynthesis, detoxification enzymes, and transport proteins were the initial targets of plant transformation.

Stress-induced gene expression can be broadly categorized into three groups: (1) genes encoding proteins with known enzymatic or structural functions, (2) proteins with as yet unknown functions, and (3) regulatory proteins.

Osmoprotectants

Osmoprotectants are proteins that probably function in stress tolerance. They are water channel proteins involved in movement of water through membranes, the enzymes required for the biosynthesis of various osmoprotectants (sugars, Pro, and Gly-betaine). Stress tolerant transgenic plants to engineer genes that encode enzymes for the synthesis of selected osmolytes (Bray, 1993) or osmoprotectants such as glycine-betaine (Sakamoto et al., 1998, 2000; Holmstrom, 2000; McNeil et al., 2000) and proline (Zhu et al., 1998; Yamada et al., 2005). Also, a number of "sugar alcohols" (mannitol, trehalose, myo-inositol and sorbitol) have been targeted for the engineering of compatible-solute overproduction, thereby protecting the membrane and protein complexes during stress (Gao et al., 2000;Zhao et al., 2000; Garg et al., 2002; Cortina and Culia´n˜ez, 2005). Transgenics were engineered for the overexpression of polyamines have also been developed (Roy and Wu, 2001; 2002; Kumria and Rajam, 2002; Waie and Rajam, 2003; Capell et al., 2004). Similarly, transgenics engineered for Genes encoding enzymes that synthesize osmotic and other protectants are *adc* (Arginine decarboxylase), *Adc* (Polyamine synthesis), *Apo-Inv* (Apoplastic invertase), *AtHAL3a* (Phosphoprotein phosphatase), *AtGolS2* Galactinol and raffinose accumulation), *AtTPS1* (trehalose-6-phosphate synthase), *beta* [Choline dehydrogenase (glycinebetaine synthesis)], *BADH-1* (Betaine aldehyde dehydrogenase), *CHIT33, CHIT42* (Endochitinase synthesis), *codA* [Choline oxidase (glycine betaine synthesis)], *COX* (Choline oxidase (glycine betaine synthesis), *CMO* [Choline monooxygenase (glycine betaine synthesis)], *Ect A...ect C* (Edtoin accumulation in chloroplasts), *GS2* (Chloroplastic glutamine synthetase), *IMT1*[Myo-inositol o-methyltransferase (D-ononitol synthesis)], *M6PR*(Mannose-6-phosphate reductase),*mt1D* [Mannitol-1-phosphate dehydrogenase (mannitol synthesis)], *mt1D* and *GutD* [Mannitol-1-phosphate dehydrogenase and glucitol-6-phosphate dehydrogenase], *Osm1 ...Osm4* [Osmotin protein accumulation], *OsP5CS2* (Highly homologous to P5CS), *otsA* [Trehalose-6-phosphate synthase (trehalose synthesis)], *otsB* [Trehalose-6-phosphate synthase (trehalose synthesis)], *P5CS* [Pyrroline carboxylate synthase(proline synthesis)], *PPO* (Polyphenol oxidases suppression), *SAMDC* [S-adenosylmethioninedecarboxylase (polyamine synthesis)], *spe1-1; spe2-1* (Spermidine non-accumulating), *SPE* (Spermidine synthase), *SST/FFT* (Fructan accumulation), *TPSP;TPS;TPS1 and TPS2* (Trehalose synthesis), *PpDHNA* (Dehydrin protein accumulation) in crops plants and they were mentioned in Table 1.

Late embryogenesis abundent (LEA) proteins

Proteins that may protect macromolecules and membranes (LEA protein, osmotin, antifreeze protein, chaperon, and mRNA binding proteins).LEA proteins represent another category of high molecular weight proteins that are abundant during late embryogenesis and accumulate during seed desiccation and in response to water stress (Galau et al., 1987). Amongst the several groups of LEA proteins, those belonging to group 3 are predicted to play a role in sequestering ions that are concentrated during cellular dehydration. These proteins have 11-mer amino acid motifs with the consensus sequence TAQAAKEKAGE repeated as many as 13 times (Dure, 1993). The group 1 LEA proteins are predicted to have enhanced water-binding capacity, while the group 5 LEA proteins are thought to sequester ions during water loss. Constitutive over expression of the HVA1, a group 3 LEA protein from barley conferred tolerance to soil water deficit and salt stress in transgenic rice plants (Xu et al., 1996). Constitutive or stress induced expression of the HVA1 gene resulted in the improvement of growth characteristics and stress tolerance in terms of cell integrity in wheat and rice under salt- and water-stress conditions (Sivamani et al., 2000; Rohilla et al., 2002). The water use efficiency (WUE) was extremely low when compared to other data reported in wheat cultigens, transgenic rice (TNG67) plants expressing a wheat LEA group 2 protein (PMA80) gene or the wheat LEA group 1 protein (PMA1959) gene resulted in increased tolerance to dehydration and salt stresses (Cheng et al. 2002).

Table 1. Gene encoding enzymes that synthesize osmotic and other protectants (www.plantstress.com); I.D. designate ID numbers in reference database).

Gene	Gene action	Species	Phenotype	References	ID
adc	Arginine decarboxylase	Rice	Reduced chlorophyll loss under drought stress	Capell et al., 1998	5607
Adc	Polyamine synthesis	Rice	Drought resistance	Capell et al., 2004	7290
Apo-Inv	Apoplastic invertase	Tobacco	Salt tolerance, high "osmotic pressure"	Fukushima et al., 2001	5202
AtHAL3a	Phosphoprotein phosphatase	*Arabidopsis*	Regulate salinity and osmotic tolerance and plant growth	Espinosa-Ruiz et al., 1999	4601
AtHAL3	Phosphoprotein phosphatase	Tobacco	Improved salt, osmotic and Lithium tolerance of cell cultures	Yonamine et al., 2004	6947
AtGolS2	Galactinol and raffinose accumulation	*Arabidopsis*	Reduced transpiration	Taji et al., 2002	5884
AtTPS1	Trehalose-6-phosphate synthase	Tobacco	Drought resistance; sustained photosyntehsis	Almeida et al., 2007	8668
BADH-1	Betaine aldehyde dehydrogenase	Tobacco	Heat tolerance in photosynthesis	Xinghong Yang, et al., 2005	7858
BADH-1	Betaine aldehyde dehydrogenase	Tomato	Maintenance of osmotic potential	Moghaieb et al., 2000	5094
BADH-1	Betaine aldehyde dehydrogenase	Carrot	Salinity tolerance	Kumar et al., 2004	7353
betA	Choline dehydrogenase (glycinebetaine synthesis)	Tobacco	Increased tolerance to salinity stress	Lilius et al., 1996	3287
betA	Choline dehydrogenase (glycinebetaine synthesis)	Maize	Drought resistance at seedling stage and high yield after drought	Ruidang et al., 2004	7409
CHIT33, CHIT42	Endochitinase synthesis	Tobacco	Salt and metal toxicity resistance (and disease)	Dana et al., 2006	8504
codA	Choline oxidase (glycine betaine synthesis)	*Brassica juncea*	Tolerance to stress induced photoinhibition	Prasad and Saradhi, 2004	7094
codA	Choline oxidase (glycine betaine synthesis)	Rice	Increased tolerance to salinity and cold	Sakamoto et al., 1998	3859
codA	Choline oxidase (glycine betaine synthesis)	Rice	Recovery from a week long salt stress	Mohanty et al., 2003	6347
codA	Choline oxidase (glycine betaine synthesis)	*Arabidopsis*	Increased stress tolerance	Huang et al., 2000	4731
codA	Choline oxidase (glycine betaine synthesis)	*Arabidopsis*	Salt tolerance in terms of reproduction	Ronan et al., 2003	6822
codA	Choline oxidase (glycine betaine synthesis)	*Arabidopsis*	Seedlings tolerant to salinity stress and increased germination under cold	Hayashi et al., 1997 ; Alia et al., 1998	4571
COX	Choline oxidase (glycine betaine synthesis)	Rice	Salt and 'stress' tolerance	Su et al., 2006	8227
CMO	Choline monooxygenase (glycine betaine synthesis)	Tobacco	Better *in vitro* growth under salinity and osmotic (PEG6000) stress	Yi-Guo et al., 2002	6285

Table 1. Contd.

Gene	Enzyme/function	Plant	Effect	Reference	
Ect A...ect C	Edtoin accumulation in chloroplasts	Tobacco	Salt and cold tolerance	Rai et al., 2006	8090
GS2	Chloroplastic glutamine synthetase	Rice	Increased salinity resistance and chilling tolerance	Hoshida et al., 2000	4792
IMT1	Myo-inositol o-methyltransferase (D-ononitol synthesis)	Tobacco	Better CO_2 fixation under salinity stress. Better recovery after drought stress.	Sheveleva et al., 1997	3660
M6PR	Mannose-6-phosphate reductase	*Arabidopsis*	Mannitol accumulation under salt stress leading to salt tolerance	Zhifang and Loescher, 2003	6343
M6PR	Mannose-6-phosphate reductase	*Arabidopsis*	Mannitol accumulation and salt tolerance due to chloroplast protection	Sickler et al., 2007	6533
mt1D	Mannitol-1-phosphate dehydrogenase (mannitol synthesis)	*Arabidopsis*	Increased germination under salinity stress	Thomas et al., 1995	5620
mt1D and GutD	Mannitol-1-phosphate dehydrogenase & glucitol-6-phosphate dehydrogenase	loblolly pine	High salt tolerance due to mannitol and glucitol accumulation	Tang et al., 2005	7614
mtlD	Mannitol-1-phosphate dehydrogenase (mannitol synthesis)	*Populus tomentosa*	Salinity tolerance	Chiang et al., 2005	7751
mt1D	Mannitol-1-phosphate dehydrogenase (mannitol synthesis)	Tobacco	Increased plant height and fresh weight under salinity stress	Hu et al., 2005	7946
mt1D	Mannitol-1-phosphate dehydrogenase (mannitol synthesis)	Tobacco	No contribution to sustained growth under salinity and drought stress.	Tarczynski et al., 1993	2383
mt1D	Mannitol-1-phosphate dehydrogenase (mannitol synthesis)	Wheat	Drought and salinity tolerance of calli and plants	Abebe et al., 2003	6533
Osm1 ...Osm4	Osmotin protein accumulation	Tobacco	Drought and salt tolerance in plant water status and proline accumulation	Barthakur et al., 2001	5560
OsP5CS2	Highly homologous to P5CS	Rice	Cold and salinity tolerance	Hur et al., 2004	7264
otsA	Trehalose-6-phosphate synthase (trehalose synthesis)	Tobacco	Increased leaf dry weight and photosynthetic activity under drought. Increased carbohydrate accumulation.	Pilon-smits et al., 1998	3101
otsB	Trehalose-6-phosphate synthase (trehalose synthesis)	Tobacco	Increased leaf dry weight and photosynthetic activity under drought. Increased carbohydrate accumulation.	Pilon-smits et al., 1998	3101
P5CS	Pyrroline carboxylate synthase (proline synthesis) (tomato)	Citrus	Osmotic adjustment and drought resistance	Molinari et al., 2004	7361
P5CS	Pyrroline carboxylate synthase (proline synthesis)	*Petunia*	Drought resistance and high proline	Yamada et al., 2005	7750
P5CS	Pyrroline carboxylate synthase (proline synthesis)	Potato	Salinity tolerance	Hmida-Sayari et al., 2005	7864
P5CS	Pyrroline carboxylate synthase (proline synthesis)	Rice	Increased biomass production under drought and salinity stress	Zhu et al., 1998	3871

Table 1. Contd.

Gene	Description	Plant	Phenotype	Reference	Number
P5CS	Pyrroline carboxylate synthase (proline synthesis)	Rice	Reduced oxidative stress under osmotic stress	Hong Zong Lie et al., 2000	5562
P5CS	Pyrroline carboxylate synthase (proline synthesis)	Rice	Resistance to water and sainity stress	Su and Wu, 2004	7034
P5CS	Pyrroline carboxylate synthase (proline synthesis)	Soybean	Resistance to osmotic stress and heat	De Ronde et al., 2001	5767
P5CS	Pyrroline carboxylate synthase (proline synthesis) (tomato)	Soybean	Drought resistance, high RWC, high proline	De Ronde et al., 2004	7383
P5CS	Pyrroline carboxylate synthase (proline synthesis) (tomato)	Sugarcane	Drought resistance via antioxidant role of proline	Molinari et al., 2007	8859
PPO	Polyphenol oxidases suppression	Tomato	Drought resistance	Thipyapong et al., 2004	7267
SAMDC	S-adenosylmethioninedecarboxylase (polyamine synthesis)	Rice	Better seedling growth under a 2 day NaCl stress	Malabika and Wu, 2002	6252
SAMDC	S-adenosylmethioninedecarboxylase (polyamine synthesis)	Tobacco	drought, salinity, Verticillium and Fusarium wilts resistance	Waie and Rajam, 2003	6538
spe1-1; spe2-1	Spermidine non-accumulating	Arabidopsis	Decreased salt tolerance	Vasuki and Astrid, 2004	7089
SPE	Spermidine synthase	Arabidopsis	Chilling, freezing, salinity, drought hyperosmosis	Kasukabe et al., 2004	7277
SST/FFT	Fructan accumulation	Potato	Reduced proline accumulation at low water status	Knipp and Honermeier, 2006	8144
TPSP	Trehalose synthesis	Rice	Drought, salt and cold tolerance expressed by chlorophyll fluorescence	In-Cheol Jang et al., 2003	6389
TPS1	Trehalose synthesis	Tomato	Drought, salt and oxidative stress tolerance	Cortina and Culiáñez-Macià, 2005	7788
TPS1 and TPS2	Trehalose synthesis	Tobacco	Maintenance of water status under drought stress	Karim et al., 2007	8913
PpDHNA	Dehydrin protein accumulation	Moss	Salt and osmotic stress tolerance	Saavedra et al., 2006	8082

Besides, protective chaperone like function of LEA pro-teins acting against cellular damage has been proposed (Vincour and Altman, 2005), indicating the role of LEA proteins in anti aggregation of enzymes under desic-cation and freezing stresses (Goyal et al., 2005). Some more genes are DQ663481 (Lea gene), HVA1 (Group 3 LEA protein gene), OsLEA3-1(Lea protein), Rab17 (LEA protein), ME-leaN4 (LEA protein) and the crops transformed are given in Table 2.

Detoxifying genes

The higher stress tolerance and the accumulation of compatible solutes may also protect plants against da-mage by scavenging of reactive oxygen species (ROS), and by their chaperone-like activities in maintaining pro-tein structures and functions (Hare et al., 1998; Bohnert and Shen, 1999; McNeil et al., 1999; Diamant et al., 2001). In most of the aerobic organisms, there is a

need to effectively eliminate reactive oxygen species (ROS) generated as a result of environmental stresses. De-pending on the nature of the ROS, some are highly toxic and need to be rapidly detoxified. In order to control the level of ROS and protect the cells from oxidative injury, plants have developed a complex antioxidant defense system to scavenge the ROS. These antioxidant systems include various enzymes and non-enzymatic metabolites that may also play a

Table 2. Late embryogenesis abundant (LEA) related genes (www.plantstress.com).

Gene	Gene action	Species	Phenotype	References	ID
DQ663481	Lea gene	Tobacco	Drought resistance via cell membrane stability	Wang et al., 2006	8510
HVA1	Group 3 LEA protein gene	Oat	Delayed wilting under drought stress	Maqbool et al., 2002	6146
HVA1	Group 3 LEA protein gene	Oat	Salinity tolerance in yield/plant	Oraby et al., 2005	7971
HVA1	Group 3 LEA protein gene	Rice	Dehydration avoidance and cell membrane stability	Babu et al., 2004	7030
HVA1	Group 3 LEA protein gene	Rice	Drought and salinity tolerance	Rohila et al. 2002	6185
HVA1	Group 3 LEA protein gene	Wheat	Increased biomass and WUE under stress	Sivamani et al. 2000	4781
HVA1	Group 3 LEA protein gene	Wheat	Improved plant water status and yield under field drought conditions	Bahieldin et al., 2005	7618
OsLEA3-1	Lea protein	Rice	Drought resistance for yield in the field	Xiao et al., 2007	8926
Rab17	LEA protein	Arabidopsis	Resistance to osmotic and salinity stress	Figueras et al., 2004	7204
ME-leaN4	LEA protein	Lettuce	Enhanced growth and delayed wilting under drought. Salt resistance	Park et al., 2005	7671
ME-leaN4	LEA protein	Chinese cabbage	Drought and salt resistance	Park et al., 2005	7794

that may also play a significant role in ROS signaling in plants (Vranova et al., 2002). A number of transgenic improvements for abiotic stress tolerance have been achieved through detoxification strategy. These include transgenic plants over expressing enzymes involved in oxidative protection, such as glutathione peroxidase, superoxide dismutase, ascorbate peroxidases and glutathione reductases (Zhu et al., 1999; Roxas et al., 1997). Transgenic tobacco over expressing SOD in the chloroplast, mitochondria and cytosol have been generated (Bowler et al., 1991; Van Camp et al., 1996) and these have been shown to enhance tolerance to oxidative stress induced by methyl viologen (MV) in leaf disc assays. Overexpression of chloroplast Cu/Zn SOD showed a dramatic improvement in the photosynthetic performance under chilling stress conditions in transgenic tobacco (Sen et al., 1993) and potato plants (Perl et al., 1993). While transgenic alfalfa (Medicago sativa) plants cv. RA3 over expressing MnSOD in chloroplasts showed lower membrane injury (McKersie et al., 1996), the tobacco transgenic plants overproducing alfalfa aldose reductase gene (MsALR) showed lower concentrations of reactive aldehydes and increased tolerance against oxidative agents and drought stress (Oberschall et al., 2000).Tobacco transgenic plants over expressing MnSOD rendered enhanced tolerance to oxidative stress only in the presence of other antioxidant enzymes and substrates (Slooten et al., 1995), thereby, showing that the genotype and the isozyme composition also have a profound effect on the relative tolerance of the transgenic plants to abiotic stress (Rubio et al., 2002). Oxidative stress related genes like ApGPX2 and AcGPX2

(Glutathione peroxidase (GPX)-like proteins), ALR (Aldose/aldehyde reductase), Apx1 (Ascorbate peroxidase) peroxidase), APX2(Ascorbate peroxidase), Apx3(Ascorbate peroxidase), Apx3(Ascorbate peroxidase), Apx (Ascorbate peroxidase),' AO' (Ascorbate oxidase), AtMDAR1 (Monodehydroascorbate reductase; Ascorbate regeneration) DHAR (regeneration of ascorbate), Gly1;gly2 (Glutathione-based detoxification of methylglyoxal), GmTP55 (Antiquitin-like protein),GST (glutathione S-transferase overexpression), GST/GPX (Glutathione S-transferase with Glutathione peroxidase), GPX (Glutathione peroxidase), katE (Escherichia coli catalase), ndhCKJ [NADPH dehydrogenase], NtPox (Gluthathione peroxidase), Nt107 (Glutathione S-transferase), parB (glutathione S-transferase), SOD(Cu, MN, Fe. Zn-SOD),SOD(Cu/Zn superoxide dismutase), SOD(Fe superoxide dismutase), SOD (Mn superoxide dismutase), vtc1;vtc2;npq1; cad2 (reactive oxygen metabolism mutants),vtc-1 (ascorbate deficient mutant) were transferred in many crops and given in Table 3.

Multifunctional genes for lipid biosynthesis

Multifunctional genes are those genes that improve photosynthesis under abiotic stress conditions through changes in the lipid biochemistry of the membranes (Grover and Minhas, 2000). Adaptation of living cells to chilling temperatures is a function of alteration in the membrane lipid composition by increased fatty acid unsaturation. Genetically engineered tobacco plants over-expressing chloroplast glycerol-3-phosphate

Table 3. Oxidative stress related genes (www.plantstress.com).

Gene	Gene action	Species	Phenotype	References	ID*
ApGPX2 and AcGPX2	Glutathione peroxidase (GPX)-like proteins	Arabidopsis	Oxidative stress, drought and salt resistance	Gaber et al., 2006	8466
ALR	Aldose/aldehyde reductase	Tobacco	Drought and UV-B tolerance	Hideg et al., 2003	6524
ALR	Aldose/aldehyde reductase	Tobacco	Cold and cadmium stress tolerance	Hegedüs et al., 2004	7098
APX2	Ascorbate peroxidase	Arabidopsis	High light and drought tolerant mutant	Rossel et al., 2006	8164
Apx3	Ascorbate peroxidase	Tobacco	Increased protection against oxidative stress	Wang et al., 1999	4531
Apx3	Ascorbate peroxidase	Tobacco	Drought resistance in photosynthesis	Juqiang Yan et al., 2003	6614
Apx	Ascorbate peroxidase	Tomato	Chilling and salt tolerance	Kornyeyev et al., 2003	6769
'AO'	Ascorbate oxidase	Tobacco and Arabidopsis	Salt sensitivity in germination, photosynthesis, and seed yield	Yamamoto et al., 2005	7744
AtMDAR1	Monodehydroascorbate reductase; Ascorbate regeneration	Tobacco	Ozone, salt and polyethylene glycol tolerance	Eltayeb et al., 2007	8814
DHAR	Regeneration of ascorbate	Tobacco	Tolerance to ozone, drought, salt, and PEG	Elsadig et al., 2006	8297
DHAR	Regeneration of ascorbate	Arabidopsis	Salt tolerance	Ushimaru et al., 2006	8492
Gly1; gly2	Glutathione-based detoxification of methylglyoxal	Tobacco	Salt tolerance	Singla-Pareek et al., 2006	8261
GmTP55	Antiquitin-like protein	Soybean Tobacco	Resistance to drought, salt and oxidative stress	Rodrigues et al., 2006	8330
GST	Glutathione S-transferase overexpression	Arabidopsis	No whole-plant salt resistance despite antioxidant activity	Katsuhara et al., 2005	7793
GST	Glutathione S-transferase overexpression	Cotton	No whole-plant salt resistance and no antioxidant activity	Light et al., 2005	8032
GST	Glutathione S-transferase overexpression	Rice	Salt and chilling resistance	Zhao and Zhang, 2006	8555
GPX	Glutathione peroxidase	Tobacco	Chilling and salt resistance	Kazuya et al., 2004	6921
katE	Escherichia coli catalase	Tobacco	Salt tolerance by hydrogen peroxide scavenging	Al-Taweel et al., 2007	9030
ndhCKJ	NAD(P)H dehydrogenase	Tobacco	Photosystem function under heat stress	Wang et al., 2006	8353
NtPox	Gluthathione peroxidase	Arabidopsis	Protect against Al toxicity and oxidative stress	Ezaki et al., 2001	5664
Nt107	Glutathione S-transferase	Tobacco	Sustained growth under cold and salinity stress	Roxas et al., 1997	5616
parB	Glutathione S-transferase	Arabidopsis	Protect against Al toxicity and oxidative stress	Ezaki et al., 2000	4728
parB	Glutathione S-transferase	Arabidopsis	Protect against Al toxicity and oxidative stress	Ezaki et al., 2001	5664
SOD	Cu, MN, Fe, Zn-SOD	Alfalfa, rye grass	Increased winter hardiness	McKersie, 2001	5614
SOD	Cu/Zn superoxide dismutase	Tobacco, Tomato	No protection seen against superoxide toxicity	Tepperman and Dunsmuir, 1990	5619

Table 3. Contd.

Gene	Protein	Organism	Effect	Reference	No.
SOD	Cu/Zn superoxide dismutase	Tobacco	Retained photosynthesis under chilling and heat stress	Gupta et al., 1993	5609
SOD	Cu/Zn superoxide dismutase	Tobacco	Enhanced tolerance to salt, water, and PEG stresses,	Badawi et al., 2004	7033
SOD	Fe superoxide dismutase	Tobacco	Protected plants from ozone damage	Van Camp et al., 1994	5621
SOD	Mn superoxide dismutase	Tobacco	Reduced cellular damage under oxidative stress	Bowler et al. 1991	5606
SOD	Mn superoxide dismutase	Alfalfa	Tolerance to freezing stress	McKersie et al., 1993	5615
SOD	Mn superoxide dismutase	Alfalfa	Tolerance to water deficit	McKersie et al., 1996	3345
SOD	Mn superoxide dismutase	Alfalfa	Winter survival	McKersie et al., 1999	3894
SOD	Mn superoxide dismutase	Tobacco	Tolerance to Mn deficiency	Yu et al., 1999	4512
SOD	Mn superoxide dismutase	Canola	Aluminum tolerance	Basu et al., 2001	5684
SOD	Mn superoxide dismutase	Arabidopsis	Salt tolerance	Wang et al., 2004	7266
SOD	Mn superoxide dismutase	Rice	Reduced injury and sustained photosynthesis under PEG stress	Wang et al., 2005	7724
SOD	Mn/Fe superoxide dismutase	Alfalfa	Background dependent increased photosynthesis under drought stress	Maria et al., 2002	6103
vtc1, vtc2, npq1, cad2	Reactive oxygen metabolism mutants	Arabidopsis	Heat tolerance/sensitivity	Larkindale et al., 2005	7783
vtc-1	Ascorbate deficient mutant	Arabidopsis	Sensitivity to salinity stress	Huang et al., 2005	7990

acyltransferase (GPAT) gene (involved in phosphatidyl glycerol fatty acid desaturation) from squash (*Cucurbita maxima*) and *A. thaliana* (Murata et al., 1992) showed an increase in the number of unsaturated fatty acids and a corresponding decrease in the chilling sensitivity. Besides transgenic tobacco plants with silenced expression of chloroplast x3-fatty acid desaturase (Fad7, which synthesizes trienoic fatty acids) were able to acclimate to high temperature as compared to the wild type (Murakami et al., 2000).

Heat shock protein genes

The heat shock response, increased transcription of a set of genes in response to heat or other toxic agent exposure is a highly conserved biological response, occurring in all organisms (Waters et al., 1996). The response is mediated by heat shock transcription factor (HSF) which is present in a monomeric, non-DNA binding form in unstressed cells and is activated by stress to a trimeric form which can bind to promoters of heat shock genes. The induction of genes encoding heat shock proteins (Hsps) is one of the most prominent responses observed at the molecular level of organisms exposed to high temperature (Kimpel and Key, 1985; Lindquist, 1986; Vierling 1991). Genetic engineering for increased thermo-tolerance by enhancing heat shock protein synthesis in plants has been achieved in a number of plant species (Malik et al., 1999; Li et al., 2003; Katiyar-Agarwal et al., 2003). There have been a few reports on positive correlations between the levels of heat shock proteins and stress tolerance (Sun et al., 2001; Wang et al., 2005). Although the precise mechanism by which these heat shock proteins confer stress tolerance is not known, a recent study demonstrated that *in vivo* function of thermoprotection of small heat shock proteins is achieved via their assembly into functional stress granules (HSGs; Miroshnichenko et al., 2005). Genes encoding for molecular chaperones are *APG6* (Chloroplast structure), *AtDjA2 and atDjA3* (J-domain molecular chaperone family), *AtMTP3* (Metal tolerance Protein), *Atsbp1* (Selenium binding protein), *atRZ-1a* (RNA chaperone protein), *BiP* [Endoplasmic reticulum binding protein (BiP)], *CaHSP26* [Chloroplast (CP)-localized small heat shock protein], *hs* (Heat shock transcription factor), *Hsp101(* Heat shock protein), *Hsp17.7(* Heat shock protein), *Hsp70* (Heat-inducible antisense HSP70), *LeHSP100/ClpB* (Chloroplast HSP), *mHSP22* (Mitochondrial small HSP),*P5CR* [Inducible heat shock promoter (IHSP)], *pBE2113/ hiC6* (Overexpressed HIC6 cryoprotective protein), *S1pt::ECS*(glutamylcysteine synthetase),*TLHS1(* Over expressed class I cytosolic small HSP), *wx(* Control amylose synthesis) were used for transformation in crop plants are given in Table 4.

Hormone regulatory genes

Many genes that respond to multiple stresses like dehydration and low temperature at the transcriptional level are also induced by ABA (Mundy and Chua 1988), which protects the cell from dehydration (Dure et al., 1989; Skriver and Mundy, 1990). In order to restore the cellular function and make plants more tolerant to stress, transferring a single gene encoding a single specific stress protein may not be sufficient to reach the required tolerance levels (Bohnert et al., 1995). To overcome such constraints, enhancing tolerance towards multiple stresses by a gene encoding a stress inducible transcription factor that regulates a number of other genes is a promising approach (Yamaguchi-Shinozaki et al., 1994; Chinnusamy et al., 2005). Therefore, a second category of genes of recent preference for crop genetic engineering are those that switch on transcription factors regulating the expression of several genes related to abiotic stresses. Another ABA-independent, stress-responsive and senescence- activated gene expression involves ERD gene, the promoter analysis of which further identified two different novel cis acting elements involved with dehydration stress induction and in dark-induced senescence (Simpson et al., 2003). Similarly, transgenic plants developed by expressing a drought-responsive AP2-type TF, SHN1-3 or WXP1, induced several wax-related genes resulting in enhanced cuticular wax accumulation and increased drought tolerance (Aharoni et al., 2004; Zhang et al., 2005). Thus, clearly, the over expression of some drought-responsive transcription factors can lead to the expression of downstream genes and the enhancement of abiotic stress tolerance in plants (Zhang et al., 2004). The regulatory genes/factors reported so far not only play a significant role in drought and salinity stresses, but also in submergence tolerance. More recently, an ethylene response-factor-like gene Sub1A, one of the cluster of three genes at the Sub1 locus have been identified in rice and the over expression of Sub1A-1 in a submergence-intolerant variety conferred enhanced submergence tolerance to the plants (Xu et al., 2006), thus confirming the role of this gene in submergence tolerance in rice. Various other hormone regulating genes are *ABI1, ABI2* (ABA regulation), *ABA2(*ABA regulation), *hab1 group*(ABA hypersensitivity), *AtNCED3* (Increased ABA synthesis), *AtPP2CA(*Reduce ABA

sensitivity), *EIN2(*Ethylene and ABA signaling pathways), *Eto 1-1(*Ethylene over-production), *CYP707A3(*Regulate ABA levels), *LLA23(*Reduced ABA sensitivity), *NTHK1(*Ethylene receptor), *PSAG12-IPT(*Over production of cytokinins), *PLD alpha* (Phospholipase D (alpha) expression), *sp12 and sp5(*ABA overproduction), *tos1* (Increased ABA sensitivity), *ZmACS6* (Ethylene synthesis) are listed in Table 5.

Transcription factors

Transcription factors an attractive target category for manipulation and gene regulation is the small group of transcription factors that have been identified to bind to promoter regulatory elements in genes that are regulated by abiotic stresses (Shinozaki and Yamaguchi-Shinozaki, 1997; Winicov and Bastola, 1997). The transcription factors activate cascades of genes that act together in enhancing tolerance towards multiple stresses. Individual members of the same family often respond differently to various stress stimuli. On the other hand, some stress responsive genes may share the same transcription factors, as indicated by the significant overlap of the geneexpression profiles that are induced in response to different stresses (Seki et al., 2001; Chen and Murata 2002). Dozens of transcription factors are involved in the plant response to drought stress (Vincour and Altman, 2005; Bartels and Sunkar, 2005). Most of these falls into several large transcription factor families, such as AP2/ERF, bZIP, NAC, MYB, MYC, Cys2His2 zinc-finger and WRKY (Umezawa et al., 2006).

A cis-acting element, dehydration responsive element (DRE) identified in *A. thaliana*, is also involved in ABA-independent gene expression under drought, low temperature and high salt stress conditions in many dehydration responsive genes like rd29A that are responsible for dehydration and cold-induced geneexpression (Yamaguchi-Shinozaki and Shinozaki, 1993; Iwasaki et al., 1997; Nordin et al., 1991). DREB1/CBFs are thought to function in cold-responsive gene expression, whereas DREB2s are involved in drought-responsive gene expres- sion. The transcriptional activation of stress-induced genes has been possible in transgenic plants over-expressing one or more transcription factors that recognize regulatory elements of these genes. Subsequently, the overexpression of DREB1A has been shown to improvethe drought- and low-temperature stress tolerance in tobacco, wheat and groundnut (Kasuga et al., 2004; Pellegrineschi et al., 2004; Behnam et al., 2006; Bhatnagar-Mathur et al., 2004, 2006). The use of stress inducible rd29A promoter minimized the negative effects on plant growth in these crop species. However, over-expression of DREB2 in transgenic plants did not improve stress tolerance, suggesting involvement of post-translational activation of DREB2 proteins (Liu et al.,

Table 4. Genes encoding for molecular chaperones   (www.plantstress.com).

Gene	Gene action	Species	Phenotype	References	ID*
APG6	Chloroplast structure	Arabidopsis	Heat tolerance	Myouga et al., 2006	8474
AtDjA2 & atDjA	J-domain molecular chaperone family	Arabidopsis	Heat tolerance	Li et al., 2007	9034
AtMTP3	Metal tolerance protein	Arabidopsis	Zinc tolerance	Arrivault et al., 2006	8307
Atsbp1	Selenium binding protein	Arabidopsis	Selenium tolerance	Agalou et al., 2005	7899
atRZ-1a	RNA chaperone protein	Arabidopsis	Cold tolerance	Kim and Kang, 2006	8338
BiP	Endoplasmic reticulum binding protein (BiP)	Tobacco	Maintenance of plant water status under drought stress and antioxidative defence	Alvim et al., 2001	5433
CaHSP26	Chloroplast (CP)-localized small heat shock protein	Tobacco	Protection of PSII and PSI during chilling	Guo et al., 2007	8673
hs	Heat shock transcription factor	Arabidopsis	Increased thermotolerance in transgenic plants	Lee et al., 1995	5612
Hsp101	Heat shock protein	Arabidopsis	Decreased Thermotolerance in Hsp101 deficient (hot1) mutant	Hong and Vierling, 2000	5363
Hsp101	Heat shock protein	Arabidopsis	Manipulated themotolerance in transgenic plants	Queitsch et al., 2000	4733
Hsp101	Heat shock protein	Rice	Heat tolerance in plant growth	Katiyar-Agarwal et al., 2003	6430
Hsp17.7	Heat shock protein	Carrot	Increased or decreased thermotolerance	Malik et al., 1999	4526
Hsp70	Heat-inducible antisense HSP70	Arabidopsis	Increased thermotolerance in transgenic plants	Lee and Schoof, 1999	5613
LeHSP100/ClpI	Chloroplast HSP	Tomato	Heat tolerance	Yang et al., 2006	8468
mHSP22	Mitochondrial small HSP	Arabidopsis	Heat tolerance (high leaf mass after heat stress)	Rhoads et al., 2005	7619
P5CR	Inducible heat shock promoter (IHSP)	Soybean	Increased proline accumulation	de Ronde et al., 2000	4936
pBE2113/ hiC6	Overexpressed HIC6 cryoprotective protein	Tobacco	Freezing tolerance; reduced membrane injury	Honjoh et al., 2001	5531
S1pt::ECS	glutamylcysteine synthetase	Arabidopsis	Metal tolerance	Li et al., 2006	8310
TLHS1	Overexpressed class I cytosolic small HSP	Tobacco	Seedling thermotolerance	Park and Hong, 2002	5811
wx	Control amylose synthesis	Rice	Increased amylose content at low temperature	Hirano and Sano, 1998	5610

1998). Recently, an active form of DREB2 was shown to transactivate target stress-inducible genes and improve drought tolerance in transgenic Arabidopsis (Sakuma et al., 2006). The DREB2 protein is expressed under normal growth conditions and activated by osmotic stress through post-translational modification in the early stages of the osmotic stress response. To date, 55 members belonging to the DREB subfamily have been isolated from Arabidopsis (Sakuma et al., 2002) and divided into six sub-groups, A-1, A-2, A-3, A-4, A-5 and A-6, based on the homology of the AP2 conserved domains (Seki et al., 2003; Sakuma et al., 2002; Ito et al., 2006). Among them, the A-1 and A-2 subgroups, harboring the DREB1-type and DREB2-type genes, respectively, were the two largest ones that are involved in two different ABA-in-dependent pathways (Liu et al., 1998; Shinwari et al., 1998; Yamaguchi-Shinozaki et al., 2002). Currently, three DRE-binding transcription factors from Gossypium hirsutum, designated GhDREB1L, GhDBP2 and GhDBP3, are isolated and classified into the A-1, A-4 and A-6 groups of DREB subfamilies (Huang and Liu 2006; Huang et al., 2007, 2008).

Various regulatory genes are AB13(Transcription factor), ABF3(Transcription factor), ADC(Arginine decarboxylase overexpression), ADH1; ADH2(alcohol dehydrogenase), ALDH3(aldehyde dehydrogenase),

Table 5. Hormone regulating genes (www.plantstress.com).

Gene	Gene action	Species	Phenotype	References	ID*
ABI1, ABI2	ABA regulation	Arabidopsis	Heat tolerance	Larkindale et al., 2005	7783
ABA2	ABA regulation	Arabidopsis	Tolerance to various prolonged stresses?	Lin et al., 2007	8734
hab1 group	ABA hypersensitivity	Arabidopsis	Dehydration avoidance	Saez et al., 2006	8435
AtNCED3	Increased ABA synthesis	Arabidopsis	Reduced transpiration and drought resistance	Iuchi et al., 2001	5527
AtPP2CA	Reduce ABA sensitivity	Arabidopsis	Induce cold sensitivity	Tahtiharju and Palva, 2001	5437
EIN2	Ethylene and ABA signaling pathways	Arabidopsis	Salt and osmotic stress responses	Wang et al., 2007	8975
Eto 1-1	Ethylene over-production	Arabidopsis	Reduced ABA sensitivity and greater transpiration	Tanaka et al., 2005	7859
CYP707A3	Regulate ABA levels	Arabidopsis	Dehydration and rehydration responses	Umezawa et al. 2006	8245
LLA23	Reduced ABA sensitivity	Arabidopsis	Drought and salt resistance	Nakajima et al., 2002	6007
NTHK1	Ethylene receptor	Arabidopsis	Salt sensitivity	Cao et al., 2007	8733
PSAG12-IPT	Over production of cytokinins	Petunia	Delayed leaf senescence (not tested under stress)	Clark et al., 2004	6904
PLD alpha	Phospholipase D (alpha) expression	Arabidopsis	Increase sensitivity to ABA and reduce transpiration	Sang et al. 2001	5635
sp12 and sp5	ABA overproduction	Tomato	High water-use efficiency, low transpiration and greater root hydraulic conductance	Thompson et al., 2007	8805
tos1	Increased ABA sensitivity	Tomato	Hypersensitive to osmotic stress and exogenous ABA	Borsani et al., 2002	6330
ZmACS6	Ethylene synthesis	Maize	Non-functional mutant expressed drought induced senescence	Todd et al., 2004	7471

ALDH3I1 and ALDH7B4(aldehyde dehydrogenase), Alx8 (High APX2 and AREB1(ABA hypersensitivity), AREB1(ABRE-dependent ABA signaling), ASR1(Undetermined), AtAOX1(alternative oxidase (AOX) pathway of plant mitochondrial], AtCBF1-3(Transcription factor), AtGluR2(Transcription factor), AtGSK1(Homologue of GSK3/shaggy-like protein kinase), Atnoa1(Impaired Nitric Oxide synthesis), AtPCS1(Phytochelatin synthesis), ATP-PRT(Free His accumulation), AtRabG3e(Intracellular vesicle trafficking), atRZ-1a(Zinc finger glycine-rich RNA-binding protein), ATTS244 and ATTS405 (FtsH protease protecting photosystem), AZF1, AZF2, AZF3, STZ(Cys2/His2-Type Zinc-Finger and Proteins), atRZ-1a(zinc finger-containing glycine-rich RNA-binding proteins (GR-RBPs)], AtSZF1 and AtSZF2(CCCH-type zinc finger proteins,

involved in salt stress responses), BNCBF5,(CBF/DREB1-like transcription factors), CAbZIP1[Plant development (dwarf phenotype)], CAP2(Transcription factor), CaPF1(Transcription factor), CaPIF1(Cys-2/His-2 zinc finger protein), CBF1(Transcription factor), CBF1; CBF3(Transcription factor), CBF3 (Transcription factor), CBF4 (Transcription factor), CBL1(Ca sensing protein), CBP20(cap binding complex), CGS(Cystathionine-synthase), CIT1(Mitochondrial citrate synthase), CpMYB10(Glucose sensitive and ABA hypersensitive), cpSL[Selenocysteine lyase (mouse)], CRYOPHYTE/ LOS4(RNA helicase), CUP1(metallothionein accumulation), Cys(Enhanced cysteine synthase activity), desC(Acyl-lipid 9-desaturase), DREB or DREB1 (Transcription factor), DREB1 factor), OsDREB1(Transcription

DREB1A(Transcription factor), DREB2A(Transcription factor), EhCaBP (Calcium binding protein), ERA1(Farnesyl transferase),FAD3 and FAD8(Increased fatty acid desaturation),FAD7(Increased fatty acid desaturation under COR15a), FLD(Flavodoxin expression in chloroplasts), Gal(Raffinose hydrolysis), Gli1(Mutant lack glycerol catabolism),GhDREB1(Transcription factor), GPAT(glycerol-3-phosphate acyltransferase of chloroplasts), HAL1(Promote K$^+$/Na$^+$ selectivity), HAL2 (Yeast)[Promote K$^+$/Na$^+$ selectivity],HOS9(Transcription factor),HOS10(Transcription factor), HOT2(Encode a chitinase-like protein), HsfA2(Heat-inducible transactivator), HvCBF4(Transcription factor), ISPS(Isoprene synthesis), JERF3(Jasmonate and ethylene-responsive factor 3), JERF1(Jasmonate and ethylene-responsive factor 1), LeGPAT(Glycerol-3-phosphate acyltransferase), lew2(Wilting allele; cellulose synthesis complex),MBF1c(Transcriptional coactivator multiprotein bridging factor),ME (NADP-malic enzyme which converts malate and NADP to pyruvate, NADPH, and CO2), NADP-ME 2(NADP-malic enzyme), MIZ1(Hydrotropism of root), MKK9(MAP Kinase), MsPRP2 (Transcription factor), NahG(salicylate hydroxylase expression), NPK1(mitogen-activated protein kinase),NtC7(Trans-membrane protein, osmotic adjustment), OsCDPK7(Transcription factor), OsCIPK01-OsCIPK30(Calcineurin B-like protein-interacting protein kinases), OsCOIN(RING finger protein), OCPI1(Transcription factor), OPBP1(Transcription factor), OsSbp(Calvin cycle enzyme sedoheptulose-1,7-bisphosphatase),OsDREB1A(Transcription factor), OsMYB3R-2(MYB homeodomain and zinc finger proteins), SIZ1(SUMO E3 ligase), SNAC1(Stomatal activity),PARP1; PARP2(Poly(ADP-ribose) polymerase),PDH45(DNA helicase 45), RGS1(Regulation of G-protein signaling),S851(Encodes 8 sphingolipid desaturase in cell membranes), SacB(Fructan synthesis), SCABP8(Interacts with SOS2), SCOF1(Transcription factor), Shn (Increased epicuticular wax),SPS(sucrose phosphate synthase),SRK2C (Protein kinase),STO(Protein binds to a Myb transcription factor),Sto1(Reduced ABA accumulation), TaPP2Ac-1(catalytic subunit (c) of protein phosphatase 2A), TaSTK(serine/threonine protein kinase), TaSrg6(Transcription factor), TERF1(ERF transcription activator), Tsi1(Transcription factor), uvi1(Transcription factor),VuNCED1(Involved in ABA biosynthesis), WXP1(Epicuticular wax accumulation), WXP1; WXP2(Epicuticular wax accumulation), ZmDR EB2A(Encodes HSP and LEA proteins), ZPT2-3(Encodes a Cys2/His2-type zinc finger protein), MtZpt2(zinc finger protein) are listed in the Table 6.

Signal transduction genes

Genes involved in stress signal sensing and a cascade of stress-signaling in A. thaliana has been of recent research interest (Winicov and Bastola, 1997; Shinozaki and Yamaguchi-Shinozaki, 1999). Components of the same signal transduction pathway may also be shared by various stress factors such as drought, salt and cold (Shinozaki and Yamaguchi-Shinozaki, 1999). Although there are multiple pathways of signal-transduction systems operating at the cellular level for gene regulation, ABA is known component acting in one of the signal transduction pathways, while others act independently of ABA. The early response genes have been known to encode transcription factors that activate downstream delayed response genes (Zhu, 2002). Although, specific branches and components exist (Lee et al., 2001), the

signaling pathways for salt, drought, and cold stresses all interact with ABA, and even converge at multiple steps (Xiong et al., 1999). Abiotic stress signalling in plants involves receptor-coupled phospho-relay, phosphoion-ositol- induced Ca^{2+} changes, mitogen activated protein kinase (MAPK) cascade, and transcriptional activation of stress responsive genes (Xiong and Zhu, 2001). A number of signaling components are associated with the plant response to high temperature, freezing, drought and anaerobic stresses (Grover et al., 2001). One of the merits for the manipulation of signaling factors is that they can control a broad range of downstream events that can result in superior tolerance for multiple aspects (Umezawa et al., 2006). Alteration of these signal transduction components is an approach to reduce the sen-sitivity of cells to stress conditions, or such that a low level of constitutive expression of stress genes is induced (Grover et al., 1999). Overexpression of functionally con-served At-DBF2 (homolog of yeast DBF2 kinase) showed striking multiple stress tolerance in Arabidopsis plants (Lee et al., 1999). Pardo et al. (1998) also achieved salt stress-tolerant transgenic plants by over expressing calcineurin (a Ca^{2+}/Calmodulin dependent protein phosphatase), a protein phosphatase known to be involved in salt-stress signal transduction in yeast. Trans-genic tobacco plants produced by altering stress sign-aling through functional reconstitution of activated yeast calcineurin not only opened-up new routes for study of stress signaling, but also for engineering transgenic crops with enhanced stress tolerance (Grover et al., 1999). Overexpression of an osmotic-stress-activated protein kinase, SRK2C resulted in a higher drought tolerance in A. thaliana, which coincided with the upregulation of stress-responsive genes (Umezawa et al., 2004). Simi-larly, a truncated tobacco mitogen-activated protein kinase kinase kinase (MAPKKK), NPK1, activated an oxidative signal cascade resulting in cold, heat, salinity and drought tolerance in transgenic plants (Kovtun et al., 2000; Shou et al., 2004). However, suppression of sign-aling factors could also effectively enhance tolerance to abiotic stress (Wang et al., 2005).

SALT TOLERANCE

Transporter genes for salt tolerance

An important strategy for achieving greater tolerance to abiotic stress is to help plants to re-establish homeostasis under stressful environments, restoring both ionic and osmotic homeostasis. This has been and continues to be a major approach to improve salt tolerance in plants through genetic engineering, where the target is to achieve Na^+ excretion out of the root, or their storage in the vacuole. A number of abiotic stress tolerant trans-genic plants have been produced by increasing the cellular levels of proteins (such as vacuolar antiporter

Table 6. Various regulatory genes (www.plantstress.com).

Gene	Gene action	Species	Phenotype	References	ID*
--	Overexpression of nicotianamine synthase	Tobacco and Arabodopsis	Heavy metal tolerance by chelation	Kim et al., 2005	8006
AB13	Transcription factor	Arabidopsis	Enhanced freezing tolerance	Tamminen et al., 2001	5217
ABF3	Transcription factor	Rice	Drought resistance	Oh et al., 2005	7780
ADC	Arginine decarboxylase overexpression	Rice	Polyamine accumulation and salt resistance in biomass accumulation	Roy and Wu, 2001	5561
ALDH3I1 & ALDH7B4	aldehyde dehydrogenase	Arabidopsis	Salt, dehydration and oxidative stress tolerance	Kotchoni et al., 2006	8303
Alx8	High APX2 and ABA	Arabidopsis	Drought resistance	Rossel et al., 2006	8164
AREB1	ABA hypersensitivity	Arabidopsis	Dehydration survival	Fujita et al., 2005	8099
AREB1	ABRE-dependent ABA signaling	Arabidopsis	Drought resistance	Fujita et al., 2006	8099
ASR1	Undetermined	Tobacco	Decreased water loss; salt tolerance	Perlson et al., 2004	7462
AtAOX1	Alternative oxidase (AOX) pathway of plant mitochondria	Arabidopsis	Cold acclimation	Fiorani et al., 2005	8085
AtCBF1-3	Transcription factor	potato	Promoter driven freezing tolerance in yield	Kim et al., 2006	8253
AtGluR2	Transcription factor	Arabidopsis	Calcium utilization under ionic stress	Kim et al., 2001	5172
AtGSK1	Homologue of GSK3/shaggy-like protein kinase	Arabidopsis	Salt tolerance in whole plant and root growth	Piao et al., 2001	5526
Atnoa1	Impaired Nitric Oxide synthesis	Arabidopsis	Salt tolerance	Zhao et al., 2007	8866
AtPCS1	Phytochelatin synthesis	Arabidopsis	Paradoxically showed hypersensitivity to Cd stress	Lee et al., 2003	6387
AtPCS1	Phytochelatin synthesis	Arabidopsis	Arsenic tolerance and cadmium hypersensitivity	Li et al., 2004	7513
AtPCS1	Phytochelatin synthesis	Tobacco	Cadmium tolerance	Pomponi et al., 2006	8092
AtPCS1	Phytochelatin synthesis	Indian mustard	As and Cd tolerance	Gasic et al., 2007a	8912
---	Phytochelatin synthesis (3 genes)	Tobacco	Cadmium tolerance	Wawrzyski et al., 2006	8372
ATP-PRT	Free His accumulation	Alyssum Indian Mustard	Nickel tolerance	Ingle et al., 2005	7812
AtPCS1	Phytochelatin synthesis	Arabidopsis	Cadmium and zinc tolerance	Gasic and Korban, 2007b	8815
AtRabG3e	Intracellular vesicle trafficking	Arabidopsis	Salt and osmotic stress tolerance	Mazel et al., 2004	6975
atRZ-1a	Zinc finger glycine-rich RNA-binding proteins	Arabidopsis	Negative effect on germination and seedling growth under salt stress	Kim et al., 2007	9026
atRZ-1a	zinc finger-containing glycine-rich RNA-binding proteins (GR-RBPs)	Arabidopsis	Freezing tolerance	Yeon-Ok et al., 2005	7773
AtSZF1 & AtSZF2	CCCH-type zinc finger proteins, involved in salt stress responses	Arabidopsis	Salt tolerance	Sun et al., 2007	9025
BNCBF5- and 17 17	CBF/DREB1-like transcription factors	Brassica napus	Freezing tolerance and photosynthetic capacity	Savitch et al., 2005	7926

Table 6. Contd.

Gene	Function	Species	Trait	Reference	No.
CAbZIP1	Plant development (dwarf phenotype)	*Arabidopsis*	Disease, drought and salt tolerance	Lee et al., 2006	8477
CAP2	Transcription factor	Tobacco	Drought and salt tolerance	Shukla et al., 2006	8470
CaPF1	Transcription factor	Virginia pine	Antioxidant activity and metal tolerance	Tang et al., 2005	8022
CaPIF1	Cys-2/His-2 zinc finger protein	Tomato	Chilling and disease resistance	Seong et al., 2007	8716
CBF1	Transcription factor	*Arabidopsis*	Cold tolerance	Jaglo-Ottosen et al., 1998	5611
DREB	Transcription factor	*Arabidopsis*	Increased tolerance to cold, drought and salinity	Kasuga et al., 1999	4534
DREB1 or OsDREB1	Transcription factor	Rice	Drought, salt and cold tolerance with reduced growth under non-stress	Ito et al., 2006	8176
DREB1A	Transcription factor	Tobacco	Drought and cold tolerance	Kasuga et al., 2004	7091
DREB1A	Transcription factor	wheat	Delayed wilting under drought stress	Pellegrineschi et al., 2004	7443
DREB2A	Transcription factor	*Arabidopsis*	Drought resistance	Sakuma et al., 2006	8302
FAD3 & FAD8	Increased fatty acid desaturation	Tobacco	Drought resistance	Meng et al., 2005	8020
GhDREB1	Transcription factor	Tobacco	Chilling tolerance, negatively regulated by gibberellic acid	Shan et al., 2007	9012
MKK9	MAP Kinase	*Arabidopsis*	Salt resistance in germination	Alzwiya et al., 2007	8979
OsDREB1A	Transcription factor	*Arabidopsis*	Drought, salt, freezing tolerance	Dubouzet et al., 2003	6429
OsMYB3R-2	MYB homeodomain, and zinc finger proteins	*Arabidopsis*	Drought, salt, freezing tolerance	Dai et al., 2007	8803
SCABP8	Interacts with SOS2	*Arabidopsis*	Salt tolerance	Quan et al., 2007	8908
TaPP2Ac-1	catalytic subunit (c) of protein phosphatase 2A	Tobacco	Drought resistance; maintain RWC and membrane stability	Xu et al., 2007	8658
ZIF1	Zn sequestration	*Arabidopsis*	Zinc tolerance	Haydon and Cobbett, 2007	8802
ZmDREB2A	Encodes HSP & LEA proteins	*Arabidopsis*	Drought and heat tolerance	Qin et al., 2007	8829
ZPT2-3	Encodes a Cys2/His2-type zinc finger protein	Petunia	Dehydration tolerance	Shoji et al., 2004	6920
MtZpt2	zinc finger protein	Medicao	Recover root growth under salt stress	Merchan et al., 2007	8911

proteins) that control the transport functions. For example, transgenic melon (Borda's et al., 1997) and tomato (Gisbert et al., 2000) plants expressing the HAL1 gene showed a certain level of salt tolerance as a result of retaining more K^+ than the control plants under salinity stress. A vacuolar chloride channel, AtCLCd gene, which is involved in cation detoxification, and AtNHXl gene which is homologous to NhxI gene of yeast have been cloned and over expressed in *Arabidopsis* to confer salt tolerance by compartmentalizing Na^+ ions in the vacuoles. Transgenic *Arabidopsis* and tomato plants that over express AtNHX1 accumulated abundant quantities of the transporter in the tonoplast and exhibited substantially enhanced salt tolerance (Apse et al., 1999; Quintero et al., 2000; Zhang and Blumwald, 2001). Salt Overly Sensitive I (SOSI) locus in *A. thaliana*, which is similar to plasma membrane Na^+/H^+ antiporter from bacteria and fungi, was

cloned and over expressed using CaMV 35S promoter. The up-regulation of SOSI gene was found to be consistent with its role in Na^+ tolerance, providing a greater proton motive force that is necessary for elevated Na^+/H^+ antiporter activities (Shi et al., 2000).

Genes encoding proton pumps, antiporters and ion trans-porters are AtMRP4 (Stomatal guard cell plasma membrane ABCC-type ABC transporter), *AtNHX1* (Vacuolar Na^+/H^+ antiporter), *AtNHX2;*

AtNHX5 (Vacuolar Na$^+$/H$^+$ antiporter), AVP1(AVP1 proton pump overexpression), GmCAX1(Cation/proton antiporter), HKT1(Potassium transporter), AtHKT1(Sodium and Potassium transporter), AtHKT1(Reduction in Sodium in root), GhNHX1(Vacuolar Na$^+$/H$^+$ antiporter), HvAACT1(Citrate transporter), HvPIP2;1(PIP2 plasma membrane aquaporin Over-expression), IRT1(Divalent cation transporter), NtAQP1(PIP1 plasma membrane aquaporin), NtPT1(Phosphate transporter), NRT2.1(Nitrate transporter) , OsNHX1 (Vacuolar Na$^+$/H$^+$ antiporter), OsSOS1(Plasma membrane Na$^+$/H$^+$ exchanger), PcSrp[Serine rich protein (enhancing ion homeostasis)], Pht1, Pht1;4(Phosphate acquisition by roots), PIP(Plasma membrane aquaporin over exression), PgTIP1(Tonoplast intrinsic protein), PIP2;2(Plasma membrane aquaporin knockout), PIP1b(Plasma membrane aquaporin over exression), PIP1bn(Plasma membrane aquaporin over exression), PIP1;4 and PIP2;5(Plasma membrane aquaporin over exression), RWC3(Aquaporin overexpression), SOS4(Involved in the synthesis of pyridoxal-5-phosphate which modulates ion transporters), SOS3(Sodium accumulation in roots), SOS1(Na$^+$-H$^+$ antiporter), SOD2(Vacuolar Na$^+$/H$^+$ antiporter), SsVP-2(Vacuolar Na$^+$/H$^+$ antiporter), SsNHX1(Vacuolar Na$^+$/H$^+$ antiporter), SULTR1;2(High affinity root sulfate transporter), TNHX1 and H$^+$-PPase TVP1(Vacuolar Na$^+$/H$^+$ antiporter), TsVP (Vacuolar Na$^+$/H$^+$ antiporter), YCF1(Sequester glutathione-chelates of heavy metals into vacuoles), ZntA(Regulation of Cd, PB and Zn pump) are listed in Table 7.

Antioxidant protection

Stress induces production of reactive oxygen species (ROS) including superoxide radicals, hydrogen peroxide (H_2O_2) and hydroxyl radicals (OH-) and these ROS cause oxidative damage to different cellular components including membrane lipids, protein and nucleic acids (Halliwell and Gutteridge, 1986). Reduction of oxidative damage could provide enhanced plant resistance to salt stress. Plants use antioxidants such reduced glutathione (GSH) and different enzymes such as superoxide dismutases (SOD), CAT, APX, glutathione-S-trans-ferases (GST) and glutathione peroxidases (GPX) to scavenge ROS. Transgenic tobacco plants over ex-pressing both GST and GPX showed improved seed germination and seedling growth under stress (Roxas et al., 1997). A major function of glutathione in protection against oxidative stress is the reduction of H_2O_2 (Foyer and Halliwell, 1976). Ruiz and Blumwald (2002) inves-tigated the enzymatic pathways leading to glutathione synthesis during the response to salt stress of wild-type and salt-tolerant B. napus L. (Canola) plants over expressing a vacuolar Na$^+$/H$^+$ antiporter (Zhang et al.,

2001).

Ion homeostasis

Plants respond to salinity using two different types of res-ponses. Salt-sensitive plants restrict the uptake of salt and adjust their osmotic pressure through the synthesis of compatible solutes (e.g. proline, glycinebetaine, so-luble sugars; Greenway and Munns, 1980). Salt-tolerant plants sequester and accumulate salt into the cell vacuoles, controlling the salt concentrations in the cytosol and maintaining a high cytosolic K$^+$/Na$^+$ ratio in their cells. The maintenance of a high cytosolic K$^+$/Na$^+$ ratio and precise regulation of ion transport is critical for salt tolerance (Glenn et al., 1999).The alteration of ion ratios in plants could result from the influx of Na$^+$ through path-ways that also function in the uptake of K$^+$ (Blumwald et al., 2000). This can be achieved by extrusion of Na$^+$ ions from the cell or vacuolar compartmentation of Na$^+$ ions. Three classes of low-affinity K$^+$ channels have been identified (Sanders, 2001), these are K$^+$ Inward rectifying channels (K IRC); K$^+$ outward rectifying channels (KORCs) and Voltage-independent cation channels (VIC). K$^+$ outward rectifying channels (KORCs) could play a role in mediating the influx of Na$^+$ into plant cells.

These channels, which open during the depolarization of the plasma membrane, could mediate the efflux of K$^+$ and the influx of Na$^+$ ions. Na$^+$ competes with K$^+$ uptake through Na$^+$ - K$^+$ co-transporters and may also block the K$^+$ specific transporters of root cells under salinity (Zhu, 2003). This could result in toxic levels of sodium as well as insufficient K$^+$ concentration for enzymatic reactions and osmotic adjustment. The influx of Na$^+$ is controlled by AtHKT1, a low affinity Na$^+$ transporter (Rus et al., 2001; Uozumi et al., 2000). The knockout mutant (hkt1) from Arabidopsis suppressed Na$^+$ accumulation and sodium hypersensitivity (Rus et al., 2001), suggesting that AtHKT1 is a salt tolerance determinant, while the efflux is Hussain et al. (2009) controlled by Salt OverlySensitive1 (SOS1), a plasma membrane Na$^+$/H$^+$ anti-porter (Shi et al., 2000). This antiporter is powered by the operation of H$^+$ -ATPase (Blumwald et al., 2000).

In addition to its role as an antiporter, the plasma membrane Na$^+$/K$^+$ SOS1 may act as a Na$^+$ sensor (Zhu, 2003). The overexpression of SOS1 improved salt tolerance in Arabidopsis (Shi et al., 2003) The compart-mentation of Na$^+$ ions in vacuoles provides an efficient and cost effective mechanism to prevent the toxic effects of Na$^+$ in the cytosol. The overexpression of AtNHX1, resulted in the generation of transgenic arabidopsis (Apse et al., 1999), tomato (Zhang and Blumwald, 2001), Brassica napus (Canola) (Zhang et al., 2001), rice (Ohta et al., 2002), tobacco (Wu et al., 2004), maize (Yin et al., 2004), tall fescue plants (Luming et al., 2006) that were not only able to grow in significantly higher salt concentration (200 mM NaCl) but could also flower and set fruit.

Table 7. Genes encoding proton pumps, antiporters and ion transporters (www.plantstress.com).

Gene	Gene action	Species	Phenotype	References	ID*
AtMRP4	Stomatal guard cell plasma membrane ABCC-type ABC transporter,	Arabidopsis	Drought susceptibility due to loss of stomatal control	Markus et al., 2004	8917
AtNHX1	Vacuolar Na$^+$/H$^+$ antiporter	Arabidopsis	Salt tolerance	Yokoi et al., 2002	8872
AtNHX2	Vacuolar Na$^+$/H$^+$ antiporter	Arabidopsis	Salt tolerance	Yokoi et al., 2002	7515
AtNHX5					
AtNHX1	Vacuolar Na$^+$/H$^+$ antiporter	Brassica napus	Salt tolerance, growth, seed yield and seed oil quality	Zhang et al., 2001	8219
AtNHX1	Vacuolar Na$^+$/H$^+$ antiporter	Cotton	Salt tolerance in photosynthesis and yield	He et al., 2005	7024
AtNHX1	Vacuolar Na$^+$/H$^+$ antiporter	Tomato	Salt tolerance, growth, fruit yield	Apse et al., 1999	7428
AtNHX1	Vacuolar Na$^+$/H$^+$ antiporter	Wheat	Salt tolerance for grain yield in the field	Xue et al., 2004	8362
AVP1	AVP1 proton pump overexpression	Arabidopsis	Salt tolerance in growth and sustained plant water status	Gaxiola et al., 2001	6970
GmCAX1	Cation/proton antiporter	Arabidopsis	Salt tolerance	Luo et al., 2005	8231
HKT1	Potassium transporter	Wheat	Salt tolerance in growth and improved K$^+$/Na$^+$ ratio	Laurie et al., 2002	8568
AtHKT1	Sodium and Potassium transporter	cells	Reduced sodium accumulation	Tomoaki et al. 2005	7252
AtHKT1	Reduction in Sodium in root	Arabidopsis	Salt tolerance	Horie et al. 2006	8981
GhNHX1	Vacuolar Na$^+$/H$^+$ antiporter	Arabidopsis (cotton)	Salt tolerance	Wu et al., 2004	6120
HvAACT1	Citrate transporter	Tobacco	Aluminum tolerance	Furukawa et al., 2007	6120
IRT1	Divalent cation transporter	Arabidopsis	Iron uptake by root and elimination of iron deficiency	Vert et al., 2002	8007
NtAQP1	PIP1 plasma membrane aquaporin	Tobacco	High root hydraulic conductance and reduced plant water deficit under drought stress	Siefritz et al., 2002	5523
NtPT1	Phosphate transporter	Rice	Phosphate acquisition	Park et al., 2007	7270
NRT2.1	Nitrate transporter	Arabidopsis	Root architecture and nitrate uptake under N stress	Remans et al., 2006	7781
OsNHX1	Vacuolar Na$^+$/H$^+$ antiporter	rice	Salt tolerance	Fukuda et al., 2004	7254
OsSOS1	Plasma membrane Na$^+$/H$^+$ exchanger	rice	Salt tolerance	Martinez-Atienza et al., 2007	7256
PcSrp	Serine rich protein (enhancing ion homeostasis?)	Finger millet	Salt tolerance	Mahalakshmi et al., 2006	8830
Pht1, Pht1;4	Phosphate acquisition by roots	Arabidopsis	Phosphate efficiency	Shin et al., 2004	5597
PIP	Plasma membrane aquaporin overexression	Soybean, lettuce	Downregulated by arbuscular mycorrhisa causing water conservation	Porcel et al., 2006	8352
PgTIP1	Tonoplast intrinsic protein	Arabidopsis	Salt tolerance; root dependant drought tolerance	Peng et al., 2007	6871

Table 7. Genes encoding

PIP2;2	Plasma membrane aquaporin knockout	*Arabidopsis*	Reduced hydraulic conductivity of root cortex cells	Javot et al., 2003	8591
PIP1b	Plasma membrane aquaporin overexression	Tobacco	No effect under salt and negative effect under drought stress	Aharon et al. 2003	6659
PIP1bn	Plasma membrane aquaporin overexression	Tobacco	Tolerance to osmotic stress	Yua et al., 2005	7863
PIP1;4 & PIP2;5	Plasma membrane aquaporin overexression	Tobacco	Excessive water loss and retarded seedling growth under drought stress	Jang et al., 2007	8974
RWC3	Aquaporin overexpression	Rice	Maintenance of leaf water potential and transpiration under 10 h PEG stress	Lian et al., 2004	7177
SOS4	Involved in the synthesis of pyridoxal-5-phosphate which modulates ion transporters	*Arabidopsis*	Salt tolerance through Na^+/K^+ homeostasis	Shi et al., 2002	5931
SOS3	Sodium accumulation in roots	*Arabidopsis*	Salt tolerance	Horie et al. 2006	8335
SOS1	Na^+-H^+ antiporter	*Arabidopsis*	Protect K^+ permeability during salt stress	Qi and Spalding, 2004	7350
SOD2	Vacuolar Na^+/H^+ antiporter	*Arabidopsis*	Salt tolerance; higher plant K/Na ratio	Gao et al., 2004	6924
SOD2	Vacuolar Na^+/H^+ antiporter	Rice	Salt tolerance	Zhao et al. 2006	8088
SsVP-2	Vacuolar Na^+/H^+ antiporter	*Arabidopsis*	Salt tolerance	Guo et al. 2006	8166
SsNHX1	Vacuolar Na^+/H^+ antiporter	Rice	Salt tolerance	Zhao et al. 2006	8216
SULTR1;2	High affinity root sulfate transporter	*Arabidopsis*	Selenate sensitivity	El Kassis et al., 2007	8800
TNHX1 and H^+-PPase TVP1	Vacuolar Na^+/H^+ antiporter	*Arabidopsis*	Salt tolerance	Brini et al. 2007	8697
TsVP	Vacuolar Na^+/H^+ antiporter	Tobacco	Salt tolerance	Gao et al., 2006	8462
YCF1	Sequester glutathione-chelates of heavy metals into vacuoles	*Arabidopsis*	Heavy metal and salt tolerance	Koh et al. 2006	8172
ZntA	Regulation of Cd, PB and Zn pump	*Arabidopsis*	Cd and Pb resistance; reduced metal accumulation	Lee et al., 2003	6824

Synthesis/over expression of compatible solutes

The cellular response of salt-tolerant organisms to both long- and short-term salinity stresses includes the syn-thesis and accumulation of a class of osmoprotective compounds known as compatible solutes. These re-latively small organic molecules are not toxic to meta-bolism and include proline, glycinebetaine, polyols, sugar alcohols, and soluble sugars. These osmolytes stabilize proteins and cellular structures and can increase the osmotic pressure of the cell (Yancey et al., 1982). This response is homeostatic for cell water status, which is perturbed in the face of soil solutions containing higher amounts of NaCl and the consequent loss of water from the cell. Glycinebetaine and trehalose act as stabilizers of quartenary structure of proteins and highly ordered states of membranes. Mannitol serves as a free radical scavenger. It also stabilizes sub cellular structures (membranes and proteins), and buffers cellular redox potential under stress. Hence these organic osmolytes are also known as osmoprotectants (Bohnert and Jensen, 1996; Chen and Murata, 2000). Genes involved in osmoprotectant biosynthesis are upregulated under salt stress and concentrations of accumulated osmo-protectants correlate with osmotic stress tolerance (Zhu, 2002). Although enhanced synthesis and accu-mulation of compatible solutes under osmotic stress is well known, little is known about the signaling cascades that regulate compatible solute biosynthesis in higher plants.

Salt tolerance of transgenic tobacco engineered to over accumulate mannitol was first demonstrated by Tarczynski et al. (1993). The other examples of compatible solute genetic engineering includes the transformation of genes for Ectoine synthesis with enzymes from the halophilic bacterium *Halomonas elongata* (Nakayama et al., 2000; Ono et al., 1999) and trehalose synthesis in potato (Yeo et al., 2000), rice (Garg et al., 2002), and sorbitol synthesis in plantago (Pommerrenig et al., 2007) (Table 1). Initial strategies aimed at engineering higher concentrations of proline began with the overexpression of genes encoding the enzymes pyrroline-5-carboxylate (P5C) synthetase (P5CS) and P5C reductase (P5CR), which catalyze the two steps between the substrate (glutamic acid) and the product (proline). P5CS overexpression in transgenic tobacco dramatically elevated free proline (Kishor et al., 1995). However there is strong evidence that free proline inhibits P5CS (Roosens et al., 1999). Hong et al. (2000) achieved a two-fold increase in free proline in tobacco plants by using a P5CS modified by site directed mutagenesis. The procedure alleviated the feedback inhibition of P5CS activity by proline and resulted in improved germination and growth of seedlings under salt stress.

In spinach and sugar beet which naturally accumulate glycinebetaine, the synthesis of this compound occurs in the chloroplast. The first oxidation to betaine aldehyde is catalyzed by choline mono-oxygenase (CMO). Betaine aldehyde oxidation to glycinebetaine is catalyzed by betaine aldehyde dehydrogenase (BADH) (Rathinasabapathi, 2000). In *Arthrobacter globiformis,* the two oxidation steps are catalyzed by one enzyme, choline oxidase (COD), which is encoded by the *codA* locus (Sakamoto and Murata, 2000). Hayashi et al. (1997) used choline oxidase of *A. globiformis* to engineer glycine-betaine synthesis in *Arabidopsis* and subsequently tolerance to salinity during germination and seedling establishment was improved markedly in the transgenic lines. Huang et al. (2000) used COX from *A. panescens*, which is homologous to the *A. globiformis* COD, to transform *arabidopsis*, *B. napus* and tobacco. In this set of experiments COX protein was directed to the cytoplasm and not to the chloroplast. Improvements in tolerance to salinity, drought and freezing were observed in some transgenics from all three species, but the tolerance was variable. The results offered the possibility that the protection offered by glycinebetaine is not only osmotic but also function as scavengers of oxygen radicals. The level of glycinebetaine production in transgenics could be limited by choline. A dramatic increase in glycinebetaine levels (to 580 mmol/g dry weight in *Arabidopsis thaliana*) was achieved when the growth medium was supplemented with choline (Huang et al.,

2000). The enhancement of glycinebetaine syntheses in target plants has received much attention (Rontein et al., 2002).

Conclusions

This review summarizes the recent efforts to improve abiotic stress tolerance in crop plants by employing some of the stress-related genes and transcription factors. There is a clear and urgent need to begin to introduce stress tolerance genes into crop plants, in addition to establishing gene stacking or gene pyramiding. Although progress in improving stress tolerance has been slow, there are a number of reasons for optimism. The use of transgenes to improve the tolerance of crops to abiotic stresses remains an attractive option. Options targeting multiple gene regulation appear better than targeting single genes. An important issue to address is how the tolerance to specific abiotic stress is assessed, and whether the achieved tolerance compares to existing tolerance. A well focused approach combining the molecular physiological and metabolic aspects of abiotic stress tolerance is required for bridging the knowledge gaps between the molecular or cellular expression of the genes and the whole plant phenotype under stress.

ACKNOWLEDGMENTS

The authors thank the Indo-US collaboration with Agricultural knowledge initiative programme in Biotechnology that is jointly funded by the Indian Council of Agricultural Research, Government of India.

REFERENCES

Abe H, Shinwari ZK, Seki M, Shinozaki K (2002). Biological plants. Plant Mol Biol., 12: 475-486.

Abebe T, Guenzi AC, Martin B, Cushman JC (2003). Tolerance of Mannitol-Accumulating Transgenic Wheat to Water Stress and Salinity. Plant Physiol., 131: 1748-1755.

Agalou A, Roussis A, Spaink HP (2005)The *Arabidopsis* selenium-binding protein confers tolerance to toxic levels of selenium. Funct. Plant Biol., 32: 881-890.

Aharon R, Shahak Y, Wininger S, Bendov R,Kapulnik Y, Galili G (2003). Overexpression of a Plasma Membrane Aquaporin in Transgenic Tobacco Improves Plant Vigor under Favorable Growth Conditions but Not under Drought or Salt Stress. Plant Cell, 15: 439-447.

Aharoni A, Dixit S, Jetter R, Thoenes E, van Arkel G, Pereira A (2004). The SHINE clade of AP2 domain transcription factors activates wax biosynthesis, alters cuticle properties, and confers drought tolerance when overexpressed in *Arabidopsis*. Plant Cell, 16: 2463–2480.

Alia KY, Sakamoto A, Nonaka H, Hayashi H, Saradhi PP, Chen THH, Murata N (1998). Enhanced tolerance to light stress of transgenic *arabidopsis* plants that express the codA gene for a bacterial choline oxidase. Plant Mol. Biol., 40: 279–288.

Almeida AM, Silva AB, Arajo SS, Cardoso LA, Santos DM, Torn JM, Silva JM, Paul MJ, Fevereiro PS (2007). Responses to water withdrawal of tobacco plants genetically engineered with the AtTPS1 gene: A special reference to photosynthetic parameters. Euphytica, 154: 113-126.

Al-Taweel K, Iwaki T, Yabuta Y, Shigeoka S, Murata N, Wadano A

(2007) A Bacterial Transgene for Catalase Protects Translation of D1 Protein during Exposure of Salt-Stressed Tobacco Leaves to Strong Light. Plant Physiol., 144: 258-265.

Alvim CFC, Cascardo JCM, Nunes CC, Martinez CA, Otoni WC, Fontes EPB (2001). Enhanced accumulation of BiP in transgenic plants confers tolerance to water stress. Plant Physiol., 126: 1042-1054.

Alzwiya IA, Morris PC (2007). A mutation in the *Arabidopsis* MAP kinase kinase 9 gene results in enhanced seedling stress tolerance. Plant Sci., 173: 302-308.

Apse MP, Aharon GS, Snedden WA, Blumwald E (1999). Salt tolerance conferred by overexpression of a vascoular Na⁺/H⁺ antiport in *Arabidopsis*. Science, 285: 1256–1258.

Araus JL, Bort J, Steduto P, Villegas D, Royo C (2003). Breeding cereals for Mediterranean conditions: Ecophysilogy clues for biotechnology application. Ann. Appl. Biol., 142: 129-141.

Arrivault S, Toralf S, Ute K (2006). The *Arabidopsis* metal tolerance protein AtMTP3 maintains metal homeostasis by mediating Zn exclusion from the shoot under Fe deficiency and Zn oversupply. Plant J., 46: 861-879.

Babu RC, Zhang J, Blum A, David HTH, Wu R, Nguyen HT (2004). HVA1, a LEA gene from barley confers dehydration tolerance in transgenic rice (Oryza sativa L.) via cell membrane protection. Plant Sci.166: 855-862.

Badawi GH, Yamauchi Y, Shimada E, Sasaki R, Kawano N, Tanaka K (2004). Enhanced tolerance to salt stress and water deficit by overexpressing superoxide dismutase in tobacco (*Nicotiana tabacum*) chloroplasts. Plant Sci., 166: 919-928.

Bahieldin A, Mahfouz HT, Eissa HF, Saleh OM, Ramadan AM, Ahmed IA, Dyer WE, El-Itriby HA, Madkour MA (2005). Field evaluation of transgenic wheat plants stably expressing the HVA1 gene for drought tolerance. Physiol. Plant, 123: 421-427.

Bartels D, Furini A, Ingram J, Salamini F (1996). Responses of plant to dehydration stress: A molecular analysis. Plant Growth Reg., 20: 111-118.

Bartels D, Sunkar R (2005). Drought and salt tolerance in plants. Crit. Rev. Plant Sci., 21: 1-36.

Barthakur S, Babu V, Bansal KC (2001). Over-expression of osmotin induces proline accumulation and confers tolerance to osmotic stress in transgenic tobacco. J. Plant Biochem. Biotechnol., 10: 31-37.

Basu U, Good AG, Taylor GJ (2001). Transgenic Brassica napus plants overexpressing aluminium-induced mitochondrial manganese superoxide dismutase cDNA are resistant to aluminium. Plant Cell Environ., 24: 1269-1278.

Behnam B, Kikuchi A, Celebi-Toprak F, Yamanaka S, Kasuga M, Yamaguchi-Shinozaki K, Watanabe KN (2006). The *Arabidopsis* DREB1A gene driven by the stress-inducible rd29A promoter increases salt-stress tolerance in proportion to its copy number in tetrasomic tetraploid potato (*Solanum tuberosum*). Plant Biotechnol., 23: 169–177.

Bhatnagar-Mathur P, Devi MJ, Reddy DS, Vadez V, Yamaguchi-Shinozaki K, Sharma KK (2006). Overexpression of *Arabidopsis* thaliana DREB1A in transgenic peanut (*Arachis hypogaea* L.) for improving tolerance to drought stress (poster presentation). In: Arthur M. Sackler Colloquia on "From Functional Genomics of Model Organisms to Crop Plants for Global Health", April 3-5, 2006. National Academy of Sciences, Washington, DC.

Bhatnagar-Mathur P, Devi MJ, Serraj RY, amaguchi-Shinozaki K, Vadez V, Sharma KK (2004). Evaluation of transgenic groundnut lines under water limited conditions. Int. Arch. Newslett., 24: 33-34.

Blum A (1988). Plant breeding for stress environments. Boca Roton, USA: CRS Press.

Blumwald E, Aharon GS, Apse MP (2000).Sodium transport in plant cells. Biochimica et Biophysica Acta, 1465: 140–151.

Bohnert HJ, Jensen RG (1996). Strategies for engineering water stress tolerance in plants. Trends Biotechnol., 14: 89-97.

Bohnert HJ, Nelson DF, Jenson RG (1995). Adaptation to environmental stresses. Plant Cell, 7: 1099-1111.

Bohnert HJ, Shen B (1999) Transformation and compatible solutes. Sci. Hortic., 78: 237-260.

Borda´s M, Montesinos C, Dabauza M, Salvador A, Roig LA, Serrano R, Moreno V (1997). Transfer of the yeast salt tolerance gene HAL1 to *Cucumis melo* L. cultivars and *in vitro* evaluation of salt tolerance.

Transgenic Res., 5: 1-10.

Borsani O, Cuartero J, Valpuesta V, Botella MA (2002).Tomato tos1 mutation identifies a gene essential for osmotic tolerance and abscisic acid sensitivity. Plant J., 32: 905-914.

Bowler C, Slooten L, Vandenbranden S, Rycke R D, Botterman J, Sybesma C, Van Montagu M, Inze D (1991). Manganese superoxide dismutase can reduce cellular damage mediated by oxygen radicals in transgenic plants. EMBO J., 10: 1723-1732.

Bray E (1993). Molecular responses to water deficit. Plant Physiol., 103: 1035-1040.

Brini F, Hanin M, Mezghani I, Gerald AB, Masmoudi K (2007). Overexpression of wheat Na⁺/H⁺ antiporter TNHX1, H⁺⁺ pyrophosphatase TVP1 improve salt- and drought-stress tolerance in *Arabidopsis* thaliana plants. J. Exp. Bot., 58: 301-308.

Cao W-H, Liu J, He X-J, Mu RL, Zhou H-L, Chen S-Y, Zhang J-S (2007). Modulation of Ethylene Responses Affects Plant Salt-Stress Responses. Plant Physiol., 143: 707-719.

Capell T, Bassie L, Christou P (2004). Modulation of the polyamine biosynthetic pathway in transgenic rice confers tolerance to drought stress. Proc. Natl. Acad. Sci. USA, 101: 9909–9914.

Capell T, Escobar C, Liu H, Burtin D, Lepri O, Christou P (1998). Over-expression of the oat arginine decarboxylase cDNA in transgenic rice (*Oryza sativa* L.) affects normal development patterns *in vitro* and results in putrescine accumulation in transgenic plants. TAG, 97: 246-254.

Ceccarelli S, Grando S (1996). Drought as a challenge for the plant breeder. Plant Growth Reg., 20: 149-155.

Chaves MM, Maroco JP, Pereira JS (2003). Understanding plant responses to drought from genes to the whole plant. Funct. Plant Biol., 30: 239-264.

Chen TH, Murata N (2002). Enhancement of tolerance of abiotic stress by metabolic engineering of betaines and other compatible solutes. Curr. Opin. Plant Biol., 5: 250–257.

Chen THH, Murata N (2000).Enhancement of tolerance of abiotic stress by metabolic engineering of betaines and other compatible solutes. Curr. Opin. Plant Biol., 5: 250–257.

Cheng WH, Endo A, Zhou L, Penney J, Chen HC, Arroyo A, Leon P, Nambara E, Asami T, Seo M (2002). A unique short-chain dehydrogenase/reductase in *Arabidopsis* glucose signaling and abscisic acid biosynthesis and functions. Plant Cell, 14: 2723–2743.

Chinnusamy V, Jagendorf A, Zhu JK (2005) Understanding and improving salt tolerance in plants. Crop Sci., 45: 437-448.

Clark DG, Dervinis C, Barrett JE, Klee H, Jones M (2004). Drought-induced Leaf Senescence and Horticultural Performance of Transgenic PSAG12-IPT Petunias. J. Am. Soc. Hortic. Sci.129: 93-99.

Cortina C, Culia n ez-Macia F (2005). Tomato abiotic stress enhanced tolerance by trehalose biosynthesis. Plant Sci., 169: 75–82.

Dai X, Xu Y, Ma O, Xu W, Wang T, Xue Y, Chong K (2007). Overexpression of an R1R2R3 MYB Gene, OsMYB3R-2, Increases Tolerance to Freezing, Drought, and Salt Stress in Transgenic *Arabidopsis*. Plant Physiol., 143: 1739-1751.

Diamant S, Eliahu N, Rosenthal D, Goloubinoff P (2001). Chemical chaperones regulate molecular chaperones *in vitro* and in cells under combined salt and heat stresses. J. Biol. Chem., 276: 39586–39591.

Dubouzet JG, Yoh S, Yusuke K, Mie D, Emilyn G, Miura S, Seki M, Shinozaki K, Yamaguchi-Shinozaki K (2003). OsDREB genes in rice, (*Oryza sativa* L.), encode transcription activators that function in drought-, high-salt- and cold-responsive gene expression. Plant J. 33: 751-763.

Dure L III (1993). A repeating 11-mer amino acid motif and plant desiccation. Plant J., 3: 363-369.

Dure L III, Crouch M, Harada J, Ho T-HD, Mundy J (1989). Common amino acid sequence domains among the LEA proteins of higher

El Hafid R, Smith DH, Karrou M, Samir K (1998). Physiological attributes associated with early-season drought resistance in spring durum wheat cultivars. Can. J. Plant Sci., 78: 227-237.

El Kassis E, Cathala N, Rouached H, Fourcroy P, Berthomieu P, Terry N, Davidian J-C (2007). Characterization of a Selenate-Resistant *Arabidopsis* Mutant. Root Growth as a Potential Target for Selenate Toxicity. Plant Physiol. 143: 1231-1241.

Elsadig EA, Naoyoshi K, Hamid KG, Hironori K, Takeshi S, Isao M,

Toshiyuki S, Shinobu I, Kiyoshi T (2006). Enhanced tolerance to ozone and drought stresses in transgenic tobacco overexpressing dehydroascorbate reductase in cytosol. Physiol. Plant, 127: 57-65.

Eltayeb AE, Kawano N, Badawi GH, Kaminaka H, Sanekata T, Shibahara T, Inanaga S, Tanaka K (2007). Overexpression of monodehydroascorbate reductase in transgenic tobacco confers enhanced tolerance to ozone, salt and polyethylene glycol stresses. Planta, 225: 1255-1264.

Espinosa-Ruiz A, Belles JM, Serrano R, Culianez-Macia FA (1999). Arabidopsis thaliana AtHAL3: A flavoprotein related to salt and osmotic tolerance and plant growth. Plant J., 20: 529-539.

Ezaki B, Gardner RC, Ezaki Y, Matsumoto H (2000). Expression of aluminum-induced genes in transgenic Arabidopsis plants can ameliorate aluminum stress and/or oxidative stress. Plant Physiol., 122: 657-665.

Ezaki B, Katsuhara M, Kawamura M, Matsumoto H (2001). Different mechanisms of four aluminum (Al)-resistant transgenes for Al toxicity in Arabidopsis. Plant Physiol., 127: 918-927.

Ezaki B, Katsuhara M, Kawamura M, Matsumoto H (2001) Different mechanisms of four aluminum (Al)-resistant transgenes for Al toxicity in Arabidopsis. Plant Physiol., 127: 918-927.

Figueras M, Pujal J, Saleh A, Savé R, Pagès M, Goday A (2004). Maize Rab17 overexpression in Arabidopsis plants promotes osmotic stress tolerance. Ann. Appl. Biol., 144: 251-257.

Fiorani F, Ann L, Umbach, James N, Siedow (2005). The Alternative Oxidase of Plant Mitochondria Is Involved in the Acclimation of Shoot Growth at Low Temperature. A Study of Arabidopsis AOX1a Transgenic Plants. Plant Physiol., 139: 1795-1805.

Foyer C, Halliwell B (1976).The presence of glutathione and glutathione reductase in chloroplasts: A proposed role in ascorbic acid metabolism. Planta, 133: 21-25.

Fujita Y, Fujita M, Satoh R, Maruyama K, Parvez MM, Seki M, Hiratsu K, Ohme-Takagi M, Shinozaki K, Yamaguchi-Shinozaki K (2005). AREB1 Is a Transcription Activator of Novel ABRE-Dependent ABA Signaling That Enhances Drought Stress Tolerance in Arabidopsis. Plant Cell, 17: 3470-3488.

Fukuda A, Nakamura A, Tagiri A, Tanaka H, Miyao A, Hirochika H,Tanaka Y (2004). Function, Intracellular Localization and the Importance in Salt Tolerance of a Vacuolar Na+/H+ Antiporter from Rice. Plant Cell Physiol., 45: 146-159.

Fukushima E, Arata Y, Endo T, Sonnewald U, Sato F (2001). Improved salt tolerance of transgenic tobacco expressing apoplastic yeast-derived invertase. Plant Cell Physiol., 42: 245-249.

Furukawa J, Yamaji N, Wang H, Mitani N, Murata Y, Sato K, Katsuhara M, Ma KTJF (2007). An Aluminum-Activated Citrate Transporter in Barley. Plant Cell Physiol., 48: 1081-1091.

Gaber, A, Yoshimura K, YamamotoT, Yabuta, Y,Takeda T, Miyasaka H, Nakano Y, Shigeoka S (2006) Glutathione peroxidase-like protein of Synechocystis PCC 6803 confers tolerance to oxidative and environmental stresses in transgenic Arabidopsis. Physiol. Plant, 128: 251-262.

Galau GA, Bijaisoradat N, Hughes DW (1987). Accumulation kinetics of cotton late embryogenesis-abundent (Lea) mRNAs and storage protein mRNAs: Coordinate regulation during embryogenesis and role of abscisic acid. Dev. Biol., 123: 198-212.

Gao F, Gao O, Duan X, Yue G, Yang A, Zhang J (2006). Cloning of an H+-PPase gene from Thellungiella halophila and its heterologous expression to improve tobacco salt tolerance. J. Exp. Bot., 57: 3259-3270.

Gao M, Sakamoto A, Miura K, Murata N, Sugiura A, Tao R (2000). Transformation of Japanese persimmon (Diospyros kaki Thunb.) with a bacterial gene for choline oxidase. Mol. Breed, 6: 501-510.

Gao X, Ren Z, Zhao Y, Zhang H (2004). Overexpression of SOD2 Increases Salt Tolerance of Arabidopsis.Plant Physiol., 133: 1873-1881.

Garg AK, Kim JK, Owens TG, Ranwala AP, Choi YC, Kochian LV, Wu RJ (2002). Trehalose accumulation in rice plants confers high tolerance levels to different abiotic stresses. Proc. Natl. Acad Sci., USA, 99: 15898–15903.

Garg AK, Kim JK, Owens TG, Ranwala AP, Do Choi Y, Kochian LV, Wu RJ (2002). Trehalose accumulation in rice plants confers high tolerance levels to different abiotic stresses. PNAS, 99: 15898-15903.

Gasic K, Korban SS (2007a). Expression of Arabidopsis phytochelatin synthase in Indian mustard (Brassica juncea) plants enhances tolerance for Cd and Zn. Planta, 225: 1277-1285.

Gasic K, Korban SS (2007b). Transgenic Indian mustard (Brassica juncea) plants expressing an Arabidopsis phytochelatin synthase (AtPCS1) exhibit enhanced As and Cd tolerance. Plant Mol. Biol., 64: 361-369.

Gaxiola RA, Soledad JL, Undurraga LM, Dang GJ, Allen SL, Alper GR (2001). Drought- and salt-tolerant plants result fro overexpression of the AVP1 H+-pump. Proc. Natl. Acad. Sci., 98: 11444-11449.

Gisbert C, Rus AM, Bolarin MC, Lopez-Coronado M, Arrillaga I, Montesinos C, Caro M, Serrano R, Moreno V (2000). The yeast HAL1 gene improves salt tolerance of transgenic tomato. Plant Physiol., 123: 393-402.

Glenn EP, Brown JJ, Blumwald E (1999).Salt tolerance and crop potential of halophytes. Crit. Rev. Plant. Sci., 18: 227-255.

Goyal K, Walton LJ, Tunnacliffe A (2005). LEA proteins prevent protein aggregation due to water stress. Biochem. J., 388: 151-157.

Greenway H, Munns R (1980). Mechanisms of salt tolerance in nonhalophytes. Annu. Rev. Plant Physiol., 31: 149-190.

Grover A, Kapoor A, Satya Lakshmi O, Agrawal S, Sahi C, Katiyar-Agarwal S, Agarwal M, Dubey H (2001). Understanding molecular alphabets of the plant abiotic stress responses. Curr. Sci., 80: 206-216.

Grover A, Minhas D (2001). Towards production of abiotic stress tolerant transgenic rice plants: issues, progress and future research needs. Proc Indian Natl Sci. Acad. B. Rev. Tracts. Biol Sci 66:13–32.

Grover A, Sahi C, Sanan N, Grover A (1999). Taming abiotic stresses in plants through genetic engineering: Current strategies and perspective. Plant Sci., 143: 101-111.

Guo H-J, Zhou H-Y, Zhang X-S, Li X-G, Meng Q-W (2007). Overexpression of CaHSP26 in transgenic tobacco alleviates photoinhibition of PSII and PSI during chilling stress under low irradiance S. J. Plant Physiol., 164: 126-136.

Guo S, Yin H, Zhang X, Zhao F, Li P, Chen S, Zhao Y, Zhang H (2006). Molecular Cloning and Characterization of a Vacuolar H+-pyrophosphatase Gene, SsVP, from the Halophyte Suaeda salsa and its Overexpression Increases Salt and Drought Tolerance of Arabidopsis. Plant Mol. Biol., 60: 41-50.

Gupta AS, Heinen JL, Holaday AS. Burke JJ, Allen RD (1993). Increased Resistance to Oxidative Stress in Transgenic Plants that Overexpress Chloroplastic Cu/Zn Superoxide Dismutase. Proc. Nat. Acad. Sci., 90: 1629-1633.

Halliwell B, Gutteridge JMC (1986). Oxygen free radicals and iron in relation to biology and medicine: Some problems and concepts. Arch. Biochem. Biophys., 246: 501-514.

Hare PD, Cress WA, Van Staden J (1998). Dissecting the roles of osmolyte accumulation during stress. Plant Cell Environ., 21: 535–553.

Hayashi H, Alia ML, Deshnium PG, Id M, Murata N (1997).Transformation of Arabidopsis thaliana with the codA gene for choline oxidase; accumulation of glycine betaine and enhanced tolerance to salt and cold stress. Plant J., 12: 133-142.

Haydon MJ, Cobbett CS (2007). A Novel Major Facilitator Superfamily Protein at the Tonoplast Influences Zinc Tolerance and Accumulation in Arabidopsis Plant Physiol., 143: 1705-1719.

He C, Yan J, Shen G, Fu L, Holaday AS, Auld D, Blumwald E, Zhang H (2005). Expression of an Arabidopsis Vacuolar Sodium/Proton Antiporter Gene in Cotton Improves Photosynthetic Performance Under Salt Conditions and Increases Fiber Yield in the Field. Plant Cell Physiol., 46: 1848-1854.

Hideg NT, Oberschall A, Dudits D, Vass I (2003) Detoxification function of aldose/aldehyde reductase during drought and ultraviolet-B (280-320 nm) stresses. Plant Cell Environ., 26: 513-522.

Hirano HY, Sano Y (1998). Enhancement of Wx gene expression and the accumulation of amylose in response to cool temperatures during seed development in rice. Plant Cell Physiol., 39: 807-812.

Hmida-Sayari A, Gargouri-Bouzid R, Bidani A, Jaoua L, Savouré A, Jaou S (2005). Overexpression of 1-pyrroline-5-carboxylate synthetase increases proline production and confers salt tolerance in transgenic potato plants. Plant Sci., 169: 746-752.

Holmstrom KO, Somersalo S, Mandal A, Palva ET, Welin B (2000).

Improved tolerance to salinity and low temperature in transgenic tobacco producing glycine betaine. J. Exp. Bot., 51: 177-185.

Hong SW, Vierling E (2000). Mutants of Arabidopsis thaliana defective in the acquisition of tolerance to high temperature stress. Proc. Natl. Acad. Sci., 97: 4392-4397.

Hong ZL, Lakkineni K, Zhang Z M, Verma DPS, Hong Z, Zhang ZM (2000). Removal of feedback inhibition of DELTA1-pyrroline-5-carboxylate synthetase results in increased proline accumulation and protection of plants from osmotic stress. Plant Physiol., 122: 1129-1136.

Hong ZL, Lakkineni K, Zhang ZM, Verma DPS (2000). Removal of feedback inhibition of delta (1)-pyrroline-5-carboxylate synthetase results in increased proline accumulation and protection of plants from osmotic stress. Plant Physiol., 122: 1129-1136.

Honjoh K, Shimizu H, Nagaishi N, Matsumoto H, Suga K, Miyamoto T, Iio M, Hatano S (2001). Improvement of freezing tolerance in transgenic tobacco leaves by expressing the hiC6 gene. Biosci. Biotechnol. Biochem., 65: 1796-1804.

Horie T, Horie R, Chan W-Y, Leung H-Y, Schroeder JI (2006). Calcium Regulation of Sodium Hypersensitivities of sos3 and athkt1 Mutants. Plant Cell Physiol., 47: 622-633.

Hoshida H, Tanaka Y, Hibino T, Hayashi Y, Tanaka A, Takabe T (2000). Enhanced tolerance to salt stress in transgenic rice that overexpresses chloroplast glutamine synthetase. Plant Mol. Biol., 43: 103-111.

Hu L, Lu H, Liu Q, Chen X, Jiang X (2005). Overexpression of mtlD gene in transgenic Populus tomentosa improves salt tolerance through accumulation of mannitol. Tree Physiol., 25: 1273-1281.

Huang B, Jin LG, Liu JY (2007). Molecular cloning and functional characterization of a DREB1/CBF-like gene (GhDREB1L) from cotton. Sci. China C Life Sci., 50(1): 7–14.

Huang B, Jin LG, Liu JY (2008). Identification and characterization of the novel gene GhDBP2 encoding a DRE-binding protein from cotton (Gossypium hirsutum). J. Plant Physiol., 165: 214–223.

Huang B, Liu JY (2006). Cloning and functional analysis of the novel gene GhDBP3 encoding a DRE-binding transcription factor from Gossypium hirsutum. Biophys. Biophys. Acta, 1759(6): 263-269.

Huang C, Wenliang He, Guo J, Chang X, Peixi Su, Lixin Z (2005) Increased sensitivity to salt stress in an ascorbate-deficient Arabidopsis mutant. J. Exp. Bot., 56: 3041-3049.

Huang J, Hirji R, Adam L, Rozwadowski KL, Hammerlindl JK, Keller WA, Selvaraj G (2000). Genetic engineering of glycinebetaine production toward enhancing stress tolerance in plants: Metabolic limitations. Plant Physiol., 122: 747–756.

Hur J, Jung K-H, Lee C-L, An G (2004). Stress-inducible OsP5CS2 gene is essential for salt and cold tolerance in rice. Plant Sci., 167: 417-426.

Hussain M, Farooq M, Jabran K , Wahid A (2009). Foliar Application of Glycinebetaine and Salicylic Acid Improves Growth, Yield and Water Productivity of Hybrid Sunflower Planted by Different Sowing Methods. J. Agron. Crop Sci., 196(2): 136 – 145.

Ingle RA, Mugford ST, Rees JD, Campbell MM, Smith JAC (2005) Constitutively High Expression of the Histidine Biosynthetic Pathway Contributes to Nickel Tolerance in Hyperaccumulator Plants. Plant Cell, 17: 2089-2106.

Ingram J, Bartels D (1996). The molecular basis of dehydration tolerance in plants. Ann. Rev. Plant Physiol. Plant Mol. Biol., 47: 377-403.

Ito Y, Katsura K, Maruyama K, Taji T, Kobayashi M, Seki M, Shinozaki K, Yamaguchi-Shinozaki K (2006). Functional analysis of rice DREB1/CBF-type transcription factors involved in cold-responsive gene expression in transgenic rice. Plant Cell Physiol., 47: 141-153.

Ito Y, Katsura K, Maruyama K, Taji T, Kobayashi M, Seki M, Shinozaki K, Kazuko Y-S (2006). Functional Analysis of Rice DREB1/CBF-type Transcription Factors Involved in Cold-responsive Gene Expression in Transgenic Rice. Plant Cell Physiol., 47: 141-153.

Iuchi S, Kobayashi M, Taji T, Naramoto M, Seki M, Kato T, Tabata S, Kakubari Y, Yamaguchi-Shinozaki K, Shinozaki K (2001). Regulation of drought tolerance by gene manipulation of 9-cis-epoxycarotenoid dioxygenase, a key enzyme in abscisic acid biosynthesis in Arabidopsis. Plant J., 27: 325-333.

Iwasaki T, Kiyosue T, Yamaguchi-Shinozaki K (1997). The dehydration-inducible rd17 (cor47) gene and its promoter region in Arabidopsis thaliana. Plant Physiol., 115: 128.

Jaglo-Ottosen KR, Gilmour SJ, Zarka DG, Schabenberger O, Thomashow MF (1998). Arabidopsis CBF1 overexpression induces COR genes and enhances freezing tolerance, Science, 280: 104-106.

Jang JY, Lee SH, Rhee JY, Chung GC, Ahn SJ, Kang H (2007). Transgenic Arabidopsis and tobacco plants overexpressing an aquaporin respond differently to various abiotic stresses. Plant Mol. Biol., 64: 621-632.

Javot H, Lauvergeat V, Santonia V, Martin-Laurent F, Josette G, Vinh J, Heyes J, Katja IF, Anton RF, David BC (2003). Role of a Single Aquaporin Isoform in Root Water Uptake. Plant Cell, 15: 509-522.

Jones HJ (1993). Drought tolerance and water-use efficiency. In: Water deficits, plant response from cell to community (Smith JAC, Griffiths H eds). Oxford, UK: BIOS Scientific Publishers, pp. 193-203.

Karim S, Aronsson H, Ericson H, Pirhonen M, Leyman B, Welin B, Palva ET, Patrick VD, Kjell-Ove H (2007). Improved drought tolerance without undesired side effects in transgenic plants producing trehalose. Plant Mol. Biol., 64: 371-386.

Kasuga M, Miura S, Shinozaki K, Kazuko Y-S (2004). A Combination of the rabidopsis DREB1A Gene and Stress-Inducible rd29A Promoter Improved Drought- and Low-Temperature Stress Tolerance in Tobacco by Gene Transfer. Plant Cell Physiol., 45: 346-350.

Kasuga M, Miura S, Shinozaki K, Yamaguchi-Shinozaki K (2004). A combination of the Arabidopsis DREB1A gene and stressinducible rd29A promoter improved drought- and low-temperature stress tolerance in tobacco by gene transfer. Plant Cell Physiol., 45: 346-350.

Kasuga M, Qiang L, Miura S, Shinozaki K Y, Shinozaki K, Liu Q (1999). Improving plant drought, salt, and freezing tolerance by gene transfer of a single stress-inducible transcription factor. Nature Biotechnol., 17: 287-291.

Katiyar-Agarwal S, Agarwal M, Grover A (2003). Heat-tolerant basmati rice engineered by over-expression of hsp101. Plant Mol. Biol., 51: 677–686.

Kim D-Y, Bovet L, Kushnir S, Noh EW, Martinoia E, Lee Y (2006). AtATM3 Is Involved in Heavy Metal Resistance in Arabidopsis. Plant Physiol., 140: 922-932.

Kim S, Takahashi M, Higuchi K, Tsunoda K, Nakanishi H, Yoshimura E, Mori S, Nishizawa NK (2005). Increased Nicotianamine Biosynthesis Confers Enhanced Tolerance of High Levels of Metals, in Particular Nickel, to Plants. Plant Cell Physiol., 46: 1809-1818.

Kim Y-O, Kang H (2006). The Role of a Zinc Finger-containing Glycine-rich RNA-binding Protein during the Cold Adaptation Process in Arabidopsis thaliana. Plant Cell Physiol., 47: 793-798.

Kim Y-O, Pan SO, Jung C-H, Kang H (2007). A Zinc Finger-Containing Glycine-Rich RNA-Binding Protein, atRZ-1a, Has a Negative Impact on Seed Germination and Seedling Growth of Arabidopsis thaliana Under Salt or Drought Stress Conditions. Plant Cell Physiol., 48: 1170-1181.

Kimpel JA, Key JL (1985). Heat shock in plants. Trends Biochem. Sci., 10: 353–357.

Kishor PBK, Hong Z, Miao GH, Hu CAA, Verma DPS (1995). Overexpression of 1-pyrroline-5-carboxylate synthetase increases proline production and confers osmotolerance in transgenic plants. Plant Physiol., 108: 1387–1394.

Knipp G, Honermeier B (2006). Effect of water stress on proline accumulation of genetically modified potatoes (Solanum tuberosum L.) generating fructans. J. Plant Physiol., 163: 392-397.

Koh E-J, Songb W-Y, Leeb Y, Kimd KH, Kima K, Chunga N, Leed K-W, Hongc S-W, Lee H (2006). Expression of yeast cadmium factor 1 (YCF1) confers salt tolerance to Arabidopsis thaliana. Plant Sci., 170: 534-541.

Kornyeyev D, Logan BA, Allen RD, Holaday AS (2003). Effect of chloroplastic overproduction of ascorbate peroxidase on photosynthesis and photoprotection in cotton leaves subjected to low temperature photoinhibition. Plant Sci.,165: 1033-1041.

Kovtun Y, Chiu WL, Tena G, Sheen J (2000) Functional analysis of oxidative stress-activated mitogen-activated protein kinase cascade in plants. Proc Natl Acad Sci USA 97:2940–2945.

Koyama ML, Levesley A, Koebner RMD, Flowers TJ, Yeo AR (2001).

Quantitative trait loci for component physiological traits determining salt tolerance in rice. Plant Physiol., 125: 406-422.

Kumar S, Dhingra A, Daniell H (2004). Plastid-Expressed Betaine Aldehyde Dehydrogenase Gene in Carrot Cultured Cells, Roots, and Leaves Confers Enhanced Salt Tolerance. Plant Physiol., 135: 2843-2854.

Kumer A, Singh DP (1998). Use of physiological indices as screening technique for drought tolerance in oilseed Brassica speicies. Ann. Bot., 81: 413-420.

Kumria R, Rajam MV (2002). Ornithine decarboxylase transgene in tobacco affects polyamines, in vitro-morphogenesis and response to salt stress. J. Plant Physiol., 159: 983-990.

Larkindale J, Hall JD, Knight MR, Vierling E (2005). Heat Stress Phenotypes of Arabidopsis Mutants Implicate Multiple Signaling Pathways in the Acquisition of Thermotolerance. Plant Physiol., 138: 882-897.

Larkindale JD, Hall J, Knight MR, Vierling E (2005). Heat Stress Phenotypes of Arabidopsis Mutants Implicate Multiple Signaling Pathways in the Acquisition of Thermotolerance. Plant Physiol., 138: 882-897.

Laurie S, Kevin A, Frans JM, Heard PJ, Brown, Leigh SJ, Roger A (2002). A role for HKT1 in sodium uptake by wheat roots. Plant J., 32: 139-149.

Lee H, Xiong L, Gong Z, Ishitani M, Stevenson B, Zhu JK (2001). The Arabidopsis HOS1 gene negatively regulates cold signal transduction and encodes a RING-finger protein that displays coldregulated nucleo-cytoplasmic partitioning. Gene Dev., 15: 912-924.

Lee J, Bae H, Jeong J, Lee J-Y, Yang Y-Y, Hwang I, Martinoia E, Lee Y (2003). Functional Expression of a Bacterial Heavy Metal Transporter in Arabidopsis Enhances Resistance to and Decreases Uptake of Heavy Metals. Plant Physiol., 133: 589-596.

Lee J, Song SD, Hwang W-Y, Youngsook IL (2004). Arabidopsis metallothioneins 2a and 3 enhance resistance to cadmium when expressed in Vicia faba guard cells. Plant Mol. Biol., 54: 805-815.

Lee JH, Hubel A, Schoffl F (1995). Depression of the activity of genetically engineered heat shock factor causes constitutive synthesis of heat shock proteins and increased thermotolerance in transgenic Arabidopsis. Plant J., 8: 603-612.

Lee JH, Schoffl F (1996). An Hsp70 antisense gene affects the expression of HSP70/HSC70, the regulation of HSF, and the acquisition of thermotolerance in transgenic Arabidopsis thaliana. Mol. Gen. Genet., 252: 11-19.

Lee JH, van Montagu M, Verbruggen N (1999). A highly conserved kinase is an essential component for stress tolerance in yeast and plant cells. Proc. Natl. Acad. Sci., USA, 96: 5873-5877.

Lee SC, Choi HW, Hwang IS, Choi DS, Hwang BK (2006). Functional roles of the pepper pathogen-induced bZIP transcription factor, CAbZIP1, in enhanced resistance to pathogen infection and environmental stresses. Planta, 224: 1209-1225.

Levitt J (1972). Responses of plants to environmental stresses. Academic Press: New York.

Levitt J (1980). A survey of cold hardness in some olive (Olea europaea L.) cultivars. ISHS. Acta Horticulturae, p. 791.

Li G-L, Chang H, Li B, Zhou W, Sun D-Y, Zhou R-Z (2007). The roles of the at DjA2 and at DjA3 molecular chaperone proteins in improving thermotolerance of Arabidopsis thaliana seedlings. Plant Sci., 173: 408-416.

Li HY, Chang CS, Lu LS, Liu CA, Chan MT, Chang YY (2003). Over-expression of Arabidopsis thaliana heat shock factor gene (AtHsfA1b) enhances chilling tolerance in transgenic tomato. Bot. Bull. Acad. Sin., 44: 129-140.

Li Y, Dankher OP, Carreira L, Smith AP, Meagher RB (2006). The Shoot-Specific Expression of -Glutamylcysteine Synthetase Directs the Long-Distance Transport of Thiol-Peptides to Roots Conferring Tolerance to Mercury and Arsenic. Plant Physiol., 141: 288-298.

Lian H-L, Yu X, Ye O, Ding X-S, Kitagawa Y, Kwak S-S, Su W-A, Tang Z-C (2004). The Role of Aquaporin RWC3 in Drought Avoidance in Rice. Plant Cell Physiol., 5: 481-489.

Light GG, Ahan JR, Roxas VP, Allen RD (2005). Transgenic cotton (Gossypium hirsutum L.) seedlings expressing a tobacco glutathione S-transferase fail to provide improved stress tolerance. Planta. 222: 346-354.

Lin PC, Hwang SG, Endo A, Okamoto M, Koshiba T, Cheng WH (2007). Expression of abscisic acid 2/glucose insensitive 1 in Arabidopsis Promotes Seed Dormancy and Stress Tolerance. Plant Physiol., 143: 745-758.

Lindquist S (1986). The heat-shock response. Annu. Rev. Biochem., 55: 1151-1191.

Luming T, Conglin H, Rong Y, Ruifang L, Zhiliang L, Lusheng Z, Yongqin W, Xiuhai Z Zhongyi W (2006). Overexpression AtNHX1 confers salt-tolerance of transgenic tall fescue. Afr. J. Biotechnol., 5(11): 1041-1044.

Luo G-Z, Wang H-W, Jian H, Tian A-G, Wang Y-J, Zhang J-S, Chen S-Y (2005). A Putative Plasma Membrane Cation/proton Antiporter from Soybean Confers Salt Tolerance in Arabidopsis. Plant Mol. Biol., 59: 809-820.

Mahalakshmi S, Christopher GBS, Reddy TP, Rao KV, Reddy VD (2006). Isolation of a cDNA clone (PcSrp) encoding serine-rich-protein from Porteresia coarctata T. and its expression in yeast and finger millet (Eleusine coracana L.) affording salt tolerance. Planta, 224: 347-359.

Malik MK, Slovin JP, Hwang CH, Zimmerman JL (1999). Modified expression of a carrot small heat-shock protein gene, Hsp17.7, results in increased or decreased thermotolerance. Plant J., 20: 89-99.

Maqbool SB, Zhong H, El-Maghraby Y, Ahmad A, Chai B, Wang W, Sabzikar R, Sticklen MB (2002). Competence of oat (Avena sativa L.) shoot apical meristems for integrative transformation, inherited expression, and osmotic tolerance of transgenic lines containing hva1. TAG, 105: 201-208.

Maria C, Rubio EM, Gonzalez FR, Minchin K, Judith W, Arrese-Igor C, Ramos J, Becana M (2002). Effects of water stress on antioxidant enzymes of leaves and nodules of transgenic alfalfa overexpressing superoxide dismutases. Physiol. Plant, 115: 531-536.

Markus, KG , Markus Suh, Su JS, Kolukisaoglu A, Louis P, Sonia C, Mark D, Richter A, Weder B, Schulz B, Martinoia E (2004). Disruption of AtMRP4, a guard cell plasma membrane ABCC-type ABC transporter, leads to deregulation of stomatal opening and increased drought susceptibility. Plant J., 39: 219-236.

Martinez-Atienza J, Jiang X, Garciadeblas B, Mendoza I, Zhu J-K, M. Pardo JM, Quintero FJ (2007). Conservation of the Salt Overly Sensitive Pathway in Rice. Plant Physiol., 143: 1001-1012.

Mazel A, Leshem Y, Tiwari BS, Levine A (2004). Induction of Salt and Osmotic Stress Tolerance by Overexpression of an Intracellular Vesicle Trafficking Protein AtRab7 (AtRabG3e). Plant Physiol., 134: 118-128.

McKersie BD (2001). Stress tolerance of transgenic plants overexpressing superoxide dismutase. J. Exp. Bot., 52: 8 (abstract).

Mckersie BD, Bowley SR, Harjanto E, Leprince O (1996). Water deficit tolerance and field performance of transgenic alfalfa overexpressing superoxide dismutase. Plant Physiol., 111: 1177-1181.

McKersie BD, Bowley SR, Jones KS, (1999) Winter survival of transgenic alfalfa overexpressing superoxide dismustase. Plant Physiol., 119: 839-847.

McKersie BD, Chen Y, Beus M de, Bowley SR, Bowler C, Inze D, D'Halluin K, Botterman J (1993). Superoxide dismutase enhances tolerance of freezing stress in transgenic alfalfa (Medicago sativa L.). Plant Physiol., 103: 1155-1163.

McNeil SD, Nuccio ML, Hanson AD (1999). Betaines and related smoprotectants. Targets for metabolic engineering of stress resistance. Plant Physiol., 120: 945-949.

McNeil SD, Nuccio ML, Rhodes D, Shachar-Hill Y, Hanson AD (2000). adiotracer and computer modeling evidence that phosphobase methylation is the main route of choline synthesis in tobacco. Plant Physiol., 123: 371–380.

Meng Z Barg R, Yin M, Gueta-Dahan Y, Leikin-Frenkel A, Salts Y, Shabtai S, Ben-Hayyim G (2005). Modulated fatty acid desaturation via overexpression of two distinct ?-3 desaturases differentially alters tolerance to various abiotic stresses in transgenic tobacco cells and plants. Plant J., 44: 361-371.

Merchan F, Lorenzo LD, Rizzo SG, Niebel A, Manyani H, Frugier F, Sousa C, Crespi M (2007). Identification of regulatory pathways involved in the reacquisition of root growth after salt stress in Medicago truncatula Plant J., 51: 1-17.

Miroshnichenko S, Tripp J, Nieden UZ, Neumann D, Conrad U, Manteuffel R (2005). Immunomodulation of function of small heat shock proteins prevents their assembly into heat stress granules and results in cell death at sublethal temperatures. Plant J., 41: 269-281.

Moghaieb REA, Tanaka N, Saneoka H, Hussein HA, Yousef SS, Ewada MAF, Aly MAM, Fujita K (2000). Expression of betaine aldehyde dehydrogenase gene in transgenic tomato hairy roots leads to the accumulation of glycine betaine and contributes to the maintenance of the osmotic potential under salt stress. Soil Sci. Plant Nutr., 46: 873-883.

Mohanty C, Kathuria H, Ferjani A, Sakamoto A, Mohanty P, Murata N, Tyagi AK (2003). Transgenics of an elite indica rice variety Pusa Basmati 1 harbouring the codA gene are highly tolerant to salt stress. TAG, 106: 51-57.

Molinari HBC, Marur CJ, Daros E, Freitas de Campos MK, Portela de Carvalho JFR, Filho JCB, Pereira LFP, Vieira LGE (2007). Evaluation of the stress-inducible production of proline in transgenic sugarcane (Saccharum spp.): Osmotic adjustment, chlorophyll fluorescence and oxidative stress. Physiol. Plant., 130: 218-229.

Molinari HBC, Marur CJ, Filho GCB, Kobayashi AK, Pileggi M, Rui PL Jnr, Luiz FPP, Luiz GEV (2004). Osmotic adjustment in transgenic citrus rootstock Carrizo citrange (Citrus sinensis Osb. x Poncirus trifoliata L. Raf.) overproducing proline. Plant Sci., 167: 1375-1381.

Moony HA, Pearcy RW, Ehleringer J (1987). Plant Physiology ecology today. BioScience, 37: 18-20.

Morgan JM (1984). Osmoregulation and water stress in higher plants. Annu. Rev. Plant Physiol. Plant Mol. Biol., 35: 299-319.

Mundy J, Chua N-H (1988). Abscisic acid and water-stress induce the expression of a novel rice gene. EMBO J., 7: 2279-2286.

Mundy J, Yamaguchi-Shinozaki K, Chua NH (1990). Nuclear proteins bind conserved elements in the abscisic acid-responsive promoter of a rice rab gene. Proc. Natl. Acad. Sci. USA, 87: 406-410.

Murakami Y, Tsuyama M, Kobayashi Y, Kodama H, Iba K (2000). Trienoic fatty acids and plant tolerance of high temperature. Science, 287: 476-479.

Murata N, Ishizaki-Nishizawa O, Higashi S, Hayashi S, Tasaka Y, Nishida I (1992). Genetically engineered alteration in the chilling sensitivity of plants. Nature, 356: 710-713.

Myouga F, Motohashi R, Kuromori T, Nagata N, Shinozak K (2006). An Arabidopsis chloroplast-targeted Hsp101 homologue, APG6, has an essential role in chloroplast development as well as heat-stress response. Plant J., 48: 249-260.

Nakajima N, Itoh T, Takikawa S, Asai N, Tamaoki M, Aono M, Kubo A, Azumi Y, Kamada H, Saji H (2002). Improvement in ozone tolerance of tobacco plants with an antisense DNA for 1-aminocyclopropane-1-carboxylate synthase. Plant Cell Environ., 25: 727-735.

Nakayama H, Yoshida K, Ono H, Murooka Y, Shinmyo A (2000). Ectoine, the compatible solute of Halomonas elongata, confers hyperosmotic tolerance in cultured tobacco cells. Plant Physiol., 122: 1239-1247.

Nanjo T, Kobayashi M, Yoshiba Y, Kakubari Y, Yamaguchi-Shinozaki K, Shinozaki K (1999). Antisense suppression of proline degradation needs. Proc. Indian Natl. Sci. Acad. B Rev. Tracts. Biol. Sci., 66: 13-32.

Nepomuceno AL, Stewart JMCD, Oosterhuis D, Turley R, Neumaier N, Farias JRB (2000). Isolation of a cotton NADP(H) oxidase homologue induced by drought stress. Pesqui.Agropecu. Bras., 35: 1407-1416.

Nordin K, Heino P, Palva ET (1991) Separate signal pathways regulate the expression of a low-temperature-induced gene in Arabidopsis thaliana (L.) Heynh. Plant Mol. Biol., 115: 875-879.

Oberschall A, Deak M, Torok K, Sass L, Vass I, Kovacs I, Feher A, Dudits D, Hovarth GV (2000). A novel aldose/aldehyde reductase protects transgenic plants against lipid peroxidation under chemical and drought stress. Plant J., 24: 422, 437-446.

Oh S-J, Song SI, Kim YS, Jang H-J, Kim SY, Kim M, Kim Y-K, Nahm BH, Ju- Kim K (2005). Arabidopsis CBF3/DREB1A and ABF3 in Transgenic Rice Increased Tolerance to Abiotic Stress without Stunting Growth. Plant Physiol., 138: 341-351.

Ohta M, Hayashi Y, Nakashima A, Hamada A, Tanaka A, Nakamura T, Hayakawa T (2002). Introduction of a Na⁺/H⁺ antiporter gene from Atriplex gmelini confers salt tolerance in rice. FEBS Letters 532: 279-282.

Oraby HF, Ransom CB, Kravchenko AN, Sticklen MB (2005). Barley HVA1 Gene Confers Salt Tolerance in R3 Transgenic Oat. Crop Sci., 45: 2218-2227.

Pardo JM, Reddy MP, Yang S (1998). Stress signaling through Ca^{2+}/Calmodulin dependent protein phosphatase calcineurin mediates salt adaptation in plants. Proc. Natl. Acad. Sci. USA, 95: 9681–9683.

Park BJ, Liu Z , Kanno A , Kamey T (2005). Genetic improvement of Chinese cabbage for salt and drought tolerance by constitutive expression of a B. napus LEA gene. Plant Sci., 169: 553-558.

Park BJ, Liu Z, Kanno A, Kamey T (2005). Increased tolerance to salt- and water-deficit stress in transgenic lettuce (Lactuca sativa L.) by constitutive expression of LEA. Plant Growth Regul., 45: 165-171.

Park MR, Baek S-H, Benildo G. Reyes DL, Yun SJ (2007). Overexpression of a high-affinity phosphate transporter gene from tobacco (NtPT1) enhances phosphate uptake and accumulation in transgenic rice plants. Plant Soil., 291: 259-269.

Park SM, Hong CB (2002). Class I small heat-shock protein gives thermotolerance in tobacco. J. Plant Physiol., 159: 25-30.

Pellegrineschi A, Reynolds M, Pacheco M, Brito RM, Almeraya R, Kazuko Y-S, Hoisington D (2004). Stress-induced expression in wheat of the Arabidopsis thaliana DREB1A gene delays water stress symptoms under greenhouse conditions. Genome, 47: 493-500.

Peng Y, Lin W, Cai W, Arora R (2007). Overexpression of a Panax ginseng tonoplast aquaporin alters salt tolerance, drought tolerance and cold acclimation ability in transgenic Arabidopsis plants. Planta, 226: 729-740.

Perl A, Perl-Treves R, Galili S, Aviv D, Shalgi E, Malkin S, Galun E (1993). Enhanced oxidative stress defense in transgenic potato expressing tomato Cu, Zn superoxide dismutases. Theor. Appl. Genet., 85: 568-576.

Perlson E, Kalifa Y, Gilad A, Konrad Z, Scolnik Pa, Bar-Zvi D (2004). Over-Expression Of The Water And Salt Stress-Regulated Asr1 Gene Confers An increased salt tolerance. Plant Cell Environ., 27: 1459-1468.

Piao HL, Lim JH, Kim SJ, Cheong GW, Hwang I (2001). Constitutive over-expression of AtGSK1 induces NaCl stress responses in the absence of NaCl stress and results in enhanced NaCl tolerance in Arabidopsis. Plant J., 27: 305-314.

Pilon-Smits EHA, Ebskamp MJM, Paul MJ, Jeuken MJW, Weisbeek PJ, Smeekens SCM (1995). Improved performance of transgenic fructan-accumulating tobacco under drought stress. Plant Physiol., 107: 125-130.

Pommerrenig B, Papini-Terzi FS, Sauer N (2007). Differential regulation of sorbitol and sucrose loading into the phloem of Plantago major in response to salt stress. Plant Physiol., pp. 1029-1038.

Pomponi M, Censi V, Girolamo VD, Paolis AD, Toppi LS, Aromolo R, Costantino P, Cardarelli M (2006). Overexpression of Arabidopsis phytochelatin synthase in tobacco plants enhances Cd^{2+} tolerance and accumulation but not translocation to the shoot. Planta, 223: 180-190.

Porcel R, Aroca R Azcn R, Juan M R-L (2006). PIP Aquaporin Gene Expression in Arbuscular Mycorrhizal Glycine max and Lactuca sativa Plants in Relation to Drought Stress Tolerance. Plant Mol. Biol., 60: 389-404.

Prasad KVSK, Saradhi PP (2004). Enhanced tolerance to photoinhibition in transgenic plants through targeting of glycinebetaine biosynthesis into the chloroplasts. Plant Sci., 166: 1197-1212.

Qi Z, Spalding EP (2004). Protection of Plasma Membrane K⁺ Transport by the Salt Overly Sensitive1 Na⁺-H⁺ Antiporter during Salinity Stress. Plant Physiol., 135: 2548-2555.

Qin F, Kakimoto M, Sakuma Y, Maruyama K, Osakabe Y, Lam-Son Phan Tran L-Sp, Shinozaki K, Yamaguchi-Shinozaki K (2007). Regulation and functional analysis of ZmDREB2A in response to drought and heat stresses in Zea mays L Plant J., 50: 54-69.

Quan R, Lin H, Mendoza I, Zhang Y, Cao W, Yang Y, Shang M, Chen S, Pardo JM, Guo Y (2007). SCABP8/CBL10, a Putative Calcium Sensor, Interacts with the Protein Kinase SOS2 to Protect Arabidopsis Shoots from Salt Stress. Plant Cell, 19: 1415-1431.

Queitsch C, Hong SW, Vierling E, Lindquist S (2000). Heat shock protein 101 plays a crucial role in thermotolerance in Arabidopsis. Plant Cell, 12: 479-492.

Quintero FJ, Blatt MR, Pardo JM (2000). Functional conservation between yeast and plant endosomal Na($^+$)/H($^+$) antiporters. FEBS Lett., 471: 224-228.

Rai M, Pal M, Sumesh KV, Jain V, Sankaranarayanan A (2006). Engineering for biosynthesis of ectoine (2-methyl 4-carboxy tetrahydro pyrimidine) in tobacco chloroplasts leads to accumulation of ectoine and enhanced salinity tolerance Plant Sci., 170: 291-306.

Rathinasabapathi B (2000). Metabolic engineering for stress tolerance: Installing osmoprotectant synthesis pathways. Ann. Bot., 86: 709-716.

Rohila JS, Jain RK, Wu R (2002). Genetic improvement of Basmati rice for salt and drought tolerance by regulated expression of a barley Hva1 cDNA. Plant Sci., 163: 525-532.

Remans T, Nacry P, Pervent M, Girin T, Tillard P, Lepetit M, Gojon A (2006). A Central Role for the Nitrate Transporter NRT2.1 in the Integrated Morphological and Physiological Responses of the Root System to Nitrogen Limitation in Arabidopsis. Plant Physiol., 140: 909-921.

Rhoads DM, Samuel JW, You Z, Mrinalini M, Elthon TE (2005). Altered gene expression in plants with constitutive expression of a mitochondrial small heat shock protein suggests the involvement of retrograde regulation in the heat stress response. Physiol. Plant, 123: 435-444.

Rodrigues SM, Andrade MO, Gomes APS, DaMatta FM, Baracat-Pereira MC, Fontes EPB (2006). Arabidopsis and tobacco plants ectopically expressing the soybean antiquitin-like ALDH7 gene display enhanced tolerance to drought, salinity, and oxidative stress. J. Exp. Bot., 57: 1909-1918.

Rohila JS, Rajinder K, Wu JR (2002). Genetic improvement of Basmati rice for salt and drought tolerance by regulated expression of a barley Hva1 cDNA. Plant Sci., 163: 525-532.

Ronan S, Hirokazu T, Hideko N, Laszlo M, Chen THH , Norio M (2003). Enhanced formation of flowers in salt-stressed Arabidopsis after genetic engineering of the synthesis of glycine betaine. Plant J., 36: 165-176.

Rontein D, Basset G, Hanson AD (2002). Metabolic engineering of osmoprotectant accumulation in plants. Metab. Eng., 4: 49-56.

Roosens NH, Willem R, Li Y, Verbruggen II, Biesemans M, Jacobs M (1999). Proline metabolism in the wild-type and in a salt-tolerant mutant of Nicotiana plumbaginifolia studied by (13)C-nuclear magnetic resonance imaging. Plant Physiol., 121: 1281-1290.

Rossel JB, Philippa BW, Luke H, Soon CW, Andrew P, Philip MM, Barry JP (2006). A Mutation Affecting Ascorbate Peroxidase 2 Gene expression reveals a link between responses to high light and drought tolerance. Plant Cell Environ., 29: 269-281.

Rossel JB, Walter PB, Hendrickson L, Chow WS, Poole A, Mullineaux PM, Pogson BJ (2006) A mutation affecting ascorbate peroxidase 2 gene expression reveals a link between responses to high light and drought tolerance. Plant Cell Environ., 29: 269-281.

Roxas VP, Smith RK Jr, Allen ER, Allen RD (1997). Overexpression of glutathione S-transferase/glutathione peroxidase enhances the growth of transgenic tobacco seedlings during stress. Nat. Biotechnol., 15: 988-991.

Roy M, Wu R (2001). Arginine decarboxylase transgene expression and analysis of environmental stress tolerance in transgenic rice. Plant Sci., 160: 869-875.

Roy M, Wu R (2002). Overexpression of S-adenosylmethionine decarboxylase gene in rice increases polyamine level and enhances sodium chloride-stress tolerance. Plant Sci., 163: 987-992.

Roy M, Wu R (2001) Arginine decarboxylase transgene expression and analysis of environmental stress tolerance in transgenic rice. Plant Sci., 160: 869-875.

Rubio MC, Gonza´lez EM, Minchin FR, Webb KJ, Arrese-Igor C, Ramos J, Becana M (2002). Effects of water stress on antioxidant enzymes of leaves and nodules of transgenic alfalfa overexpressing superoxide dismutases. Physiol. Plant, 115: 531-540.

Ruidang Q, Mei S, Hui Z, Yanxiu Z, Juren Z (2004). Engineering of enhanced glycine betaine synthesis improves drought tolerance in maize. Plant Biotechnol. J., 2: 477-486.

Ruiz JM, Blumwald, E (2002). Salinity-induced glutathione synthesis in Brassica napus. Planta, 214: 965-969.

Rus A, Yokoi S, Sharkhuu A, Reddy M, Lee B, Matsumoto TK, Koiwa H,

Zhu JK, Bressan RA, Hasegawa PM (2001). AtHKT1 is a salt tolerance determinant that controls Na$^+$ entry into plant roots. PNAS, 98: 14150-14155.

Saavedra L, Svensson J, Carballo V, Izmendi D, Welin B, Vidal S (2006). A dehydrin gene in Physcomitrella patens is required for salt and osmotic stress tolerance. Plant J., 45: 237-249.

Saez A, Robert N, Maktabi MH, Schroeder JI, Serrano R, Rodriguez PL (2006). Enhancement of Abscisic Acid Sensitivity and Reduction of Water Consumption in Arabidopsis by Combined Inactivation of the Protein Phosphatases Type 2C ABI1 and HAB11. Plant Physiol., 141: 1389-1399.

Sakamoto A, Murata A, Murata N (1998). Metabolic engineering of rice leading to biosynthesis of glycinebetaine and tolerance to salt and cold. Plant Mol. Biol., 38: 1011-1019.

Sakamoto A, Murata N (2000). Genetic engineering of glycinebetaine synthesis in plants: current status and implications for enhancement of stress tolerance. J. Exp. Bot., 51: 81-88.

Sakuma Y, Maruyama K, Osakabe Y, Qin F, Seki M, Shinozaki K, Yamaguchi-Shinozaki K (2006). Functional analysis of an Arabidopsis transcription factor, DREB2A, involved in droughtresponsive gene expression. Plant Cell, 18: 1292-1309.

Sakuma Y, Maruyama K, Osakabe Y, Qina F, Seki M, Shinozaki K, Kazuko Y-S (2006). Functional Analysis of an Arabidopsis Transcription Factor, DREB2A, Involved in Drought-Responsive Gene Expression. Plant Cell, 18: 1292-1309.

Sakuma Yoh, Qiang Liu, Joseph G, Dubouzet HA, Kazuo SK, Yamaguchi S (2002). DNA-Binding Specificity of the ERF/AP2 Domain of Arabidopsis DREBs, Transcription Factors Involved in Dehydration- and Cold-Inducible Gene Expression. Biochem. Biophys. Res. Comm., 290(3): 998-1009.

Sanders (2001). Phylogenetic Relationships within Cation Transporter Families of Arabidopsis. Plant Physiol., 126: 1646-1667.

Sang YM, Zheng SQ, Li WQ, Huang BR, Wang XM (2001). Regulation of plant water loss by manipulating the expression of phospholipase D alpha. Plant J., 28: 135-144.

Savitch LV, Allard G, Seki M, Robert LS, Tinker NA, Huner NPA, Shinozaki K, Singh J (2005). The Effect of Overexpression of Two Brassica CBF/DREB1-like Transcription Factors on Photosynthetic Capacity and Freezing Tolerance in Brassica napus, Plant Cell Physiol., 46: 1525-1539.

Seki M, Narusaka M, Abe H, Kasuga M, Yamaguchi-Shinozaki K, Carninci P, Hayashizaki Y, Shinozaki K (2001). Monitoring the expression pattern of 1300 Arabidopsis genes under drought and cold stresses by using a full-length cDNA Microarray. Plant Cell, 13: 61-72.

Sen Gupta A, Heinen JL, Holady AS, Burke JJ, Allen RD (1993). Increased resistance to oxidative stress in transgenic plants that over-express chloroplastic Cu/Zn superoxide dismutase. Proc. Natl. Acad. Sci. USA, 90: 1629–1633.

Shan D-P, Huang J-G, Yang Y-T, Guo Y-H, Wu C-A, Yang G-D, Gao Z, Zheng C-C (2007). Cotton GhDREB1 increases plant tolerance to low temperature and is negatively regulated by gibberellic acid. New Phytol., 175: 70-81.

Sheveleva E, Chmara W, Bohnert HJ, Jensen RG (1997). Increased salt and drought tolerance by d-ononitol production in transgenic Nicotiana tabacum l. Plant Physiol., 115: 1211-1219.

Shi H, Ishitani M, Kim C, Zhu JK (2000). The Arabidopsis thaliana salt tolerance gene SOS1 encodes a putative Na$^+$/H$^+$ antiporter. Proc. Natl. Acad. Sci. USA, 97: 6896-6901.

Shi H, Lee BH, Wu SJ, Zhu JK (2003). Overexpression of a plasma membrane Na$^+$/H$^+$ antiporter gene improves salt tolerance in Arabidopsis thaliana. Nat. Biotechnol., 21: 81-85.

Shi HZ, Xiong LM, Stevenson BM, Lu TG, Zhu JK (2002). The Arabidopsis salt overly sensitive 4 mutants uncover a critical role for vitamin B6 in plant salt tolerance. Plant Cell, 14: 575-588.

Shi WM, Muramoto Y, Ueda A, Takabe T (2001). Cloning of peroxisomal ascorbate peroxidase gene from barley and enhanced thermotolerance by overexpressing in Arabidopsis thaliana. Gene, 273: 23-27.

Shin H, Shin H-S, Dewbre GR, Harrison MJ (2004). Phosphate transport in Arabidopsis: Pht1;1 and Pht1;4 play a major role in phosphate acquisition from both low- and high-phosphate environments. Plant J., 39: 629-642.

Shinozaki K, Yamaguchi-Shinozaki K (1997). Gene expression and signal transduction in water-stress response. Plant Physiol., 115: 327-334.

Shinozaki K, Yamaguchi-Shinozaki K (1999). Molecular responses todrought stress. In: Shinozaki K, Yamaguchi-Shinozaki K (eds) Molecular responses to cold, drought, heat and salt stress in higher plants. R.G. Landes Co., Austin, pp. 11-28.

Shinwari ZK, Nakashima K, Miura S, Kasuga M, Seki M, Yamaguchi-Shinozaki K, Shinozaki K (1998). An *Arabidopsis* gene family encoding DRE/CRT binding proteins involved in low-temperature-responsive gene expression. Biochem. Biophys. Res. Commun., 250: 161-170.

Shoji S, Hironori K, Catala RZ, Salinas R, Matsui J, Ohme-Takagi K, Hiroshi M-T (2004). Stress-responsive zinc finger gene ZPT2-3 plays a role in drought tolerance in petunia. Plant J., 36: 830-841.

Shou H, Bordallo P, Wang K (2004). Expression of the *Nicotiana* protein kinase (NPK1) enhanced drought tolerance in transgenic maize. J. Exp. Bot., 55: 1013-1019.

Shukla RK, Raha S, Tripathi V, Chattopadhyay D (2006). Expression of CAP2, an APETALA2-Family Transcription Factor from Chickpea, Enhances Growth and Tolerance to Dehydration and Salt Stress in Transgenic Tobacco. Plant Physiol., 142: 113-123.

Siefritz F, Tyree MT, Lovisolo C, Schubert A, Kaldenhoff R (2002). PIP1 Plasma Membrane Aquaporins in Tobacco: From Cellular Effects to Function in Plants. Plant Cell, 14: 869-876.

Simpson SD, Nakashima K, Narusaka Y, Seki M, Shinozaki K, Yamaguhci-Shinozaki K (2003). Two different novel cis-acting elements of erd1, a clpA homologous *Arabidopsis* gene function in induction by dehydration stress and dark-induced senescence. Plant J., 33: 259-270.

Singla-Pareek SL, Reddy MK, Sopory SK (2003) Genetic engineering of the glyoxalase pathway in tobacco leads to enhanced salinity tolerance. PNAS, 100: 14672-14677.

Sivamani E, Bahieldin A, Wraith JM, Al-Niemi T, Dyer WE, Ho THD, Qu RD (2000). Improved biomass productivity and water use efficiency under water deficit conditions in transgenic wheat constitutively expressing the barley HVA1 gene. Plant Sci., 155: 1-9.

Sivamani E, Bahieldin A, Wraith JM, Al-Niemi T, Dyer WE, Ho THD, Qu R (2000). Improved biomass productivity and water use efficiency under water deficit conditions in transgenic wheat constitutively expressing the barley HVA1 gene. Plant Sci., 155: 1-9.

Skriver K, Mundy J (1990). Gene expression in response to abscisic acid and osmotic stress. Plant Cell, 5: 503-512.

Slooten L, Capiau K, Van Camp W, Montagu MV, Sybesma C, Inze′ D (1995). Factors affecting the enhancement of oxidative stress tolerance in transgenic tobacco overexpressing manganese superoxide dismutase in the chloroplasts. Plant Physiol., 107: 737-775.

Su J, Hirji R, Zhang L, He C, Selvaraj G, Wu R (2006). Evaluation of the stress-inducible production of choline oxidase in transgenic rice as a strategy for producing the stress-protectant glycine betaine. J. Exp. Bot., 57: 1129-1135.

Sun J, Jiang H, Xu Y, Li H, Wu X, Xie Q, Li C (2007). The CCCH-Type Zinc Finger Proteins AtSZF1 and AtSZF2 Regulate Salt Stress Responses in *Arabidopsis*. Plant Cell Physiol., 48: 1148-1158.

Sun W, Bernard C, van de Cotte B, Montagu MV, Verbruggen N (2001). At-HSP17.6A, encoding a small heat-shock protein in *Arabidopsis*, can enhance osmotolerance upon overexpression. Plant J., 27: 407-415.

Tahtiharju S, Palva T (2001). Antisense inhibition of protein phosphatase 2C accelerates cold acclimation in *Arabidopsis* thaliana. Plant J., 26: 461-470.

Tamminen I, Makela P, Heino P, Palva ET (2001). Ectopic expression of ABI3 gene enhances freezing tolerance in response to abscisic acid and low temperature in *Arabidopsis* thaliana. Plant J., 25: 1-8.

Tanaka Y, Sano T, Tamaoki M, Nakajima N, Kondo N, Hasezawa S (2005). Ethylene Inhibits Abscisic Acid-Induced Stomatal Closure in *Arabidopsis*. Plant Physiol., 138: 2337-2343.

Tang W, Peng X, Newton RJ (2005). Enhanced tolerance to salt stress in transgenic loblolly pine simultaneously expressing two genes encoding mannitol-1-phosphate dehydrogenase and glucitol-6-

phosphate dehydrogenase. Plant Physiol. Biochem., 43: 139-146.

Tangpremsri T, Fukai S, Fischer KS (1995). Growth and yield of sorghum lines extracted from a population for differences in osmotic adjustment. Aust. J. Agric. Res., 46: 61-74.

Tarczynski MC, Jensen RG, Bohnert HJ (1993). Stress protection of transgenic tobacco by production of the osmolyte mannitol. Science, 259: 508-510.

Tepperman JM, Dunsmuir P (1990). Transformed plants with elevated levels of chloroplasts SOD are not more resistant to superoxide toxicity. Plant Mol. Biol., 14: 501-511.

Thomas JC, Sepahi M, Arendall B, Bohnert HJ (1995). Enhancement of seed germination in high salinity by engineering mannitol expression in *Arabidopsis* thaliana. Plant Cell Environ., 18: 801-806.

Thompson AJ, Andrews J, Mulholland BJ, McKee JMT, Hilton HW, Horridge JS, Farquhar GD, Smeeton RC, Smillie IRA, Black CR, Taylor IB (2007). Overproduction of Abscisic Acid in Tomato Increases Transpiration Efficiency and Root Hydraulic Conductivity and Influences Leaf Expansion.

Todd EY, Robert BM, Daniel RG (2004). ACC synthase expression regulates leaf performance and drought tolerance in maize. Plant J., 40: 813-825.

Tomoaki H, Jo m, Masahiro K, Hua Y, Kinya Y, Rie H, Yin CW, Yin LH, Kazumi H, Mami K, Masako O, Mutsumi Y, Schroeder Julian I, Nobuyuki U (2005). Enhanced salt tolerance mediated by AtHKT1 transporter-induced Na$^+$ unloading from xylem vessels to xylem parenchyma cells. Plant J., 44: 928-938.

Turner NC (1986) Crop water deficits: a decade of progress. Adances in Agronomy 39: 1-51.

Umezawa T, Fujita M, Fujita Y, Yamaguchi-Shinozaki K, Shinozaki K (2006). Engineering drought tolerance in plants: Discovering and tailoring genes to unlock the future. Curr. Opin. Biotechnol., 17: 113-122.

Umezawa T, Okamoto M, Kushiro T, Nambara E, Oono Y, Seki M, Kobayashi M, Koshiba T, Kamiya Y, Shinozaki K (2006). CYP707A3, a major ABA 8'-hydroxylase involved in dehydration and rehydration response in *Arabidopsis* thaliana. Plant J., 46: 171-182.

Umezawa T, Yoshida R, Maruyama K, Yamaguchi-Shinozaki K, Shinozaki K (2004). SRK2C, a SNF1-related protein kinase 2, improves drought tolerance by controlling stress-responsive gene expression in *Arabidopsis* thaliana. Proc. Natl. Acad. Sci. USA, 101: 17306-17311.

Uozumi N, Kim EJ, Rubio F, Yamaguchi T, Muto S, Tsuboi A, Bakker EP, Nakamura T, Scroeder JI (2000). The *Arabidopsis* HKT1 gene homolog mediates inward Na$^+$ currents in *Xenopus laevis* oocytes and Na$^+$ uptake in *Saccaharomyces cerevisiae*. Plant Physiol., 122: 1249-1259.

Ushimaru T, Nakagawa T, Fujioka Y, Daicho K, Naito M, Yamauchi Y, Nonaka H, Amako K, Yamawaki K, Murata N (2006) Transgenic *Arabidopsis* plants expressing the rice dehydroascorbate reductase gene are resistant to salt stress. J. Plant Physiol., 163: 1179-1184.

Van Camp W, Capiau K, Van Montagu M, Inze′ D, Slooten L (1996). Enhancement of oxidative stress tolerance in transgenic tobacco plants overproducing Fe-superoxide dismutase in chloroplasts. Plant Physiol., 112: 1703-1714.

Vert G, Fabienne NG, Champ D, Gaymard F, Guerinot ML, Briata J-F, Curie C (2002). IRT1, an *Arabidopsis* Transporter Essential for Iron Uptake from the Soil and for Plant Growth. Plant Cell, 14: 1223-1233.

Vierling E (1991). The roles of heat shock proteins in plants. Annu. Rev. Plant Physiol. Plant Mol. Biol., 42: 579-620.

Vinocur B, Altman A (2005). Recent advances in engineering plant tolerance to abiotic stress: achievements and limitations. Curr. Opin. Biotechnol., 16: 123-132.

Virginia P, Roxas R K, Smith Jr, Allen ER, Allen RD (1997) Overexpression of glutathione S-transferase/glutathione peroxidase enhances the growth of transgenic tobacco seedlings during stress . Nat. Biotechnol., 15: 988.

Vranova E, Inze D, Van Breusegem F (2002). Signal transduction during oxidative stress. J. Exp. Bot., 53: 1227-1236.

Waie B, Rajam MV (2003). Effect of increased polyamine biosynthesis on stress responses in transgenic tobacco by introduction of human S-adenosylmethionine gene. Plant Sci., 164: 727-734.

Wang J, Zhang H, Allen RD, Wang J, Zhang H(1999). Overexpression of an *Arabidopsis* peroxisomal ascorbate peroxidase gene in tobacco increases protection against oxidative stress. Plant Cell Physiol., 40: 725-732.

Wang P, Duan W, Takabayashi A, Endo T, Shikanai T, Ji-Yu Ye, Hualing Mi (2006). Chloroplastic NAD(P)H Dehydrogenase in Tobacco Leaves Functions in Alleviation of Oxidative Damage Caused by Temperature Stress. Plant Physiol., 141: 465-474.

Wang Y, Jiang J, Zhao X, Liu G , Yang C , Zhan L (2006). A novel LEA gene from Tamarix and Rossowii confers drought tolerance in transgenic tobacco. Plant Sci., 171: 655-662.

Wang Y, Liu C, Li K, Sun F, Hu H, Xia Li, Zhao Y, Han C, Zhang W, Duan Y, Liu M, Li X (2007). *Arabidopsis* EIN2 modulates stress response through abscisic acid response pathway. Plant Mol. Biol., 64: 633-644.

Wang Y, Ying J, Kuzma M, Chalifoux M, Sample A, McArthur C, Uchacz T, Sarvas C, Wan J, Dennis DT et al (2005). Molecular tailoring of farnesylation for plant drought tolerance and yield protection. Plant J., 43: 413-424.

Wang Y, Ying Y, Chen J, Wang X (2004). Transgenic *Arabidopsis* overexpressing Mn-SOD enhanced salt-tolerance. Plant Sci., 167: 671-677.

Waters ER, Lee GJ, Vierling E (1996). Evolution, structure and function of the small heat shock proteins in plants. J. Exp. Bot., 47: 325-338.

Wilson JR, Ludlow MM, Fisher MJ, Schulze ED (1980). Adapttion to water stress of the leaf water relation of four tropical forage species. Aust. J. Plant Physiol., 7: 207-220.

Wawrzyski A, Kopera E, Wawrzyska A, Kamiska J, Bal W, Sirko A (2006). Effects of simultaneous expression of heterologous genes involved in phytochelatin biosynthesis on thiol content and cadmium accumulation in tobacco plants. J. Exp. Bot., 57: 2173-2182.

Winicov I, Bastola DR (1997). Salt tolerance in crop plants: New approaches through tissue culture and gene regulation. Acta Physiol. Plant, 19: 435-449.

Winicov I, Bastola DR (1999). Transgenic overexpression of the transcription factor Alfin1 enhances expression of the endogenous MsPRP2 gene in alfalfa and improves salinity tolerance of the plants. Plant Physiol., 120: 473-480.

Wu C-A, Yang G-D, Meng O-W, Zheng C-C (2004). The Cotton GhNHX1 Gene Encoding a Novel Putative Tonoplast Na$^+$/H$^+$ Antiporter Plays an Important Role in Salt Stress. Plant Cell Physiol., 5: 600-607.

Wu CA, Yang GD, Meng QW, Zheng CC (2004).The cotton GhNHX1 gene encoding a novel putative tonoplast Na$^+$/K$^+$ antiporter plays an important role in salt stress. Plant Cell Physiol., 45: 600-607.

Xiao B, Huang Y, Tang N, Xiong L (2007). Over-expression of a LEA gene in rice improves drought resistance under the field conditions. TAG, 115: 35-46.

Xinghong Y, Zheng L, Congming L (2005). Genetic Engineering of the Biosynthesis of Glycinebetaine Enhances Photosynthesis against HighTemperature Stress in Transgenic Tobacco Plants. Plant Physiol., 138: 2299-2309.

Xiong L, Ishitani M, Zhu J-K (1999). Interaction of osmotic stress, temperature, and abscisic acid in the regulation of gene expression in *Arabidopsis*. Plant Physiol., 119: 205-211.

Xiong L, Zhu JK (2001). Plant abiotic stress signal transduction: molecular and genetic perspectives. Physiol. Plant, 112: 152-166.

Xu C, Jing R, Mao X, Jia X, Chang X (2007). A Wheat (*Triticum aestivum*) Protein Phosphatase 2A Catalytic Subunit Gene Provides Enhanced Drought Tolerance in Tobacco. Ann. Bot., 99: 439-450.

Xu D, Duan X, Wang B, Hong B, Ho T-HD, Wu R (1996). Expression of a late embryogenesis abundant protein gene, HVA1, from barley confers tolerance to water deficit and salt stress in transgenic rice. Plant Physiol., 110: 249-257.

Xu K, Xu X, Fukao T, Canlas P, Maghirang-Rodriguez R (2006). Sub1A is an ethylene-response-factor–like gene that confers submergence tolerance to rice. Nature, 442: 705-708.

Xue Z-Y, Zhi D-Y, Xue G-P, Zhang H, Zhao Y-X, Xi G-M (2004) Enhanced salt tolerance of transgenic wheat (*Triticum aestivum* L.) expressing a vacuolar Na$^+$/H$^+$ antiporter gene with improved grain yields in saline soils in the field and a reduced level of leaf Na$^+$. Plant Sci., 167: 849-859.

Yamada M, Morishita H, Urano K, Shiozaki N, Kazuko Y-S, Shinozaki

K, Yoshiba Y (2005). Effects of free proline accumulation in petunias under drought stress. J. Exp. Bot., 56: 1975-1981.

Yamada M, Morishita H, Urano K, Shiozaki N, Yamaguchi-Shinozaki K, Shinozaki K, Yoshiba Y (2005). Effects of free proline accumulation in petunias under drought stress. J. Exp. Bot., 56: 1975-1981.

Yamaguchi-Shinozaki K, Kasuga M, Liu Q, Nakashima K, Sakuma Y, Abe H, Shinwari ZK, Seki M, Shinozaki K (2002) Biological mechanisms of drought stress response. JIRCAS Working Rep. 45:1–8.

Yamaguchi-Shinozaki K, Shinozaki K (1993). Characterisation of the expression of a dessication-responsive rd29 gene of *Arabidopsis* thaliana and analysis of its promoter in transgenic plants. Mol. Gen. Genet., 236: 331-340.

Yamaguchi-Shinozaki K, Urao T, Iwasaki T, Kiyosue T, Shinozaki K (1994). Function and regulation of genes that are induced by dehydration stress in *Arabidopsis* thaliana. JIRCAS J., 1: 69-79.

Yancey PH, Clark ME, Hand SC, Bowlus RD, Somero GN (1982). Living with water stress: Evolution of osmolyte systems. Science, 217: 1214-1222.

Yang J-V, Sun Y, Sun A-G, Yi S-Y, Qin J, Li M-H, Liu J (2006). The involvement of chloroplast HSP100/ClpB in the acquired thermotolerance in tomato. Plant Mol. Biol., 62: 385-395.

Yano M, Sasaki T (1997). Genetic and molecular dissection of quantitative traits in rice. Plant Mol. Biol., 35: 145-153.

Yeo ET, Kwon HB, Han SE, Lee JT, Ryu JC, Byu MO (2000). Genetic engineering of drought resistant potato plants by introduction of the trehalose-6-phosphate synthase (TPS1) gene from *Saccharomyces cerevisiae*. Mol. Cells, 10: 263-268.

Yi-Guo S, Bao-Xing D, Wan-Ke Z, Jin-Song Z, Shou-Yi C (2002). AhCMO, regulated by stresses in *Atriplex hortensis*, can improve drought tolerance in transgenic tobacco.TAG, 105: 815-821.

Yin XY, Yang AF, Zhang KW, Zhang JR (2004). Production and analysis of transgenic maize with improved salt tolerance by the introduction of AtNHX1 gene. Acta Bot. Sin., 46: 854-861.

Yokoi S, Cubero FJQB, Ruiz MT, Bressan RA, Hasegawa PM, Pardo JM (2002). Differential expression and function of *Arabidopsis* thaliana NHX Na$^+$/H$^+$ antiporters in the salt stress response. Plant J., 30: 529-536.

Yokoi S, CuberoFJQB, Ruiz MT, Bressan RA, Hasegawa PM, Pardo JM (2002). Differential expression and function of *Arabidopsis* thaliana NHX Na$^+$/H$^+$ antiporters in the salt stress response. Plant J., 30: 529.

Yoshimura K, Miyao K, Gaber A, Takeda T, Kanaboshi H, Miyasaka H, Shigeoka S (2004). Enhancement of stress tolerance in transgenic tobacco plants overexpressing Chlamydomonas glutathione peroxidase in chloroplasts or cytosol. Plant J., 37: 21-33.

Yu Q, Osborne LD, Rengel Z (1999). Increased tolerance to Mn deficiency in transgenic tobacco overproducing superoxide dismutase. Ann. Bot., 84: 543-547.

Yua Q, Hua Y, Lib J, Wua Q, Lin Z,(2005). Sense and antisense expression of plasma membrane aquaporin BnPIP1 from *Brassica napus* in tobacco and its effects on plant drought resistance. Plant Sci., 169: 647-656.

Zhang H-X , Hodson JN, Williams JP, Blumwald E (2001). Engineering salt-tolerant *Brassica* plants: Characterization of yield and seed oil quality in transgenic plants with increased vacuolar sodium accumulation. Proc. Natl. Acad.Sci., 98: 6896-6901, 12832-12836.

Zhang H-X, Blumwald E (2001). Transgenic salt-tolerant tomato plants accumulate salt in foliage but not in fruit. Nat. Biotechnol., 19: 765-768.

Zhang JY, Broeckling CD, Blancaflor EB, Sledge MK, Sumner LW, Wang ZY (2005). Overexpression of WXP1, a putative *Medicago truncatula* AP2 domain-containing transcription factor gene, increases cuticular wax accumulation and enhances drought tolerance in transgenic alfalfa (*Medicago sativa*). Plant J., 42: 689-707.

Zhang JZ, Creelman RA, Zhu JK (2004). From laboratory to field. Using information from *Arabidopsis* to engineer salt, cold, and drought tolerance in crops. Plant Physiol., 135: 615-621.

Zhao F, Guo S, Zhang H, Zhao Y (2006). Expression of yeast SOD2 in transgenic rice results in increased salt tolerance. Plant Sci., 170: 216-224.

Zhao F, Wang Z, Zhang Q, Zhao Y, Zhang H (2006). Analysis of the physiological mechanism of salt-tolerant transgenic rice carrying a vacuolar Na$^+$/H$^+$ antiporter gene from *Suaeda salsa*. J. Plant Res., 119: 95-104.

Zhao F, Zhang H (2006). Expression of *Suaeda salsa* glutathione S-transferase in transgenic rice resulted in a different level of abiotic stress resistance. J. Agric. Sci., 144: 547-554.

Zhao HW, Chen YJ, Hu YL, Gao Y, Lin ZP (2000). Construction of a trehalose-6-phosphate synthase gene driven by drought responsive promoter and expression of drought-resistance in transgenic tobacco. Acta Bot. Sin., 42: 616-619.

Zhao M-G, Tian Q-Y, Zhang W-H (2007). Nitric Oxide Synthase-Dependent Nitric Oxide Production Is Associated with Salt Tolerance in *Arabidopsis*. Plant Physiol., 144: 206-217.

Zhifang G, Loescher WH (2003). Expression of a celery mannose 6-phosphate reductase in *Arabidopsis* thaliana enhances salt tolerance and induces biosynthesis of both mannitol and a glucosyl-mannitol dimer. Plant Cell Environ., 26: 275-283.

Zhu BC, Su J, Chan MC, Verma DPS, Fan YL, Wu R (1998). Overexpression of a Delta (1)-pyrroline-5-carboxylate synthetase gene and analysis of tolerance to water- and salt-stress in transgenic rice. Plant Sci. 139: 41-48.

Zhu JK (2001). Cell signaling under salt, water and cold stress. Curr. Opin. Plant Biol., 4: 401-406.

Zhu JK (2002). Salt and drought stress signal transduction in plants. Annu. Rev. Plant Biol., 53: 247-273.

Zhu JK (2003). Regulation of ion homeostasis under salt stress. Curr. Opin. Plant Biol., 6: 441-445.

Zhu L, Tang GS, Hazen SP, Kim HS, Ward RW (1999). RFLP-based genetic diversity and its development in Shaanxi wheat lines. Acta Bot. Boreali Occident Sin., 19: 13.

14

Review on application of biomimetics in the design of agricultural implements

Benard Chirende[1] and Jianqiao Li[2]*

Key Laboratory for Terrain-Machine Bionic Engineering (Ministry of Education), Jilin University, Changchun 130022, P.R. China.

This paper aims at reviewing the application of biomimetics in design of agricultural implements. Most of the biomimetic works done were aimed at investigating the effect of non-smooth surfaces on soil resistance based on soil burrowing animals. The characteristics of soil-burrowing animals for improved soil scouring and their mechanism for reducing soil adhesion and friction are discussed. From past research works, it can be concluded that non-smooth surfaces can generally reduce soil resistance however the extent of reduction is still a gray area. The main factors affecting soil adhesion like the nature and properties of the soil, the properties of the soil-engaging component surfaces and the experimental conditions which are difficult to replicate, could be the explanation for inconsistencies in the extent of soil resistance reduction. Generally, when applying the concept of non-smooth surfaces in biomimetic implement design, general factors considered in arranging non-smooth structures are distribution of normal stresses, choice of non-smooth type and material, soil motion tracks during operation and choice of non-smooth convex parameters.

Key words: Biomimetics, anti-friction, anti-adhesion, soil resistance, burrowing animals.

INTRODUCTION

The anti-adhesion and anti-friction functions of a soil burrowing animal body surface against soil are an inevitable outcome of evolution and adaptation over millions of years. The body surface morphology of these animals have non-smooth units such as convex domes, con-cave dips, ridges or wavy structures, which play important roles in their anti-soil adhesion and anti-friction functions. The soil-burrowing animals' soil adhesion techniques have led to some improvement of conventional methods for reducing soil adhesion like in the design of implement surface shapes, selection of surface materials for soil-engaging components and application of electro-osmosis, magnetic fields, vibration and lubrication in implement design. These soil burrowing animals prevent soil from sticking to their bodies because of evolution of their biolo-gical systems through exchange of matter, energy and information with soil over centuries. They can comfortably move in even clay soil without soil sticking to their bodies. Soil-engaging tools have been designed based on these features of living organisms which are efficient in bio-mimetic anti-adhesion, anti-friction and anti-abrasion aga-inst soil (Tong et al., 2004). Different biomimetic designs were found to have different effects on improving imple-ment performance against soil resistance.

Characteristics of soil-burrowing animals

Soil-burrowing animals include animals such as dung beetle, ground beetle and mole cricket living in soil and also those which dig burrows in earth without necessarily living in it such as house mouse, yellow mouse and pangolin. The living surroundings of these soil animals are very different from those of animals living on land and in water. It is generally more difficult for animals to move in soil especially when the soil is moist and this has led to

---

*Corresponding author. E-mail: jqli@jlu.edu.cn.

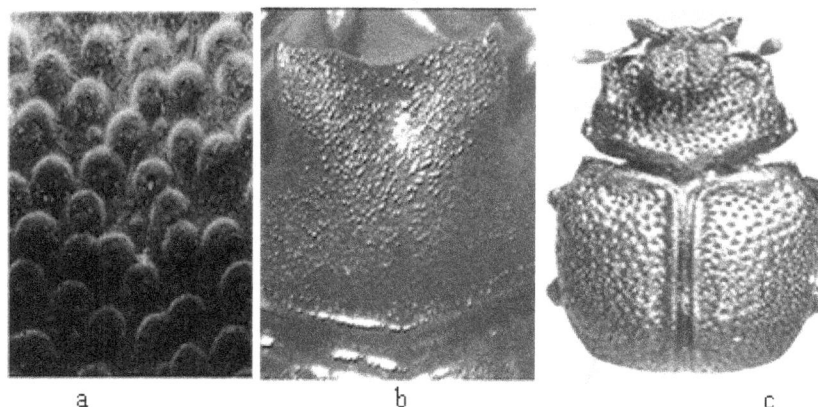

Figure. 1 The morphological surfaces of dung beetle *Copris ochus Motschulsky*.
(a) Stereoscopyimage of the clypeus and pronotum of a female dung beetle *Copris ochu Motschulsky*.
(b) Stereoscopy image of the pronotum of a male dung beetle *Copris ochus Motschulsky*.
(c) Scanning electron microscopy image of the convex domes on pronotum cuticle surface of *Chlamydopsinae* (Moayad, 2004).

their natural adaptation to suit the difficult conditions. The gradual adaptation is evident in their anti-adhesion and anti-friction behaviour. The body surface morphologies, chemical composition, bioelectricity, secretion and flexibility of cuticle of soil-burrowing animals are the main features helping in achieving anti-adhesion and anti-friction functions.

Soil-burrowing animals have geometrically textured structures on their body surfaces. For example, there exist varied textured structures on the clypeus, pronotum, elytra, abdomen and legs of all the *Lamellicornia* beetles (Liu et al., 1997). In addition, the geometrical surface morphologies of earthworm (*Lumbricidae*), centipede (*Chilopoda*), dung beetle (*Scarabaeidae*), ground beetle (*Carabidae*), ant (*Formicidae*), mole cricket (*Gryllotalpidae*) were examined and non-smooth structures were seen on the body surface (Tong et al., 2004). The different morphological features of the body surfaces exist in different species of soil animals as well as in different segments of the same animal.

The geometrically textured surfaces include the embossed morphology with small convex domes, dimpled morphology with small concave hollows, wavy morphology, scaly morphology, corrugated morphology with ridges, and seta-covered morphology. For a dung beetle such as *Scarabaeus typhoon Fischer, Gymnopleurus mopsus Pallas, Sisyphus schaefferi Linnaeus*, the clypeus has a curved shape surface (Moayad, 2004). Figure 1 to 3 illustrate photographs of some beetles and their surface morphologies.

According to Tong et al. (2005), the body surfaces of soil animals have a strong intrinsic hydrophobic nature,

which implies that the force of attraction between the body surface materials and water molecules is very small. From the pronotum surface of dung beetle *(Copris ochus Motschulsky)*, it was demonstrated that the apparent contact angles of water on the surface were 91 to 106.5° and the average contact angle was 97.2°, a figure which represents its hydrophobic property (Figure 4). The non-smooth structures on the body surface of these soil animals help to enhance their hydrophobicity. The combination of geometrically textured surface and the hydrophobic nature prevents soil from sticking to the body surfaces of soil animals.

The locomotion of some organisms like caterpillar and earthworm is crawling by reversing the direction of normal peristaltic wave. Brackenbury (1999) conducted some researches on the crawling movement characteristics of such animals. He described the reverse gaits available to caterpillars and Figure 5 gives the forward movement procedure. The caterpillar and earthworm movement inspired Yao et al. (2001) into designing a push-pull air-cushioned platform vehicle. Two symmetrical air-cushioned platforms with a crawling mechanism were designed. The two air-cushioned sub-platforms with a grabbing mechanism were linked with one hydraulic cylinder. The alternating movement can be produced through the pushing and pulling operation of the hydraulic cylinder. This movement is discontinuous. The push-pull air-cushioned platform vehicle can turn around forward, backward, left or right (Moayad, 2004).

The legs and tarsal of the mole cricket and the claws and toes of the house mouse and yellow mouse have such functions as grasping, walking, clinging, and

Figure 2. Photographs of four beetles showing the textured surface structures.
(a) *Trachypachusslevini* (location: Oregon, USA).
(b) *Eustra japonica* (Location, Japan).
(c) *Omoglymmius hamatus* (wrinkled bark beetles. Location, California, USA).
(d) *Arrowina anguliceps* (Location: South India).

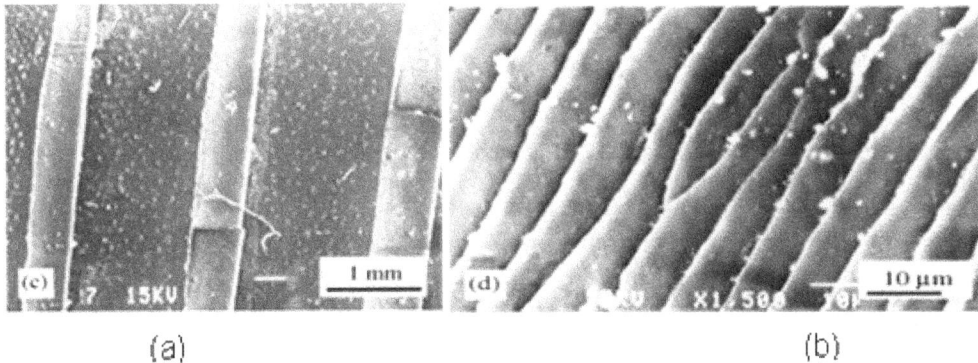

Figure 3. The textured morphologies with
(a) Ridged surface structure on the abdomen of a ground beetle.
(b) Stepped surface structure on the head of a black ant (Tong et al., 1994).

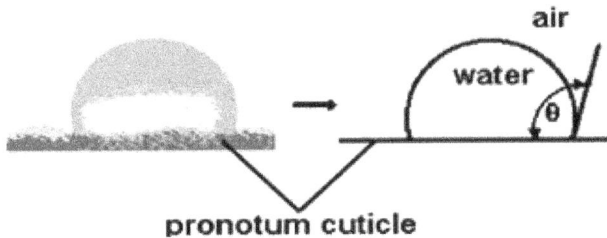

Figure 4. Hydrophobic nature of the pronotum cuticle surface of the dung beetle *(Copris ochus Motschulsky)* (Tong et al., 2004).

particularly, digging. The house mouse and yellow mouse are soil-burrowing animals with strong digging claws. House mouse and yellow mouse can use their toes to excavate. A quantitative understanding of the curvature characteristics of digging tarsal or toes of soil-burrowing animals is very useful to the biomimetic designs of soil-cutting tools like subsoilers, moldboard ploughs and furrow openers. The curvature variation of the soil-contacting surfaces of the soil-engaging components is an important factor affecting the forward resistance and working quality of the components (Tong et al., 2004).

Soil-tool interface

During soil-tool interaction, soil on the tool surface results in the pressure transmitted across the interface and the reaction force is closely related to the weight and type of soil. If the interface is not horizontal, then the reaction force is made up of many components. The main forces

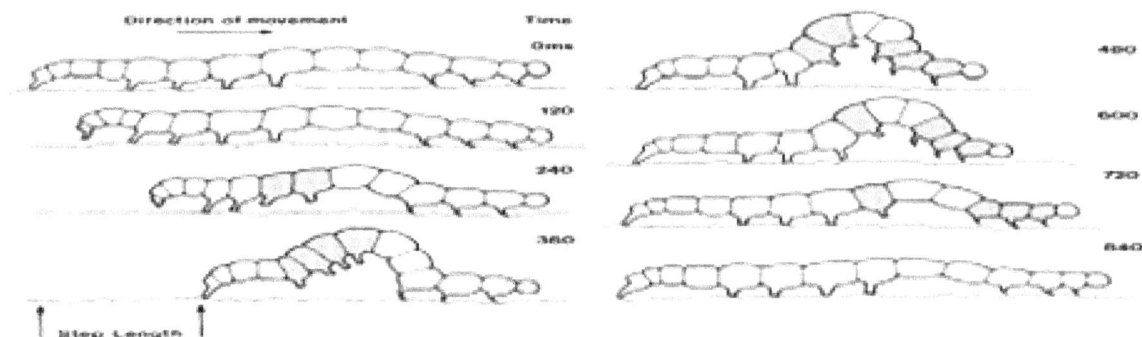

Figure 5. Forward movement in early instar of *Cucullia verbasc.* (Brackenbury, 1999)

at the interface include adhesion and friction forces which result in wear of the tool. In some cases these effects (adhesion, friction and wear) are very large, quite complex and greatly influence the interaction between soil and tool (Rabinowicz, 1995). Greenwood and Johnson (1998) suggested models for contact and adhesion of rubber. They concluded that there was a linear relationship between adhesion and size of contact area. Ren et al. (2001) investigated the effect of moisture on the adhesion forces. He found that when the moisture content of soil or its normal stress was high, the contact interface was filled with water and the soil was linked to the solid surface by a continuous water film. Moreover, soil adhesion increased with soil water tension and adhesion was also highest between the plastic limit and liquid limit. Jia (2004) concluded that the adhesion force of soil to solid materials mainly consists of intermolecular force between soil and solid, and the attraction force of the water film, which depends on the interface state of the soil and solid material. In order to see the effect of different animal body surfaces on adhesion, many researchers scanned burrowing animals using scanning electron microscope and their characteristics which enable them to overcome these problems were investigated.

Application of non-smooth structures in implement design

Non-smooth structures have been the most commonly used biomimetic anti-adhesion and anti-friction technique compared to others such as biomimetic electro-osmosis and biomimetic flexible structures. This could be explained by its simplicity in application compared to the other techniques.

Ren et al. (1995a) applied the concept of pseudo-variable approximation D-optimum theory to design biomimetic non-smooth surfaces of bulldozer plates based on the angle of cut, depth of cut, forward speed and soil particle size distribution. He imitated the surface morphology of the head of a dung beetle *(Ohthophagus lenzii harold).* On average, the sliding resistance was reduced by 13.2% compared to conventional plate. 18.02% was the maximum reached. The parameters used for designing the convex domes used in the best biomimetic plate were 7 mm in height, 25 mm in base diameter, a quantity of 45, and parallelogram arrangement. Ren et al. (1995b) used a statistical distribution derived by Li et al. (1995), which said that statistical position distribution of domes on a dung beetle followed a uniform statistical distribution governed by the equation $N_L = -0.243 + 0.1692\,L$, and the base diameter of the domes which ranged from 0.033 to 0.749 mm followed a Gaussian distribution on the basis of the χ^2 test. Ren et al. (1995b) then used plain carbon steel to make the convex dooms arranged on a bulldozer blade, which were tested on clay soil with a moisture content of 27% d.b and a speed of 13.33 - 58.82 mm/s. The resistance was lowered at all the speeds, but a more significant reduction was recorded at the highest speeds. Ren et al. (2003) tested different biomimetic blades with small convex domes that were different in quantity, base diameter, height and arrangement. The soil moisture content was 28.25% (d.b.) and the plastic limit and liquid limit were 22.62 and 36.33% respectively. He found that the sample with the largest convex dome base was the most effective in reducing soil resistance, 32.9% was the highest reached (sample number 5 in Figure 6). Sample 5 had 16 regularly arranged convex domes with base diameter 40 mm, height 4 mm and a distance between convex dome centres of 50 mm. The draft force of the smooth sample increased significantly as the experimental times increased, but the draft force of the non-smooth sample increased steadily, signifying that the soil which stuck on the smooth bulldozer plate helped increase the draft power (Figure 7).

Vander Straeten et al. (2004) used 10 plates with domes and dips ranging from 3 to 32 mm in height and 3 to 8 mm in depth respectively, and the arrangement was either hexagonal or parallelogram pattern covering

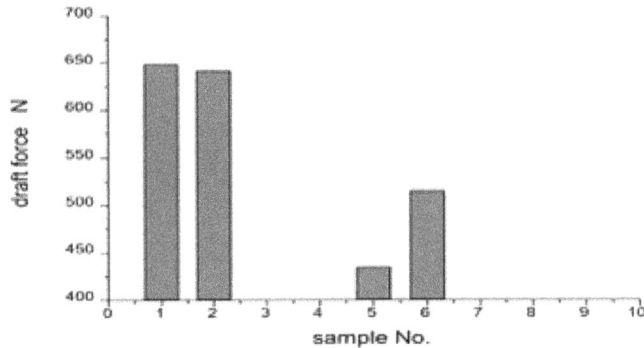

Figure 6. Effect of convex dome base diameter on draft force (Ren L et al, 2003).

*Diameters are as follows: Sample 1 was smooth, sample 2 = 30 mm, sample, sample 5 = 40 mm, sample 6 = 20 mm.

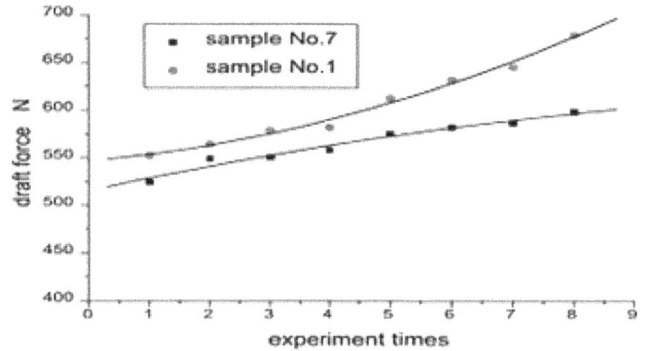

Figure 7. Relationship between draft forces and experimental times for two samples (Ren L et al, 2003).

*Sample number 1 was smooth and sample number 7 was non-smooth.

between 43 and 66% of the plate area. He found that hexagonal distribution of convex domes reduced the work per surface unit by 18.5%, twice the reduction of the parallelogram distribution. He even went on to conclude that a hexagonal distribution pattern of small diameter domes and hollows at high density can substantially reduce the sliding resistance. Although he concluded that the sliding resistance of soil is mainly influenced by type of soil, type and arrangement of non smooth structures, he also confirmed that overall his results were in disagreement with already established results especially by Ren et al. (1995a). The disagreement was mainly in the best design of the non-smooth structure and extent to which sliding resistance was reduced. However, Vander Straeten et al. (2004) used loamy soils with a moisture content of 5% and a particle distribution of 20% clay, 70% silt and 10% sand whilst Ren et al. (1995a) used clay soil with an average moisture content of 27.8% (db). This makes simple comparison of any results very difficult.

Work on bulldozer blades was also done by Qaisrani et al. (1993). He used steel-45 and ultra high molecular weight polyethylene (UHMWPE) to design biomimetic blades with convex domes. Six arrangement patterns used are shown in Figure 8. The parameters used were depth of cut 15 mm, speeds of cut 0.01, 0.02 and 0.06 m/s, angle of cut 35°. From the results, he concluded that UHMWPE was more superior to steel in reducing soil resistance with the best results of 34.0 and 15.55% at the highest speed of 0.06 m/s respectively. The resistance of the other biomimetic blades made of steel-45 convex domes was even higher than that of the conventional blade. To explain behavior of UHMWPE compared to steel, Tong et al. (1999) suggested that most polymer materials, such as ultra high molecular weight polyethylene (UHMWPE) and poly tetrafluoro ethylene (PTFE,

Teflon), possess lower adhesion force and friction force in soil because of their lower surface energy and higher hydrophobicity compared to steel. However, polymers have lower abrasive wear resistance especially against sandy soil hence they are not commonly used in making soil engaging equipment compared to steel.

Explanation for inconsistency in results

The main factors affecting soil adhesion include the nature and properties of the soil, the properties of the soil-engaging component surfaces and the experimental conditions. Soil factors affecting the soil adhesion include the soil texture, moisture content, water tension, porosity and organic matter content. The soil adhesion tends to increase as the proportion of clay particles in the soil increases and is highest when the soil moisture content is between the plastic limit and the liquid limit. An increase in the soil water tension elevates the soil adhesion. The geometry and material of the domes and dips are also critical in determining the accuracy of the results. All these factors make comparison very difficult and could be the cause for inconsistency of the results cited above.

It should be emphasized that when applying the concept of non-smooth surfaces in bionic implement design, the general factors considered in arranging non-smooth structures are distribution of normal stresses, choice of material for non-smooth structures, soil motion tracks and type of non-smooth structures (Ren et al., 2004). All this is captured by carrying out the following crucial analyses (Wilson and Andrea, 2004)

Functional analysis: The study of the natural systems physiology, including the functional mechanisms of the

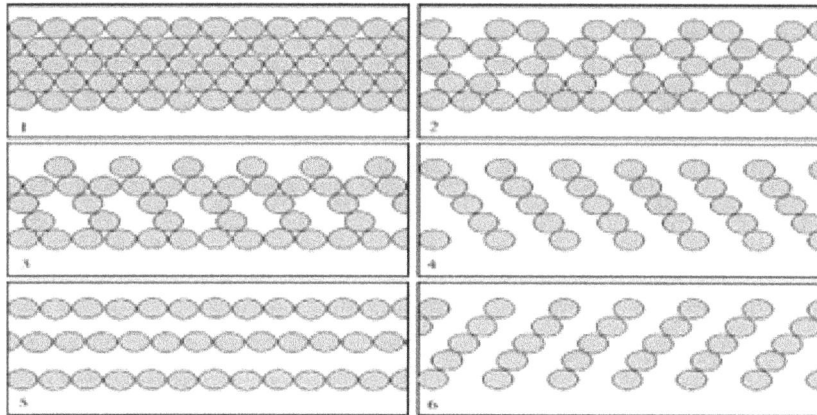

Figure 8. Schematic diagrams showing six regular distribution patterns of convex domes on biomimetic embossed non-smooth bulldozing blades (Qaisrani, 1993).

natural element and the principles that trigger its bio-mechanics. Among the relevant questions are: what is the function? What is it for? How does its functional system work?

Morphological analysis: The goal is to understand why the sample has a specific form, study the existence of geometric relationships and to observe and comprehend the texture of the sample.

Structural analysis: Aims at studying the organization of the natural element, its constituent parts and its capacity of undergoing stress, verifying its architecture and its natural growth.

Analysis of viability: Aims at studying the possibility of applying the observed characteristics into the project, and carefully evaluating all the observed aspects.

Conclusion

From past researches, it can be concluded that indeed non-smooth surfaces can reduce soil resistance however the inconsistency in the results simply means that comparison is very difficult when experiments are done under different conditions. In addition, the function or the mechanism of operation of different implements are sometimes very different from those of soil burrowing animals which means that different implements have to be uniquely designed to achieve maximum soil resistance reduction based on biomimetics and their use.

ACKNOWLEDGEMENTS

The authors are grateful for the financial support by scientific and technological development plan of Jilin Province, China (No. 20050539), and the National Natural Science Foundation of China (No. 50175045).

REFERENCES

Brackenbury J (1999). Fast locomotion in caterpillars. J. Insect Physiol. 45 (6): 525-533.

Greenwood JA, Johnson KL (1998). An alternative to the Maugismodel of adhesion between elastic spheres. J. Phys. D: Appl. Phy., 31: 3279-3290.

Jia X (2004). Theoretical analysis of adhesion force of soil to solid materials. Biosystem Engineering 87(4): 489-493.

Li J, Ren L, Chen B (1995). The statistics analysis and mathematical analogy of the unsmoothed geometrical units on the surface of soil animal. Transactions of the Chinese Society of Agricultural Engineering 11(2): 1-5

Liu G, Zhang Y, Wang R (1997). The color illustrated of common Lamellicornia beetle of Northern China. China Forestry Press.

Moayad BZ (2004). Biomimetic design and experimental research of Resistance-reducing surfaces of soil-tool systems. PhD dissertation, Jilin University, Changchun, China.

Qaisrani AR, Tong J, Ren L, Chen B (1993). The effects of unsmoothed surfaces on soil adhesion and draft of bulldozing plates. Transactions of the Chinese Society of Agricultural Engineering 9(1): 7-13.

Rabinowicz E (1995). Surface Interactions, Friction and Wear of Materials. Second Edition, A Wiley-Interscience Publication.

Ren L, Deng S, Wang J, Han Z (2004). Design principles of the non-smooth surface of bionic plow moldboard. J. Bionics Eng. 1(1): 9-19.

Ren L, Han Z, Li J, Tong J (2003). Effects of non-smooth characteristics on bionic bulldozer blades in resistance reduction against soil. J. Terramechanics 40(4): 221-230.

Ren L, Li J, Chen B (1995b). Unsmoothed surface on reducing resistance by bionics. Chinese Sci. Bull. 40(13): 1077-1080.

Ren L, Tong J, Zhang S, Chen B (1995a). Reducing sliding resistance of soil against bulldozing plates by unsmoothed bionics surfaces. J. Terramechanics 32(6): 303-309.

Ren LQ, Tong J, Li JQ, Chen BC (2001). Soil adhesion and biomimeyic of soil-engaging components: A Review. J. Agric. Eng. Res. 79(3): 239-263.

Tong J, Moayad BZ, Ren L, Chen B (2004). Biomimetic in soft terrain machines: A Review. Int. Agric. Eng. J. 13(3): 71-86.

Tong J, Ren L, Chen B (1995). Chemical constitution and abrasive wear behavior of pangolin scales. J. Mater. Sci. Lett. 14(20): 1468-1470.

Tong J, Sun J, Chen D, Zhang S (2005). Geometrical features and wettability of dung beetles and potential biomimetic engineering applicationsin tillage implements. Soil Tillage Res. 80(1-2): 1-12.

Vander Straeten P, Destain MF, Verbrugge JC (2004). Bionic improvement of soil bulldozing plates. J. Design and nature II, 545-553.

Wilson KJ, Andrea SG (2004). Methodology for product design based on the study of bionics. Materials and Design, 26 (2): 149-155.

Yao S, Zhang L, Chen J, Chen B, Zhang L (2001). Study on the structural design and walking principle of push-pull air cushioned platform vehicle. Natural Sci. J. Jilin Univ. Technol. 31(1): 1-5.

15

Homocysteine-A potent modulator

Janani Kumar*, Sowmiya Jayaraman and Nandhitha Muralidharan

Vellore Institute of Technology, VIT Vellore, India.

Homocysteine is an amino acid and is an intermediate metabolite of methionine metabolism. It is metabolized by two pathways, the trans-methylation and trans-sulphuration. These processes rely on an adequate supply of vitamin B_{12} and B_6 and folic acid. Deficiency of vitamin B_{12}, B_6 and folic acid can build up homocysteine level in blood stream. High homocysteine levels has been implicated in a variety of clinical conditions and is widely accepted, alongside smoking, obesity, hypertension and dyslipidemia as being an independent risk factor for cardiovascular disease. Homocysteine promotes artery problems in more than one way. Homocysteine got a bad rap from its cozy relationship with heart attacks and stroke. Researchers have repeatedly demonstrated that if they give a person a drink containing methionine, homocysteine will shoot up, and blood flow will shrink up. If they give the person a gram of vitamin C before they give the methionine, blood flow will be maintained. This indicates that antioxidant (vitamin C) prevents the formation of free radicals from homocysteine, which interferes with the ability of the blood vessels and causes cardio vascular diseases. Hence, further research is required to confirm whether antioxidant rich diet can prevent the homocysteine formation in the body or not.

Key words: Homocysteine, vitamins, methonine, disease.

INTRODUCTION

More over than cholesterol, there is a new molecule in the list. Last 10 years, there has been an explosion of interest. A sulfur containing amino acid that occupies a central location in the metabolic pathways–its homocysteine a major risk factor in hyperhomocysteinemia. Homocysteine is not found in our diet, and thus the answer to why it represents such a problem lies in examining how it is produced in our body. Homocysteine is formed in methionine metabolism and the imbalance between the rate of production of homocysteine through methylation, remethylation reactions in methionine metabolism can result in increase in the release of homocysteine to the extracellular medium and ultimately the plasma and urine. High levels of homocysteine have been linked to cardiovascular disease. Recent research shows that elevated levels of homocysteine is one of the risk factor for many vascular diseases, neurodegenerative diseases like Alzheimer's

and congenital heart defects.

In this article we review the various biochemical and genetic link of homocysteine and its associated risk factors and also emphasizing on the simple ways to prevent the elevation of homocysteine which may aid in the prevention of many disorders associated with it.

HISTORY OF HOMOCYSTEINE

The importance of homocysteine to human health first came to light in the 1960's by Kilmer McCully who was interested in homocystinuria, a rare disease that results in high homocysteine levels in the blood. It turned out that people with this condition often suffered from heart disease and strokes, even at a very young age. Several children with homocystinuria had died of heart attacks, despite their age and lack of fat deposits. Dr. McCully noticed that they had blood vessels damaged by atherosclerotic plaques. He suggested that maybe the homocysteine had something to do with how these cholesterol deposits were formed inside the arteries. He

*Corresponding author. E-mail: januteddy@gmail.com.

$$^+H_3N - \underset{\underset{CH_2 - CH_2 - SH}{|}}{\overset{\overset{H}{|}}{C}} - COO^-$$

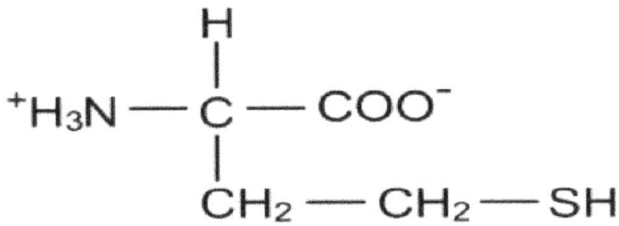

Figure 1. Structure of homocysteine.

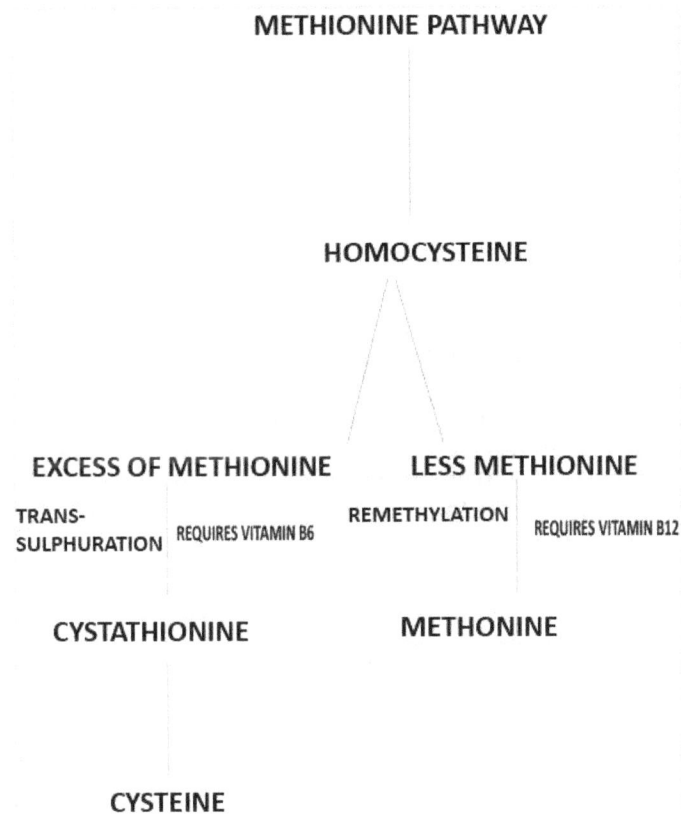

METHIONINE PATHWAY

HOMOCYSTEINE

EXCESS OF METHIONINE   **LESS METHIONINE**

TRANS-SULPHURATION REQUIRES VITAMIN B6   REMETHYLATION REQUIRES VITAMIN B12

CYSTATHIONINE   **METHONINE**

CYSTEINE

Figure 2. Homocysteine metabolism.

suggested that maybe the homocysteine had something to do with how these cholesterol deposits were formed inside the arteries (McCully, 1969), however whether to take homocysteine levels as a clinical determinant for heart disease or as other researches suggest while to take it as a marker rather than a cause is still controversial. However now researchers have also discovered that homocysteine plays a part in much more diverse health concerns like osteoporosis, depression, Alzheimer's and some conditions in pregnancy. To understand how one molecule can have influence in so many diseases, we need to take a closer look at this compound and its metabolic pathway.

Homocysteine metabolism

Homocysteine is non-protein sulphur containing amino acid and a normal intermediate in methionine metabolism (Figure 1). Proteins are metabolized, broken down into individual amino acid, including the sulfur-containing amino acid Methionine. Homocysteine formed during the methionine metabolism can be removed from the system only by two ways: when there is excess methionine, it will be converted into cysteine by transsulphuration reaction (Figure 2).

Secondly when there is low level of methionine it will be remade into methionine by remethylation both the reaction requires vitamins and folic acid as a major cofactor. Thus if a person ingests lots of protein, and there is not enough folic acid, B_6 and B_{12} available to help digest it, homocysteine levels can build up in the blood stream. Indeed, studies have shown that oral folic acid supplements are effective in bringing homocysteine levels down. In a seesaw effect, as folic acid levels rise in the blood stream, levels of homocysteine drop (Stam et al., 2005).

Genetic link

Severe homocysteinemia with homocystinuria was first identified in cases of rare inborn errors of metabolism characterized by marked elevations of plasma and urine homocysteine concentrations. The most common of these is the deficiency of CBS, the homozygous form of which occurs in approximately 1 in 2, 00,000 live births. Associated with fasting plasma homocysteine concentrations of up to 200 mol/L. Clinical manifestations include:

1) Mental retardation
2) Thromboembolism
3) Seizures
4) Premature atherosclerosis
5) Skeletal deformities

The heterozygote state is estimated to occur in 1 to 2% of the population of 3 lakhs. These patients have mild elevations of fasting homocysteine. Recent epidemiologic studies suggest that they are at increased risk for premature atherosclerosis. Homozygous deficiency of N5, N10 methylenetetrahydrofolate reductase (MTHFR) is rare and results in severe hyper homocysteinemia and early death, which is mainly due to the transition in the MTHFR Gene in the 677 codon with a change in the amino acid from valine to alanine (C677T).

Patients homozygous for the C677T mutation have slight elevations in homocysteine levels and are at increased risk for premature vascular disease. Recent observations suggest that patients with the MTHFR genotype have higher folate requirements than individuals with a normal genotype.

HOMOCYSTEINE AND VARIOUS HEALTH DISORDERS

Cardiovascular diseases

Homocysteine is believed to have a causal influence on the development of cardiovascular disease. It damages the walls of the arteries. As a result they become clogged, thickened and less flexible. The blood clots more easily so it is more likely to form a blockage and cause a heart attack. Epidemiological evidence now exists to conclude that moderately elevated homocysteine increases the risk of cardiovascular events (Lievers et al., 2003).

Artherosclerosis and stroke

Strokes affect the arteries leading to and from the brain and can be caused by either a blood clot or a blood vessel rupturing. As cerebrovascular events are similar in many ways to cardiovascular events, it should not be surprising that homocysteine is also an independent risk factor for ischemic stroke.

Alzheimer's diseases

Alzheimer's diseases destroys brain cells, causing problems with memory, thinking and behavior severe enough to affect work, lifelong hobbies or social life. Gets worse over time, and it is fatal. Today it is the seventh-leading cause of death. Recently the New England journal of medicine suggest that elevated homocysteine levels could be a strong risk factor associated with poor cognition and dementia.

Renal disorders

Normal kidney metabolism and filtration plays a prominent role in removing homocysteine from the blood interestingly patients with renal disease have unusually high rate of cardiovascular morbidity and death. Likewise, renal transplant recipients typically have elevated homocysteine levels (Friedman et al., 2001).

Other complications related to elevated homocysteine level

High homocysteine level during maternity will increase the chance of miscarriage and serious pregnancy complications, which will be fatal for both mother and the child. Moreover homocysteine seems to be inappropriately stimulate some nerve cell receptors, which can interfere brain function.it is also been proposed that homocysteine maybe a causative agent got osteoporosis condition which weakens the bone (Bostom et al., 1999b).

Factors that cause elevated levels of homocysteine

Lifestyle factors

a) Smoking
b) High consumption of coffee
c) Alcohol consumption (moderate beer intake may be beneficial)

Diet

a) Low consumption of fruits and vegetables
b) No consumption of multivitamins
c) Low intake of folic acid, vitamin B_6, vitamin B_{12}
d) High intake of methionine-containing proteins

Diseases or inherited causes

a) Cystathionine beta-synthase deficiency
b) 5MTHFR errors
c) Methionine synthase deficiencies
d) Chronic renal failure

Drugs that increase homocysteine

a) Some antiepileptic drugs (phenobarbital, valproate, phenytoin etc.)
b) Diuretic therapy
c) Methotrexate
d) Nitrous oxide
e) Estrogen-containing oral contraceptives
f) Metformin
g) Niacin

VITAMIN AND HOMOCYSTEINE METABOLISM

It is evident from homocysteine metabolism that folic acid vitamin B_6 and vitamin B_{12} are important cofactors, which metabolizes homocysteine into methionine and cysteine. Based on this fact several researchers suggest that vitamin supplementation can lower the homocysteine levels in turn the related complications. It is also been proposed that vitamin therapy with folic acid alone or complemented with vitamin B_6 and B_{12} with cereal grain products contain folate can lower plasma homocysteine (Bostom et al., 1999a). According to WHO report vitamin supplementation during pregnancy can prevent birth defects and miscarriages. Furthermore, increased dietary intake of folic acid and vitamin B_{12} (but not B_6) is inversely related to reduction in stroke risk.

How to reduce the level of homocysteine or how to keep it under control?

There are 10 commandments to lower your homocysteine levels namely:

1) Eat less fatty meat, more fish and vegetable protein
2) Eat your greens
3) Have a clove of garlic a day
4) Do not add salt to your food
5) Cut back on tea and coffee
6) Limit your alcohol
7) Reduce your stress
8) Stop smoking
9) Supplement a high-strength multivitamin
10) Proper well balanced diet with more intake of water

CONCLUSION

Because folic acid supplements do significantly lower homocysteine levels in the blood stream, such proposals to add folic acid to foods have been made. Proponents note that the addition of vitamin B6 to the food supply has resulted in a gradual drop in death from cardiovascular causes since the 1960s. However, such plans are controversial, for one because folic acid supplements can mask the symptoms of pernicious anemia, a vitamin B_{12} deficiency that hampers the bone marrow's ability to make blood, which can cause irreversible nerve damage.

Thankfully, in the absence of metabolic defects that keep homocysteine levels abnormally high, all the risks associated with high homocysteine levels seem to be avoidable with good nutrition. Leafy green vegetables, orange juice and beans are good sources of folic acid. Vitamin B6 is found in starchy foods such as whole grains, potatoes, bananas, as well as turkey and tuna, and vitamin B_{12} is found in meat, seafood and dairy products. While high homocysteine levels are found in tandem with many ailments, proving a real biochemical connection between homocysteine and disease is important before serious recommendations can be made about diet and vitamin supplementation.

REFERENCES

Bostom AG, Gohh RY, Bausserman L, Hakas D, Jacques PF, Selhub J (1999a). Serum cystatin C as a determinant of fasting total homocysteine levels in renal transplant recipients with a normal serum creatinine. J. Am. Soc. Nephrol., 10: 164-166.

Bostom AG, Rosenberg IH, Silbershatz H, Jacques PF, Selhub J, D'Agostino RB (1999b). Nonfasting plasma total homocysteine levels and stroke incidence in elderly persons: The Framingham Study. Ann. Intern. Med., 131: 352-355.

Friedman AN, Bostom AG, Selhub J, Levey AS, Rosenberg IH (2001). The kidney and homocysteine metabolism. J. Am. Soc. Nephrol., 12: 2181-2189.

Lievers KJ, Kluijtmans LA, Blom HJ (2003). Genetics of hyperhomocysteinaemia in cardiovascular disease. Ann. Clin. Biochem., 40: 46-59.

McCully KS (1969). Vascular pathology of homocysteinemia: implications for the pathogenesis of arteriosclerosis. Am. J. Pathol., 56: 111-128.

Stam F, Smulders YM, van Guldener C, Jakobs C, Stehouwer CD, de Meer K (2005). Folic acid treatment increases homocysteine remethylation and methionine transmethylation in healthy subjects. Clin. Sci. (London), 108: 449-456.

16

Urokinase - A strong plasminogen activator

Adinarayana Kunamneni[1, 2]*, Bhavani Devi Ravuri[1], Poluri Ellaiah[1], Taadimalla Prabhakhar[1] and Vinjamuri Saisha[1]

[1]Departamento de Biocatálisis, Instituto de Catálisis y Petroleoquímica, CSIC, C/ Marie Curie 2, Cantoblanco, 28049—Madrid, Spain.
[2]Pharmaceutical Biotechnology Division, University College of Pharmaceutical Sciences, Andhra University, Visakhapatnam - 530 003, India.

Urokinase (UK) is a serine protease, which specifically cleaves the proenzyme/zymogen plasminogen to form the active enzyme plasmin. It specifically catalyzes the cleavage of the Arg-Val bond in plasminogen. The active plasmin is then able to break down the fibrin polymers of blood clots. Clinically, UK is given to patients suffering from thrombolytic disorders. Among the plasminogen activators, UK provides a superior alternative for the simple reasons of its being more potent as compared to tissue-plasminogen activator and non-antigenic by virtue of its human origin unlike streptokinase. Based on these observations, UK is a strong plasminogen activator. Hence, UK, as one of the most potent plasminogen activators is attracting a great deal of attention. The mechanism of action, physico-chemical properties, *in vitro* production, cloning and expression, and clinical applications of UK are reviewed in this paper.

Key words: Urokinase (UK), plasminogen activators, fibrinolysis, strong plasminogen activator, production, cloning and expression, physico-chemical properties and clinical applications.

INTRODUCTION

Urokinase (UK) is given to patients suffering from thrombolytic disorders like deep vein thrombosis, thrombosis of the eye, pulmonary embolism, and myocardial infarction. This enzyme is a strong plasminogen activator which specifically cleaves the proenzyme/zymogen plasminogen to form the active enzyme plasmin (Lesuk et al., 1967; Kunamneni et al., 2008). It specifically catalyzes the cleavage of the arg–val bond in plasminogen. The active plasmin is then able to break down the fibrin polymers of blood clots (Bennart and Francis, 1976). The initial main source of UK was urine as described by Williams (1951). Later UK was prepared from cultures of human embryonic kidney cells (Bernick and Kwaan, 1967).

In recent years, thrombolytic therapy has revolutionized the treatment of these diverse circulatory disorders such as pulmonary embolism, myocardial infarction, deep vein thrombosis, thrombosis of the eye etc., (Collen et al.,

1988). Therapy in thrombosis has been directed towards interference with coagulation mechanism, activation of fibrinolytic system, interference with platelet aggregation or combination of these. In addition, surgical intervention to prevent embolization or to remove thrombi and restore blood flow is of critical importance.

Anticoagulant therapy includes the use of either coumarin drugs, heparin, or a combination of both. The coumarin drugs (warfarin, coumarin and dicoumarol) act by antagonizing vitamin K and result in depression of the concentration of clotting factors [prothrombin (factor II), factors VII, IX and X].

Thrombolytics are the drugs used to lyse thrombi to recanalyze occluded blood vessels. They are curative rather than prophylactic. They work by activating the natural fibrinolytic system.

The plasma fibrinolytic system

Fibrinolysis

Fibrinolysis refers to the dissolution of the fibrin blood clot by an enzyme system present in the blood of all mamma-

---

*Corresponding author. E-mail: adikunamneni@rediffmail.com.

lian species (Castillino 1981). The fibrinolytic system consists of the plasma zymogen, plasminogen; its activated product, the proteolytic enzyme, plasmin; plasminogen activators (PAs); inhibitors of both plasmin and PAs and fibrinogen and fibrin.

The basic reaction of the plasma fibrinolytic system is the conversion of a plasminogen to the active proteolytic enzyme plasmin, by a limited proteolytic cleavage mediated by different PAs (Castellino, 1984). The PAs are synthesized and released from endothelial cells and other tissues. Plasmin has the capacity to hydrolyse fibrin and various plasma coagulation proteins, including fibrinogen. The activity of the fibrinolytic system is modulated by inhibitors that inhibit both PAs and the proteolytic effect of plasmin. The main players of the fibrinolytic system are plasminogen itself, the zymogen of a trypsin like serine protease, two activators of plasminogen and three protease inhibitors (Francis and Marder, 1994).

The major two activators that occur in the circulating blood are: the tissue type plasminogen activator (t-PA) and the urinary type plasminogen activator (u-PA) also called urokinase (UK).

Fibrinolytic system

Extensive studies have been made over the last 25 years to understand the physiology of the fibrin-clot formation (Wu and Thaigarajan, 1996).

The fibrinolytic system and its constituents directly responsible for the dissolution of the fibrin clot are briefly described in Figure 1 (Bick, 1992).

Plasminogen-plasmin system

Plasminogen is a glycoprotein of molecular weight 90-KDa, which is synthesized in the liver. It is converted enzymatically by PAs to the fibrinolytic enzyme, plasmin. cuts away its covalently cross-linked α- chain protuberances. The rather open mesh like structure of a blood clot gives plasmin relatively free access to the polymerized fibrin molecules thereby facilitating clot lysis.

Plasmin, a plasma serine protease that specifically cleaves fibrins tripple stranded coiled coil segment and Plasmin is formed through the proteolytic cleavage of the 86-KDa zymogen plasminogen, a protein that is homologous to the zymogens of the blood-clotting cascade.
PA is present in the tissue (t-PA), in plasma and in urine (UK). t-PA is localized in the vascular endothelium of veins, capillaries and pulmonary arteries and in the microsomal fraction of cells. t-PA is released into the blood stream in response to a number of stimuli, includeing ischaemia, vasoactive drugs and exercise. Released activator is inactivated rapidly in the blood stream by complexing to tissue plasminogen activator inhibitors (PAIs) and has a half-life of about five minutes.

The major tissue PAIs are synthesized in the liver and in vascular endothelium, but about 30% of the total is pro-

probably megakaryocyte-derived and is stored in platelet and granules.

The activator in urine, UK, differs structurally from t-PA and is produced primarily in the kidneys and excreted into the urine where it may help to maintain urinary tract potency. Endothelial UK probably contributes a small proportion of plasma activator activity. Factor XIIa not only intimates coagulation, but also accelerates the conversion of plasminogen to plasmin via a proactivator, kallikrein.

When clotting occurs, a small amount of plasminogen is trapped in the fibrin strands. PA released locally from the vascular endothelium or traumatized tissues, binds to the fibrin of the thrombus and converts plasminogen to plasmin, itself bound to its surface fibrin and in this conformation protected from its otherwise highly effective inhibitor α_2 antiplasmin. Fibrin is then digested. There is little or no plasma fibrinolytic activity because plasmin that is formed in the blood stream from activation of plasma plasminogen is rapidly inactivated by circulating α_2 antiplasmin. The α_2 macroglobulin also acts as a secondary plasmin inhibitor in the presence of excess plasmin.

Fibrinolytics

As in the inflammation reaction, the clotting of blood is an example of a defense mechanism, which can overreact and require therapeutic intervention. It is now known that in normal tissues, there is a constant dynamic equilibrium between blood coagulation (clotting) and fibrinolysis (the process of dissolving the clotted blood). The maintenance of proper balance in this equilibrium is extremely important. If fibrinolysis is increased by a pathological cause, a predisposal to excessive bleeding results. On the other hand, if fibrinolysis is weakened so that clot formation is favoured, conditions occur that are called thromboses (clots formed in and remaining in blood vessels) and embolisms (sudden blockages of blood vessels caused by circulating fragments of clots). These can be life threatening. In chronic cases, cholesterol and other fatty materials may aggregate around clotted deposits in blood vessels. When this pathological condition is well established, it is called atherosclerosis, hardening of the arteries.

Although there are chemical anti-coagulants (such as heparin) available as well as corrective surgical techniques, acute thrombosis and embolisms (which are lumped together as acute thromboembolic vascular diseases) are still the largest single cause of death and disease in the middle-aged and elderly populations of the Western World.

The mechanism used by the body for controlling the equilibrium between clot formation and dissolution is a complex one involving a series of enzymes, proenzymes, activators, and proactivators. A brief outline of the highlights of the process will be helpful in understanding the

Extrinsic   Intrinsic

Streptokinase (SK)   Activators   Factors XIIa

Urokinase (scu-PA)   Kallikrein

rt-PA   t-PA

                                                            Fibrin (insoluble)

Plasminogen ───────────▶ Plasmin ───────────▶
(Pro fibrinolysin)

                                                            Fibrin fragments (soluble)

Extrinsic   Intrinsic

Trahexaemic acid   Inhibitors   α2-Antiplasmin

Aprotnin   α2-macroglobulin

EACA

                                                            Fibrin (insoluble)

Plasminogen ──────✕──────▶ Plasmin ──────✕──────▶
(Pro fibrinolysin)

                                                            Fibrin fragments (soluble)

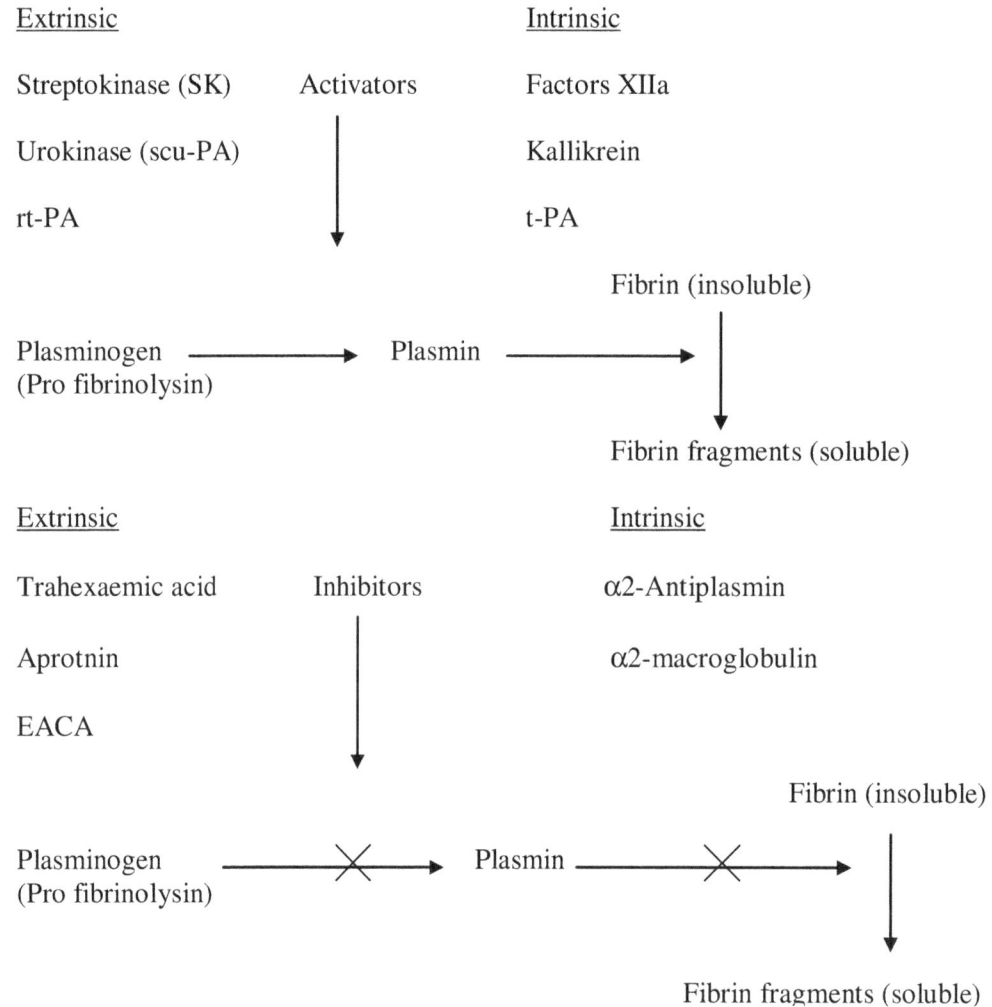

Figure 1. Relationship of components of the fibrinolytic system with activators.

therapeutic method. In forming a clot, the plasma protein, fibrinogen, is converted to insoluble fibrin by the enzyme thrombin. Fibrin forms the clot. The enzyme plasmin, which dissolves the clot, exists in the blood as the pro-enzyme plasminogen. Activators convert plasminogen to plasmin for dissolving the clot.

Just as nature uses enzymes in maintaining this crucial balance, so is man learning how to use enzymes to restore the balance once it is lost. Clinical studies have shown that the best approach to therapeutic thrombolysis (dissolution of clots) is an intravenous injection of an enzyme capable of converting plasminogen to plasmin, the enzyme which dissolves the clot. This type of therapy is known as thrombolytic (thrombus or clot-splitting) or fibrinolytic (fibrin-splitting) therapy. The enzymes most frequently used for this are streptokinase (SK) [from bacteria] and UK (from human urine). Three "new" thrombolytic enzymes which have seen some therapeutic use in the last 20 years are Arvin (from a Malayan pit viper), reptilase (from a South American snake), and brinase (from the mould *Aspergillus oryzae*).

Urokinase (UK)

UK is an endogenous substance, which is involved, in many physiological processes. It is a life saving, therapeutically important fibrinolytic enzyme used in the treatment of many disorders requiring dissolution of blood clots.

The main source of UK earlier was its isolation from urine which is a tedious and lengthy procedure and highly expensive also. Further the yield obtained is very less of the order of 6CTA (Committee on Thrombolytic Agents) units/ L urine. This value shows that minimum of 1500 L of urine are required to produce one clinical dose of this enzyme [7500-8000 CTA units].

In contrast UK can be harvested at concentrations of order 800 CTA units/ml from confluent cultures of human kidney cells. As such in recent years human cell cultures are being used for UK production and isolation.

Below is a list of cell lines being used currently for UK production and isolation:

• Human embryo kidney cells

- Human lung edenoma carcinoma cell line.
- Normal human umbilical vein endothelial cells.
- Human fibroblasts.
- Human myeloma cells.
- BHK-21/N cells.

Several forms of UK with different molecular weights were described. A single chain UK-type PA called Saruplase had been prepared. In Great Britain, potency of UK is expressed in arbitrary units known as Plough units. One plough unit is approximately equivalent to 1.5 International Units. In USA, potency is expressed in CTA units. One CTA unit is approximately equivalent to one IU.

UK is secreted from cells as a single chain proenzyme (scu-PA) from which the active two chain enzymatic plasminogen activator (tcu-PA) is derived by proteolysis, the two chains remaining linked by a disulphide bond. The UK that is used clinically is tcu-PA type. UK is enzymic and acts directly as a plasminogen activator and it is not antigenic.

The amino terminal fragment (ATF) of urokinase-type plasminogen activator (u-PA) is 130 amino acid residues long. It consists of $2-\alpha$-helices and two antiparallel β-strands. The ATF contains 2 domains; a growth factor domain and a Kringle domain. It binds fucose (magenta) as a ligand. Fucose is a deoxy-sugar, a monosaccharide with one or more hydroxyl (-OH) groups replaced by a hydrogen (-H). The formula for fucose is $C_6H_{12}O_6$. The ATF also contains 6 disulfide bonds and are as follows: Cys_{11}-Cys_{19}, Cys_{13}–Cys_{31}, Cys_{33}-Cys_{42}, Cys_{50}-Cys_{131}, Cys_{71}-Cys_{113} and Cys_{102} –Cys_{126}.

Physical properties

It had been known UK could be found in multiple molecular sizes. There appeared to be two major forms: low molecular weight (LMW, 33-KDa) and high molecular weight (HMW, 57-KDa). It was found that urinary UK contained predominantly the LMW form. It was suggested that the HMW form is the native structure and that it is converted to the LMW form by cellular proteases. However, under certain conditions, HMW form was found to be more active when assayed with standard enzyme kinetic methods.

The enzyme is moderately stable, showing no appreciable loss in activity over years in lyophilized form or over months in sterile solutions at 1 mg/ml or more at refrigerator temperatures. Stability is decreased at salt concentrations below 0.03 M sodium chloride and precipitation with loss in activity occurs at very low salt concentrations.

In diluting to the levels of activity measured in the fibrinolytic assay, it is advisable to add a protein such as human serum albumin fraction V or gelatin to prevent surface denaturation.

Mechanism of action

u-PA is a strong plasminogen activator which specifically cleaves the proenzyme/zymogen plasminogen to form the active enzyme plasmin. It is a trypsin like enzyme capable of hydrolysing L-lysine and L-arginine esters. It specifically catalyzes the cleavage of the Arg-Val bond in plasminogen by first order reaction. The active plasmin is then able to break down the fibrin polymers of blood clots.

Contraindications

Active internal bleeding, history of cerebrovascular accident, recent trauma of any kind including surgery, intracranial neoplasm, aneurysm, bleeding diathesis, severe uncontrolled hypertension are contra-indicated for UK application.

Interactions

- Platelet inhibitors such as aspirin, indomethacin, etc. can potentiate the action of UK and cause haemorrhage.
- Heparin and oral anticoagulants will increase risk of bleeding.

Adverse reactions

- Superficial bleeding at sites of venous cut down, vascular punctures, etc. may occur.
- More seriously internal bleeding could occur in any internal organ or site.
- Relatively mild allergic reactions such as bronchospasm or rash may occur occasionally.
- Other side effects include chills, blood pressure changes, dyspnoea, palpitation, cyanosis and hypoxemia. Aspirin should not be given for fever.

Assaying UK

Since 1954, when the activity of plasmin and thrombin on synthetic substrates were first demonstrated (Troll et al., 1954; Sherry and Troll, 1954) synthetic substrates have played an important role in the study and characterization of various enzymes involved in blood coagulation and fibrinolysis (Abloundi and Hagan, 1957; Kline and Fishman 1961). Activation of plasminogen to plasmin involves an enzymatic proteolytic step and the activators mediating the action shared the ability of plasmin to split arginine and lysine esters (Alkjaersig et al., 1958). In the past few years, partially purified preparations of UK, the naturally occurring PA have been available and this coupled with a fairly simple assay procedure has stimulated studies with UK relative to its mechanism of action (Alkjaersig et al., 1958), physiological role (Fletcher et al., 1962), significance in disease (Sherry and Troll, 1954), and potential for thrombolytic therapy (Fletcher et al., 1962).

Existing assay methods can be divided into three major groups:

(1). The first group consists of indirect assay of UK activity with protein as substrates eg., the fibrin plate method (Haverkate and Bradman, 1975), clot lysis method (Lassen 1958) and caseinolytic methods (Kline, 1971). PAs cannot be assayed directly by these methods but only via their activating action on plasminogen present in or added to the substrate. With the fibrin plate method extremely low levels of enzymatic activity can be determined but long incubation times are required, and the responses found depend on the quality of the fibrin substrate (Haverkate and Bradman, 1975).

(2) A second group of assays, suitable for kinetic studies is based on the determination of hydrolysis of synthetic peptide esters (Bell et al., 1974). The disadvantage of the use of these esters is that esterolytic rather than amidolytic activity of a proteolytic enzyme is measured.

(3) In the third group of synthetic substrates e.g., Acetyl L-lysine p-nitroanilines, amidolysis of the peptide amides is measured (Petkov et al., 1973).

Fibrinolytic assays

The standard system used for measuring PA in cells is an indirect, two-step assay in which plasminogen is incubated with a source of PA and the plasmin activity generated is quantitated by using fibrin, casein or protamine as substrates (Unkeless et al., 1974; Goldberg, 1974; Kessner and Troll, 1976).

Marsh and Gaffney (1977) developed a rapid fibrin plate method for plasminogen activator assay. Their study was carried out to investigate plasminogen-enrichment as a means of shortening the incubation period, which is associated with the fibrin plate method. Fibrin plates were made up to contain 2 casein units of added plasminogen. Each was opaque, firm, did not lyse spontaneously and yielded biometrically valid parallel-line assays for SK and UK.

Jespersen and Astrup (1983) described the reproducebility, precision and required conditions of the firbin plate method for determination of fibrinolytic agents.

Under optimal conditions the assay is sensitive and precise method for the quantitative determination of firbinolytic agents.

Millar and Smith (1983) compared the rapid and highly sensitive solid phase assay with the fibrin plate method for the measurement of UK, SK and the PA in human euglobulin fractions. The solid phase assay was run using Glu- or Lys- plasminogen, and significant differences were observed in the activation of the plasminogen by UK and SK. Very good agreement was obtained between the fibrin plate and solid phase methods in all cases.

Fossum and Hoem (1996) developed a firbin micro-

plate method for the estimation of UK and non-UK fibrinolytic activity in protease-inhibitor-deprived plasma. In this method fibrin clots, with a suitable dye incorporated, were formed in wells of standard high adsorption micro-titer plates.

Roche et al. (1983) presented a rapid and highly sensitive solid-phase radio assay for the measurement of PAs. The method employs a convenient and stable ^{125}I – fibrinogen – latex bead product and can reproducibly detect 0.25 milli PU/ml of UK. This represents a 100-fold increase in sensitivity of UK over radio isotopic solid-phase technique and a 120-fold increase over the sensitivity of the fibrin plate method.

Esterolytic assays

Sherry et al. (1966) investigated the ability of UK to hydrolyze a variety of alpha amino substituted Arg and Lys esters [acetyl -arg methyl ester, benzoyl –arg methyl ester, tosyl –arg methyl ester, lys methyl ester, acetyl –lys methyl ester, benzoyl –lys methyl ester and tosyl –lys methyl ester]. Their observations indicated that UK catalyses a more rapid hydrolysis of lysine esters and its derivatives than the corresponding esters of arginine. Substitution of alpha amino group of lysine methyl esters increases the sensitivity of the ester to hydrolysis. They further reported that acetyl -lys methyl ester is the most sensitive substrate among the various esters tested.

A convenient and highly sensitive colorimetric assay for various proteases such as trypsin, chymotrypsin, plasmin, thrombin and UK was reported (Ninobe et al., 1980). The substrates used were naphthyl ester derivatives of N-tosyl L-lysine, N-acetyl glycyl L-lysine and N-acetyl L-tyrosine. Activity was assayed by colorimetric determination of naphthol released. They reported that this method was more sensitive than the use of corresponding methyl or ethyl ester derivatives.

Barlow and Marder (1980) reported the use of a chromogenic substrate L–pyroglutamyl glycyl L-arginine p-nitroanilide [S-2444] for assay of plasma UK levels of patients treated with urinary source or tissue culture source of UK. The p-nitroaniline released was measured in a spectrophotometer at 405nm. A linear response relationship between UK concentration and optical activity was obtained, indicating that the method detects UK in quantitative manner.

Kulseth and Helgeland (1993) developed a simple and highly sensitive chromogenic microplate assay for quantification of rat and human plasminogen in plasma samples and subcellular fractions. The assay is based on a conversion of plasminogen to plasmin, using UK as an activator, and a subsequent cleavage of chromogenic plasmin substrate D-alanyl-L-cyclohexylanyl-L-lysine-p-nitroanilide-dihydroacetate.

p-Nitroaniline being released by the cleavage is then measured at 410 nm with a microplate reader. The assay includes an acidification step to make plasminogen more readily activated to plasmin. The method is suitable for analyses of a large number of samples, measuring plasminogen in the nanogram range (0.5 - 50ng/50µl of sample).

Fluorimetric assays

Kessner and Troll (1976) reported a new method for determining plasminogen activator levels. The assay is based on the digestion of N-terminal blocked protamine and subsequent measurement of the exposed amino groups using the flurogenic amine reagent, Fluram.

Nieuwenhuizen et al. (1978) reported fluorigenic substrates for sensitive and differential estimation of UK and t-PA. Two fluorigenic peptide amides have been synthesized, i.e. Boc L-valyl-glycyl-L-arginine-L-naphthylamide and L-valyl-glycyl-L-arginine-2-naphthylamide. The kinetic parameters of plasmin, UK and human uterine tissue plasminogen activator on substrates 1 and 11 have been determined.

Zimmerman et al. (1978) developed a simple, sensitive, direct assay that allowed both rapid measurement and kinetic analysis of PA, independent of plasmin generation. The method employed a synthetic flurogenic peptide substrate 7-(N-CbZ-glycylglycyl argininamido)- 4-methyl coumarin trifluoro acetate. The assay correlated well with the standard ^{125}I-labeled fibrin plate assay using highly purified UK.

In vitro production of UK

The presence in urine of an activator substance capable of effecting transformation of plasminogen to plasmin was first described by Williams (1951) and in the following year by Sobel et al. (1952). The latter group has assigned the name UROKINASE to this activator.

Pepper et al. (1992) demonstrated that fibroblast – conditioned medium induces Madin-Darby canine kidney (MDCK) epithelial cells to form branching tubules when grown in three dimensional collagen or fibrin gels, and that this morphogenetic effect is mediated by hepatocyte growth factor (HGF), also known as scatter factor. In fibrin gels, this effect is inhibited by addition of exogenous serine protease inhibitors, which suggests a role for PAs in the matrix remodeling required for thrombogenesis. They investigated the effect of fibroblast-conditioned medium (CM) and HGF on the production of PAs by MDCK cells. They found that u-PA activity and mRNA were increased 4.9 fold by CM from human Detroit-550 fibroblasts, which lacks thrombogenic activity. The u-PA inductive property of MRC-5 CM was completely inhibited by preincubation with antibodies to recombinant human HGF (rhHGF). Exogenously added rhHGF also increased u-PA activity and mRNA 5.9 fold in MDCK cell, with an optimal effect at approximately 10 mg/ml. MRC-5 CM also increased u-PA receptor mRNA 34.9 fold in MDCK cells, an effect which was inhibited by 71% by preincubat-

ing the CM with antibodies to rhHGF, and which was mimicked by exogenously added rhHGF (31.3 fold increase). These results demonstrated that HGF, which induces thrombogenesis by MDCK cells in vitro, also increases u-PA and u-PA receptor expression in these cells. This suggests that the resulting increase in extracellular proteolysis, appropriately localized to the cell surface, is required for epithelial morphogenesis.

By the combined use of zymographies on tissue secretions *in situ* hybridizations, Sappino et al. (1991) explained the cellular distribution of u-PA and t-PA and of their mRNAs in developing adult mouse kidneys. In 17.5-day-old embryos, renal tubules synthesized u-PA while S-shaped bodies produced t-PA. In the adult kidney, u-PA is synthesized and released in urine by the epithelial cells during lining the straight parts of both proximal and distal tubules. In contrast t-PA is produced by glomerular cells and by epithelial cells lining the distal parts of collecting ducts. The precise segmental distribution of PAs suggested that both enzymes might be implicated in the maintenance of tubular potency, by catalyzing extracellular proteolysis to prevent or circumvent protein precipitation.

Valinskey et al. (1981) studied the association of controlled extracellular proteolysis mediated by PA with embryonic tissue remodeling and cell migration in the developing *Bursa fabricius* of quail and chick embryos.

Wojta et al. (1989) stated that vascular origin determines PA expression in human endothelial cells and renal endothelial cells produce large amounts of scu-PA.

Roychoudhury et al. (1999) carried out studies in T-flasks and bioreactor to produce UK using HT 1080 human kidney cell line. While growing the cell line it has been observed that the lag phase is reduced considerably in the bioreactor as compared to T-flask culture. The HT 1080 cell adhesion rate and UK production were observed to be the function of serum concentration in the medium. The maximum UK activity of 3.1×10^{-4} PU/ml was achieved in the bioreactor at around 65 h of batch culture. Since HT 1080 is an anchorage dependent cell line, therefore, the hydrodynamic effects on the cell line were investigated.

Podorolskaia et al. (1999) studied correlative intercomnections between PA activity (fibrin plate method) and level of UK antigen (Ag uAP) and tissue PA antigen (Ag tAP) in urine and blood (ELISA) in 60 patients with chronic glomerulonephritis (CGN) and 38 patients with amyloidosis. High degree of correlation r = +0.84 and P <0.001 was found between blood Ag uAP and urine Ag tAP in amyloidosis only.

Iwamoto et al. (1990) studied the effects of thrombin interleukin (IL-1), tumor necrosis factor (TNF) and gamma interferon (gamma-IFN) on the release of PA and inhibitor (PAI) using cultivated human glomerular epithetlial cells (GEC's). Their findings indicate that the GEC's participate in the regulation of extracapillary fibrinolysis in the glomerular environment being modulated by thrombin and cytokines IL-1 and TNF.

Lee et al. (1993) presented a method for determining the plasminogen activation rate by UK via a cascade enzymatic reaction system. A procedure of parameter estimation has been proposed for the determination of the activity of UK and the kinetic constants.

Schnyder et al. (1992) developed a spectrophotometric method to quantify and discriminate UK and t-PAs.

Leprince et al. (1989) developed a colorimetric assay for the simultaneous measurement of PAs and PAIs in serum-free conditioned media from cultured cells.

Ambrus et al. (1979) reported that Plasminogen-rich and plasminogen-poor radiolabeled human fibrin clots were inserted into large veins of baboons and stump-tailed monkeys. The thrombolytic effects of PAs (UK, SK), and plasmin preparations with activator activity (SK-activated human plasmin) and without activator activity (trypsin-activated porcine plasmin, Lysofibrin) were studied. Plasminogen-free and plasminogen-rich clots lysed at equal rates. Preparations with and without activator activity were equally effective as thrombolytic agents. Endogenous activation of plasminogen in the clot thus appears not to be the essential mechanism of thrombolysis. The exogenous pathway of enzyme adsorption to fibrin fibers seems to represent an important thrombolytic mechanism. Clot lysis was achieved with doses of fibrinolytic enzymes, which produced little or no significant hematologic changes including hypofibrinogenemia and decreases of other blood coagulation factor levels.

Jamet et al. (1978) reported that UK and SK transform plasminogen into plasmin by rupture of a Arg-Val bond and the liberation of a peptide with a molecular weight of 6 to 8-KDa. UK is a physiological activator with a direct action. By contrast, SK is an enzyme of bacterial origin and two hypotheses may be advanced to explain its mechanism of action: the formation of a SK-plasminogen complex capable of activating new molecules of plasminogen or the formation of a SK-plasminogen complex within which plasminogen is transformed to plasmin.

Kang et al. (1990) reported better pro-UK production at 5% serum as compared to 10% serum supplemented medium from a human kidney CAKI-1 cell line when cultured with cytodex microcarriers in a perfusion bioreactor. The medium can be supplemented with BSA, insulin, transferrin and selenium for preservation of viability in low serum for prolonged culture duration. Therefore the optimal serum concentration for UK production depends on cell type and media additives.

Khaparde and Roychoudhury (2004) reported that a 1% of serum is optimum for UK production as well as for viability of human kidney HT 1080 cells. Similarly, Tao et al. (1987) have used 1% of serum as optimum for UK production.

Lacroix and Fritz (1982) reported that the rate of production and release of UK were greatly influenced by a variety of factors including cell density, presence of hormones, incubation temperature and duration of culture.

Suzuki et al. (1989) reported the enhanced UK productivity of 1956 IU/ml when compared to 294 IU/ml in the controls by human normal diploid fibroblasts which were cultured in serum free medium containing phospholipase-A2, phospholipase-C, bradikinin, coenzyme-A and phytohaemagglutinin as inducers.

Chen et al. (1996) have formulated serum-free media by using orthogonal experiments for the growth of genetically engineered Chinese hamster ovary (CHO) cell line 11G and reported an increase of 80% in UK production.

Deutheux et al. (1997) examined the production of UK by culturing human diploid fibroblasts in a serum-free medium supplemented with peptones (protease peptone) and K^+ ions for efficient expression of human gene in *Escherichia coli*. In cultures of 3T3 fibroblasts, the UK activity increased by 20 fold when compared to control cells within 24 h when 100 µM of sulphur mustard (SM) was added. Also ryanodine (10 µM) amplified the UK upregulation by two fold and dexamethasone (1 µM) added directly after SM treatment almost completely prevented the induction of UK at both the protein and mRNA levels.

Jo et al. (1998) developed a serum-free perfusion culture for the production of UK. The cell-growth profile showed a continuous increase in cell density, reaching 5.1×10^7 cells/ml and the production of UK remained stable throughout the culture (1586 ± 247 IU/ml).

Gomes et al. (2000) reported a 10-fold increase in UK activity with simultaneous supplementation of three amino acids, based on their repeated occurrence in the normal pro-UK produced by human kidney cell line HT 1080 culture.

Recently Chen et al. (2004) developed a serum-free medium for the production of UK by adding insulin, a trace element mixture, a lipid mixture, ascorbic acid and pluronic F68 to dulbecco´s modified eagle´s medium (DMEM)/F12 (1:1, v/v).

Usually UK production by mammalian cells depends on the following factors: (i) regulation of UK expression (ii) supply of amino acid building blocks for UK synthesis. Moreover some amino acids like glycine have been known to bring about stabilization of proteins (Chainiotakis, 2004), while arginine is known to induce UK by acting as precursor of nitric oxide, which induces UK production (Ziche et al., 1997).

For UK induction, preferable compounds are saccharides such as glucose, inositol, ribose and deoxyribose, hormones such as adrenaline (Bansal and Roychoudhury, 2006; Bansal et al., 2007).

The enhanced production of PA activity was shown to be a characteristic of many malignant cell types. The intracellular and extracellular levels of PA were demonstrated to be substantially elevated in malignant cells (Christman et al., 1975), cells treated with a tumor promoter (Wigler and Weinstein, 1976), activated macrophages (Unkeless et al., 1974), established cell lines (Mott et al., 1974; Rifkun and Pollack, 1977), granulosa

Table 1. Production of UK using different recombinant cell lines.

Recombinant cell type	UK activity	Reference
Mouse cells LB6	0.8 mg/L/day	Nolli et al. [1989]
Chinese hamster ovary cell line	1000 PU/ml	Avgerinors et al. [1990]
Saccharomyces cerevisiae cells	1863 PU/ml	Turner et al. [1991]
Namalwa KJM-1	3 μg/10^6 cells/day	Satoh et al. [1993]
E. coli cells	1500 PU/ml	Tang et al. [1997]
Chinese hamster ovary cell line	860 PU/ml	Jo et al. [1998]

cells during ovulation (Strickland and Beers, 1976), embryonic cells during differentiation (Topp et al., 1976) and hormone treated cells (Katz et al., 1977).

Recombinant studies

Human UK can be used to treat acute thromboembolic events such as venous and arterial thrombosis, pulmonary embolism, intracardiac thrombosis, and systemic embolism. However, the high cost of isolation of UK from either tissue culture cells or urine limits the use of this enzyme as a therapeutic agent. If UK could be obtained from microorganisms by recombinant DNA technology, one might have a more economical method of production.

In order to improve UK production, UK has been expressed in bacteria (Tang et al., 1997; Fahey and Chaudhuri, 2000; Sun et al. 2003; Zhong et al., 2007; Beaton et al., 2005; Gurskii et al., 2005; Ratzkin et al., 1981; Deutheux et al., 1997), fungi (Hiramatsu et al., 1991), yeast (Wang et al., 2000; Hiramatsu et al., 1989, 1991; Turner et al., 1991), mammalian cells (Nelles et al., 1987; Hu et al., 2006; Innis and Scott, 2001), insect cells (Innis and Scott, 2001) and in plants (Oishi and Zhou, 2000). Different recombinant cell lines that have been used for production of UK are listed in Table 1. Large-scale culture of bacteria or fungi is relatively easy but the main drawback of using prokaryotic system for UK production is the absence of post-translational modification machinery in these organisms. Therefore, the non-glycosylated UK so obtained does not have the same efficacy and pharmacodynamic properties as that of native UK. Mammalian cell lines are hence, preferred for production of UK. A considerable number of mammalian cell lines have been reported to date for UK production.

Tang et al. (1997) developed a system to produce recombinant urokinase-type plasminogen activator (ru-PA) in *E. coli*. The u-PA was produced with 6^x His-tag at the C-terminus, which was shown to have the same activity, after refolding, as the wild- type protein.

Kohno et al. (1984) reported the establishment of a permanent cell line, TCL-598, which produces and secretes UK into the medium in smaller quantities.

Hiramatsu et al. (1991) described the production of human pro-UK and its deletion mutants in yeast. They succeeded in producing large amounts of human pro-UK and its mutants

as the core-glycosylated forms, which were mainly accumulated within yeast endoplasmic reticulum (6667 PU/ml of culture medium). But these accumulated pro-UKs were inactive in their native state and needed to be converted to a biologically active form by a denaturation-refolding procedure.

CHO cells are considered ideal hosts for recombinant UK production (Warner, 1999). These cells offer the advantages that they can be easily genetically manipulated, can be adapted for large-scale suspension culture and it can give rise to proteins with glycans which are similar, although not identical and to those found on human glycoproteins.

Interestingly, Kim and Swartz (2004) have illustrated that UK could be efficiently synthesized in *E. coli* based cell-free systems using glutathione redox buffer coupled with the disulphide isomerase to facilitate formation of disulphide bonds. Recently, Roychoudhury et al. (2006) reported that the CHO cell line is known for its unstable karyotype. Loss of recombinant gene copy number and appearance of non-producing populations of cells were predominant causes for instability of production.

Immobilization and bioreactor studies

Wagner et al. (1990) studied the production of pro-UK by a human kidney tumor cell line in long term cultures. Cells were grown on microcarriers, which were retained inside the reactor by sedimentation, or with a spin filter.

Two modes of operation were compared: feed harvest at an average medium exchange rate of 0.3 per day and continuous perfusion at a higher dilution rate of 1.5 per day. In the two systems a stable production of pro-UK could be maintained for more than 400 h. Continuous perfusion yielded a higher cell density than feed harvest resulting in a 2-fold increase in the reactor productivity. But higher final enzyme activities were obtained with harvest recovered medium than in the perfusion medium. The cumulative medium consumption for mass of product was the same in the repeated batch and in the continuous operation mode.

Kang et al. (1990) investigated kinetics of formation of UK from pro-UK in CM and worked for a possibility of lowering the conversion to UK cultivating human cell lines under perfusion operations. A human kidney cell line,

CAKI-1, was cultivated in DMEM with FBS, glutamine and gentamicin without Ca^{2+} to prevent clumping. Cells were then inoculated into a 2L bioreactor with microcarriers when cell density reached 1×10^6 viable cells/ml. It was observed that better production of pro-UK was obtained with 5% serum containing media than 10% or serum free medium on cytodex II microcarrier under perfusion chemostat operations. Conversion of pro-UK was reduced in the serum containing media.

Senatore and Bernath (1986) had immobilized UK to the inner surfaces of fibrocollagenous tubes (FCT) in an attempt to develop a fibrinolytic biomaterial, which may be suitable for use as a small diameter vascular prosthesis. The enzyme was bound by adsorption followed by glutaraldehyde cross-linking. An in vitro kinetic study of immobilized UK was conducted by employing the tubular material as a flow through reactor operated in a batch recycle mode in which the esterolysis of the model substrate, N-α-acetyl-L-lysine methyl ester (ALME), was monitored as a function of substrate concentration, recycle flow rate, and temperature. Results were compared with data from the soluble enzyme reaction, which was conducted in the presence and absence of 10% swine skin gelatin, in order to identify the specific effects of a collagenous microenvironment. Observed rates for the UK-FCT catalyzed reaction were observed to be dependent on recycle flow rates below 12 ml/min (Re = 107). Apparent Michaelis-Menton rate parameters were determined by nonlinear search technique for two flow rates: one above the critical point for external diffusion effects (Re = 282) and one with in the mass-transfer-limited region (Re = 71). When the later data were corrected for external diffusion by applying the Graetz correlation for laminar flow in tubes to estimate the mass transfer coefficient, the corrected K_m of 6.45 ± 0.38 mM agreed very closely with the diffusion free parameter (that is, 6.13 ± 0.63). Furthermore, this value was observed to be an order of magnitude higher than that of the soluble enzyme but approximately equal to the K_m of the soluble enzyme in a 10% gelatin environment (8.13 ± 1.53 mM). It is postulated that the difference in kinetic parameters between soluble and collagen immobilized UK is due to an inherent interaction between collagen and enzyme rather than to mass transfer effects. Such an interaction is supported by the effects of collagen on thermal stability and energy of activation.

Human fetal cells (HF) from explants of neonatal fore skins were cultured in DMEM containing calf serum and antibiotics. Microcarrier cultures of these cells were prepared and seeded into microcarrier beads and incubated with medium to grow to confluence (9 - 15 days). Serum free cultures were prepared by rinsing the beads with PBS, and adding medium containing ovalbumin and epidermal growth factor. After 48 h, phorbol myristate acetate was added and the cultures were incubated for an additional 24 h. The culture medium was then collected and fresh medium containing 10% fetal calf serum was

added (Eaton et al. 1984).

One of main problems in the cultivation of human and animal cells is the fact that most of these except a few malignant continuous cell lines are anchorage dependent for their growth in vitro. This means that they require for growth in vitro a suitable solid surface to which they can attach and spread. Hence standard fermentors cannot be applied for cultivation of these cells. Production of a variety of biomolecules required anchorage-dependent cells such as primary cells or diploid cell lines. Van Wezel (1967) discussed the various cultivation systems for these anchorage-dependent cells with special attention to the microcarrier culture system. In these systems, cells are grown on small particles suspended in culture medium by stirring. Cells attach and spread upon carriers and grow out gradually to a confluent monolayer. DEAE sephadex was used as a microcarrier.

Smidsrod and Skjak-Braek (1990) stated that in recent years, entrapment of cells within spheres of Ca^{2+} alginate has become the most widely used technique for immobilizing living cells. This versatile method includes applications ranging from immobilization of living or dead cells in bioreactors, immobilization of plant protoplasts for micropropagation and immobilization of hybridoma cells for production of monoclonal antibodies, to entrapment of animal cells for implantation of artificial organs. This review evaluates the potential of this method on the basis of the current knowledge of structural and functional relationships in alginate gels.

Avgerinors et al. (1990) used a 20 liter stirred tank fermentor, equipped with a 127 mesh ethylene tetraflouro ethylene rotating screen for cell recycle, for the continuous production of recombinant single chain urokinase-type plasminogen activator (rscu-PA) from CHO cells. Viable cell densities between 60 and 74 millions per ml were maintained at medium perfusion rates of 3 to 4 fermentor volumes per day. Cells were retained by the 120-microns nominal opening filter through the formation of "clumped" cell aggregates of 200 to 600 microns in size, which did not foul the filter. The rscu-PA produced over the course of this continuous culture was purified and characterized both in vitro and in vivo and shown to be comparable to natural scu-PA produced from the transformed human kidney cell line, TCH-598.

Tamponnet et al. (1992) immobilized primary cultivated rabbit articular chondrocytes in calcium alginate beads. Both free and entrapped cells were allowed to grow under normal conditions. After long-term immobilization, the cells still exhibited metabolic activities, patterns of division, synthesis and secretion of extracellular matrix macromolecules such as type II collagen and proteoglycans. After 38 days, immobilized rabbit articular chondrocytes predominantly expressed type II but not type I collagen. Thus, they maintained their cartilage phenotype. After bead lysis, harvested cells showed normal growth patterns when resuspended in culture medium. On the basis of these results, long-duration storage and large-scale production of extracellular matrix components are being investigated.

Kuo and Bjornsson (1993) developed a simple and sensitive method for the simultaneous determination of free t-PA and u-PA concentration in biological fluids using a solid-phase immuno assay. Microtiter plates were coated with polyclonal goat antibodies and incubated with PA standards or unknown samples. The absorbed PA's were then assayed by incubation with a mixture of plasminogen, poly-L-lysine, and the chromogenic substrate H-D-norleucylherahydrotyrisyllysine-P-nitroanilide. Free t-PA and free uPA were detectable in human plasma and urine and in conditioned media from different endothelial cell cultures.

Recently Kumar et al. (2006a) developed a novel type of cell culture device based on supermacroporous poly-acrylamide cryogel support with immobilized gelatin for continuous production of UK from human fibro sarcoma HT 1080 and human colon cancer HCT 116 cell lines. The anchorage dependent cells attached to the matrix within 4 - 6 h of inoculation and grew as a tissue sheet inside the cryogel matrix. Continuous UK secretion into the circulating medium was monitored as a parameter of growth and viability of cells inside the bioreactor. A high yield of viable cell count (3.0×10^9 cells/ml) was obtained after continuously running the cell culture reactor for 32 days, during which 152,600 PU of UK was obtained from 500 ml of culture medium. No morphological changes were observed on the cells eluted from the gelatin-cryogel support andre-cultured in normal cell culture flasks. While the cryogel matrix itself is biocompatible with cells, coupling of gelatin makes it particularly suitable for growing anchorage dependent cell lines.

The potential of a cryogel bioreactor as a tool for process development of mammalian cell culture has many advantages. By using cryogel scaffolds as disposables, one can get rid of the problem of contamination and additional expenditures on securing sterile safe guards at considerable extra costs, particularly for scale up processes for continuous runs.

Integrated production and purification of UK from bioreactor

One of the most successful approaches in improving the economy of bioprocesses is to reduce the number of steps involved, as each successive step amounts to considerable loss of the product.

Khaparde and Roychoudhury (2005) successfully developed a two-step integrated process for producing UK from HT 1080 cell line in a hollow fiber reactor to which a sterile 50 ml benzamidine-Sepharose affinity column was coupled for on-line separation of UK. Approximately 4.5×10^4 PU of UK was harvested per day from the integrated set-up continuously for more than 20 days.

Similarly, Kumar et al. (2006b) developed an integrated set-up of gelatin-pAAm cryogel bioreactor was further

connected to a pAAm cryogel column carrying Cu(II)-iminodiacetic acid (Cu(II)-IDA-pAAm cryogel), which had been optimized for the capture of UK from the conditioned medium of the cell lines. Thus an automated system was built, which integrated the features of a hollow fibre reactor with a chromatographic protein separation system. The UK was continuously captured by the Cu (II)-IDA-pAAm cryogel column and periodically recovered through elution cycles. The UK activity increased from 280 PU/mg in the culture fluid to 1300 PU/mg after recovery from the capture column which gave about 4.5 fold purification of the enzyme. The integrated bioreactor system operated continuously for 32 days during which no backpressure was observed because of the porous structure of the cryogel matrix. The enzyme eluted from the Cu(II)-IDA-pAAm cryogel capture column was further purified on benzamidine-Sepharose affinity column which gave a preparation of different forms of UK with activity of 13550 PU/mg of protein. This is one of the very few successful UK production strategies using mammalian cell lines.

Clinical applications of UK

UK is used clinically as a thrombolytic agent in the treatment of severe or massive deep venous thrombosis, pumonary embolism, myocardial infarction, and occluded intravenous or dialysis cannulas. Recently, Alteplase has replaced UK as a thrombolytic drug in infarctation.

The clinical use of u-PA is as a thrombolytic agent. However, it is also a prognostic marker for tumors and its structure is the basis for the design of new anti-cancer drugs and is used for the targeting of cytotoxic agents to u-PA receptor expressing cells (Alfano et al., 2005; Blasi and Carmeliet, 2002; Dano et al., 2005; Sidenius and Blasi, 2003).

Conclusions

Thrombolytic diseases are today a major cause of morbidity and mortality world wide. Fibrinolytic enzymes have apparent significance in thrombosis therapy. Therefore great attention has been directed towards a search for thrombolytic agents of various origins with particular reference to agents with more specificity and less toxicity. UK is one such enzyme without immunogenicity and cross reactivity when given to patients. Because of its clinical importance, UK has eluded the interest of researchers as obvious from the lack of reports relating to any kind of technological advancements for its production from in vitro cell culture. Developments in cell lines and bioprocess technology have made it possible to produce UK from in vitro cell culture. The main source of UK is from cell cultures of lung, heart and kidney tissues. These tissues are normally derived from mammalian source. Cloning of the UK genes followed by heterologous expression provide higher enzyme yields. Therefore, improving the productivity and reducing the production cost are the major goals for the current studies on the UK production.

New bioreactor designs based on supermacroporous cryogel matrices promise major advantages. In the integrated set-up the coupling of purification column with the production bioreactor provides efficient strategy for the production and purification of UK by reducing the process steps and leading to significant improvement in production.

ACKNOWLEDGMENTS

Indian Council of Medical Research (ICMR) is thanked for the senior research fellowship of Dr. R. Bhavani Devi and for the Council of Scientific and Industrial Research (CSIR), India of Drs. V. Saisha and A. Kunamneni.

REFERENCES

Abloundi FB, Hagan JJ (1957). Comparison of certain properties of human plasminogen and proactivator. Proc. Soc. Exper. Biol. 95: 195-200.

Alfano D, Franco P, Vocca I, Gambi N, Pisa V, Mancini A, Caputi M, Carriero MV, Iaccarino I, Stoppelli MP (2005). The urokinase plasminogen activator and its receptor: Role in cell growth and apoptosis. Thromb. Haemost. 93: 190–191.

Alkjaersig N, Fletcher AP, Sherry S (1958). The activation of human plasminogen. II. A kinetic study of activation with trypsin, urokinase, and streptokinase. J. Biol. Chem. 233: 86-90.

Ambrus JL, Weber FJ, Ambrus CM (1979). Mechanism of action of fibrinolytic enzymes in vivo. J. Med. 10: 99-119.

Avgerinors GC, Drapeau D, Socolow JS, Mao J, Hsiao K, Broeze RJ (1990). Spin filter perfusion system for high density cell culture: production of recombinant urinary type plasminogen activator in CHO cells. Biotechnol. 8: 54-58.

Bansal V, Roychoudhury PK, Ashok KA (2007). Urokinase separation from cell culture broth of a human kidney cell line. Int. J. Biol. Sci. 3: 64-70.

Bansal V, Roychoudhury PK (2006). Production and purification of urokinase. Protein Exp. Purif. 45: 1-14.

Barlow GH, Marder VJ (1980). Plasma urokinase levels measured by chromogenic assay after infusions of tissue culture or urinary source material. Throm. Res. 18: 431-437.

Beaton M, Mark S, Pineda-Lucena A (2005). Purification and refolding of human recombinant urokinase -type plasminogen activator for structure-based inhibitor design by protein NMR using a redox pair-containing refolding buffer. PCT International Application WO2005087917 A2.

Bell PH, Dziobkowski CT, Englert ME (1974). A sensitive fluorometric assay for plasminogen, plasmin and streptokinase. Anal. Biochem. 61: 200-208.

Bennart NV, Francis JC (1976). Mechanism of the urokinase catalyzed activation of human plasminogen. J. Biol. Chem. 251: 3906-3912.

Bernick MB, Kwaan HC (1967). Origin of fibrinolytic activity in cultures of human kidney. J. Lab. Clin. Med. 70: 650-661.

Bick RL (1992). In: Disorders of thrombosis and hemostasis: Clinical and Laboratory Practice. ASCP Press, USA.

Blasi F, Carmeliet P (2002). uPAR: A versatile signalling orchestrator. Nat. Rev. Mol. Cell Biol. 3: 932–943.

Castellino FJ (1984). Biochemistry of human plasminogen. Semin. Thromb. Hemost. 10: 18-23.

Castillino FJ (1981). Recent advances in the chemistry of the fibrinolytic system. Chem. Rev. 81: 431-446.Chainiotakis NA (2004). Enzyme stabilization strategies based on electrolytes and polyelectrolytes for biosensor applications. Anal. Bioanal. Chem. 378: 89–95.

Chen Z, Wu B, Jia X, Xiao C (1996). Study on serum-free media for genetically engineered CHO cells producing prourokinase. Chin. J. Biotechnol. 12: 169–175.

Chen Z, Wu B, Liu H, Liu X, Huang P (2004). Temperature shift as a process optimization step for the production of pro-urokinase by a recombinant Chinese hamster ovary cell line in high-density perfusion culture. J. Biosci. Bioeng. 97: 239–243.

Christman JK, Sulgi S, Newcomb EW, Silverstein SC (1975). Correlated suppression by 5-bromodeoxyuridine of tumorigenicity and plasminogen activator in mouse melanoma cells. Proc. Natl .Acad. Sci. USA 72: 47-50.

Collen D, Stump JM, Gold HK (1988). Thrombolytic therapy. Ann. Rev. Med. 39: 405-423.

Dano K, Behrendt N, Hoyer-Hansen G, Johnsen M, Lund LR, Ploug M, Romer J (2005). Plasminogen activation and cancer. Thromb. Haemost. 93: 676–681.

Deutheux M, Jijaksi H, Lisson D (1997). Effect of sulfur mustard on the expression of urokinase in cultured 3T3 fibroblasts. Arch. Toxicol. 71: 243-249.

Eaton DL, Scott RW, Baker JB (1984). Purification of human fibroblast urokinase proenzyme and analysis of its regulation by proteases and protease nexin. J. Biol. Chem. 259: 6241-6247.

Fahey EM, Chaudhuri PJB (2000). Binding, refolding and purification of a urokinase plasminogen activator fragment by chromatography. J. Chromatogr. B. 737: 225-235.

Fletcher AP, Alkjaersig N, Sherry S (1962). Fibrinolytic mechanisms and the development of thrombolytic therapy. Am. J. Med. 33: 738-752.

Fossum S, Hoem NO (1996). Urokinase and non-urokinase fibrinolytic activity in protease-inhibitor-deprived plasma, assayed by a fibrin micro-plate method. Immuno. Pharmacol. 32: 119-121.

Francis CW, Marder VJ (1994). Physiological regulation and pathologic disorders of fibrinolysis, Hemostasis and Thrombosis. In R.W. Colman, J. Hirsh, V.J. Marder, W. Edwin, B. Salzman (eds) J.B. Lippincott Company, Philadelphia, p. 1077.

Goldberg AR (1974). Increased protease levels in transformed cells: a casein overlay assay for the detection of plasminogen activator production. Cell 2: 95-102.

Gomes G, Gayatri, Roychoudhury PK (2000). Urokinase production using human kidney cell line—effect of amino acid supplementation. Chemcon, p. 101–104 BIO.

Gurskii YG, Belogurov AA, Bibilashvili RS, Grigoreva NV, Delver EP, Minashkin MM (2005). Cloning and recombinant expression of a human urokinase -type plasminogen activator variant with reduced PAI-1 sensitivity for the use as a thrombolytic. Russian Patent RU2247777 C2.

Haverkate F, Bradman P (1975). In, Progress in chemical fibrinolysis and thrombolysis, Vol. 1, P. 151, ed. Raven Press, NY.

Hiramatsu R, Horinouchi S, Beppu T (1991). Isolation and characterization of human pro-urokinase and its mutants accumulated within the yeast secretory pathway. Gene 99: 235-241.

Hiramatsu T, Kasai S, Hirose M, Tsukada M, Tanabe T, Kamimura Y, Kawabe H, Arimura H (1989). Purification of human urokinase analog from cultures of recombinant Saccharomyces cerevisiae. Japanese Patent JP 01252283 A.

Hu X, Gao L, Chen H, Pan S, Yin L, Yang B, Zhang Z (2006). Human low-molecular weight urokinase mutant and its expression vector. Chinese Patent CN1880449 A.

Innis M, Scott E (2001). Transfection system and expression vectors used for recombinant production of heterologous proteins (such as CAB-2, CAB-4, uPAR, VEGF-D or viral glycoprotein) in mammalian and/or insect cells. PCT International Application WO2001032901 A1.

Iwamoto T, Nakshima Y, Suseishi K (1990). Secretion of plasminogen activator and its inhibitor by glomerular epithelial cells. Kidney Int. 37: 1466-1476.

Jamet M, Granthil C, Levy G (1978). Enzymatic fibrinolytic agents. Ann. Anesthesiol. Fr.19: 693-696.

Jespersen J, Astrup T (1983). A study of the fibrin plate assay of fibrinolytic agents. Optimal conditions, reproducibility and precision. Haemost. 13: 301–315.

Jo EC, Yun JW, Jung KH, Chung SI, Kim JH (1998). Performance study of perfusion cultures for the production of single-chain urokinase-type plasminogen activator (sc-uPA) in a 2.5 L spin-filter bioreactor. Bioprocess Eng. 19: 363-372.

Kang JK, Choi SK, Lee HY (1990). Kinetics of producing pro-urokinase in perfusion cultures. Biotechnol. Letts. 12: 173-178.

Katz J, Troll W, Adler SW, Levitz M (1977). Antipain and leupeptin restrict uterine DNA synthesis and function in mice. Proc. Natl. Acad. Sci. 74: 3754-3757.

Kessner A, Troll W (1976). Fluorometric microassay of plasminogen activators. Arch. Biochem. Biophys. 176: 411-416.

Khaparde SS, Roychoudhury PK (2005). Effect of temperature shift on urokinase production in hollow fiber bioreactor. Indian Chemical Engineering Conference. Tech. Ses. Transcript. 2: 255.

Khaparde SS, Roychoudhury PK (2004). Production of urokinase using hollow fiber bioreactor, in: Indian Chemical Engineering Conference. Tech. Ses. Transcript. 6: 11.

Kim DM, Swartz JR (2004). Efficient production of a bioactive, multiple disulfide-bonded protein using modified extracts of Escherichia coli. Biotechnol. Bioeng. 85: 122-129.

Kline DI, Fishman JB (1961). Improved procedure for the isolation of human plasminogen. J. Biol. Chem. 236: 3232-3234.

Kline DL (1971). In, Thrombosis and bleeding disorders, p. 358., ed. Academic Press, NY.

Kohno T, Hopper P, Lillquist JS, Suddith RL, Greenlee R, Moir DT (1984). Kidney plasminogen activator: A precursor form of human urokinase with high fibrin affinity. Biotechnol. 2: 628-634.

Kulseth MA, Helgeland LA (1993). Highly sensitive chromogenic microplate assay for quantification of rat and human plasminógen. Anal. Biochem. 210: 314-317.

Kumar A, Bansal V, Andersson J, Roychoudhury PK, Mattiasson B (2006b). Supermacroporous cryogel matrix for integrated protein isolation: immobilized metal affinity chromatography purification of urokinase from cell culture broth of a human kidney cell line. J. Chromatogr. A. 1103: 35–42.

Kumar A, Bansal V, Nandakumar KS, Galaev I, Roychoudhury PK, Holmdahl R, Mattiasson B (2006a). Integrated bioprocess for the production and isolation of urokinase from animal cell culture using supermacroporous cryogels. Biotechnol. Bioeng. 93: 636-646.

Kunamneni A, Ravuri BD, Saisha V, Ellaiah P, Prabhakhar T (2008). Urokinase- a very popular cardiovascular agent. Recent Patent Cardiovasc Drug Disc 3: 45-58.

Kuo BS, Bjornsson TD (1993). Simultaneous determination of free tissue-type and free urokinase-type plasminogen activators in biological fluids by a solid-phase immunoassay. Anal. Biochem. 209: 70-78.

Lacroix M, Fritz IB (1982). The control of the synthesis and secretion of plasminogen activator by rat sertoli cells in culture. Mol. Cell Endocrinol. 26: 247-258.

Lassen M (1958). The estimation of fibrinolytic components by means of the lysis time method. Scand J. Clin. Lab. Invest. 10: 384-389.

Lee WC, Chuang CY, Chang RM (1993). Plasminogen activation activity of urokinase determined via a cascade enzymatic reaction system. Biochem. Mol. Biol. Int. 29: 1039-1046.

Leprince P, Rogister B, Moonen G (1989). A colorimetric assay for the simultaneous measurement of plasminogen activators and plasminogen activator inhibitors in serum-free conditioned media from cultured cells. Anal. Biochem. 177: 341-346.

Lesuk A, Terminicko L, Traver JH, Groff JL (1967). Biochemical and biophysical studies on human urokinase. Thromb. Diath. Hemorrh. 18: 293-302.

Marsh NA, Gaffney NJ (1977). The rapid fibrin plate-a method for plasminogen activator assay. Thromb. Haemostat. 38: 545-551.

Millar WT, Smith JF (1983). The comparison of solid phase and fibrin plate methods for the measurement of plasminogen activators. Thromb. Res. 30: 431-439.

Mott DM, Tabisch PN, Sanc BP, Sorof S (1974). Lack of correlation between fibrinolysis and the transformed state of cultured mammalian cells. Biochem. Biophy. Res. Comm. 61: 621-627.

Nelles L, Lijnen HR, Collen D, Holmes WE (1987). Characterization of recombinant human single chain urokinase-type plasminogen activator mutants produced by site-specific mutagenesis of lysine. J. Biol. Chem. 262: 5682-5689.

Nieuwenhuizen W, Wijngaards G, Groeneverd E (1978). Fluorogenic substrates for sensitive and differential estimation of urokinase and

tissue plasminogen activator. Haemost 7: 146-149.

Ninobe M, Hitomi Y, Fujii S (1980). A sensitive colorimetric assay for various proteases using naphthyl ester derivatives as substrates. J. Biochem. 87: 779-783.

Oishi KK, Zhou D (2000). Production of biological active recombinant urokinase in plant expression systems. PCT International Application WO2000000624 A1.

Pepper MS, Metsumoto K, Nakamura T, Orci L, montesano R, Montesano R (1992). Hepatocyte growth factor increases urokinase-type plasminogen activator (u-PA) and u-PA receptor expression in Madin-Darby canine kidney epithelial cells. J. Biol. Chem. 267: 20493 –20496.

Petkov D, Christova E, Karadjova M (1973). Amidase activity of urokinase. I. Hydrolysis of alpha-N-acetyl-L-lysine p-nitroanilide. Thromb. Diath. Haemorrh. 29: 276-285.

Podorolskaia LV, Andreenko GV, Poliantsva LR, Bumblite ID (1999). Fibrinolytic activity of the urine during chronic glomerulonephritis and amyloidosis. Vopr. Med. Khim. 45: 158-164.

Ratzkin B, Lee SG, Schrenk WJ, Roychoudhury R, Chen M, Hamilton TA, Hung PP (1981). Expression in Escherichia coli of biologically active enzyme by a DNA sequence coding for the human plasminogen activator urokinase. PNAS 78: 3313-3317.

Rifkun DB, Pollack B (1977). Production of plasminogen activator by established cell lines of mouse origin. J. Cell Biol. 73: 47-55.

Roche PL, Compeau JD, Schaw ST (1983). A rapid and highly sensitive solid-phase radioassay for plasminogen activators. Thromb. Res. 31: 269-277.

Roychoudhury PK, Gomes J, Bhattacharyay SK, Abdullah N (1999). Production of urokinase by HT1080 human kidney cell line. Artif. Cells Blood Substit. Immobil. Biotechnol. 27: 399-402.

Roychoudhury PK, Kharparde SS, Mattiasson B, Kumar A (2006). Synthesis, regulation and production of urokinase using mammalian cell culture: A comprehensive review. Biotechnol. Adv. 24: 514-528.

Sappino AP, Huarte J, Vassalli JD, Belin D (1991). Sites of synthesis of urokinase and tissue-type plasminogen activators in the murine kidney. J. Clinical Invest. 87: 962-970.

Schnyder J, Marti R, Cooper PH, Payne TG (1992). Spectrophotometric method to quantify and discriminate urokinase and tissue-type plasminogen activators. Anal. Biochem. 200: 156-162.

Senatore FF, Bernath FR (1986). Urokinase bound to fibrocollagenous tubes: an in vitro kinetic study. Biotechnol. Bioengg. 28: 58-63.

Sherry S, Troll W (1954). The action of thrombin on synthetic substrates. J. Biol. Chem. 208: 95-105.

Sherry S, Alkjaersig N, Fletcher AP (1966). Activity of plasmin and streptokinase-activator on substituted arginine and lysine esters. Throm. Diath. Haemorrh. 16: 18-31.

Sidenius N, Blasi F (2003). The urokinase plasminogen activator system in cancer: Recent advances and implication for prognosis and therapy. Cancer Metast. Rev. 22: 205–222.

Smidsrod O, Skjak-Braek G (1990). Alginate as immobilization matrix for cells. Trends Biotechnol. 8: 71-78.

Sobel GW, Mohler SR, Jones WW, Dawdy AB, Guest MM (1952). Urokinase: an activator of plasma profibrinolysin extracted from urine. Am. J. Physiol. 171: 768-769.

Strickland S, Beers WH (1976). Studies on the role of plasminogen activator on ovulation: in vitro response of granulose cells to gonadotropins, cyclic nucleotides and prostaglandins. J. Biol. Chem. 251: 5694-5702.

Sun S, Yan H, Chen R (2003). Chimeric protein mA5UKB comprising annexin A5 mutant mAnxA5 and urokinase B chain for use as bifunctional anticoagulant and thrombolytic agent. Chinese Patent CN1401662 A.

Tamponnet C, Ramdi H, Guyot JB, Lievremont M (1992). Rabbit articular chondrocytes in alginate gel: characterisation of immobilized preparations and potential applications. Appl. Microbiol. Biotechnol. 37: 311-315.

Tang W, Sun Z, Pannell R, Gurewich V, Liu JN (1997). An efficient system for production of recombinant urokinase-type plasminogen Activator. Protein Exp. Purif. 11: 279-283.

Tao TY, Bohn MA, Ji GY, Einsele A, Hu WS (1987). Kinetics of prourokinase production by human kidney cells in culture. J. Biotechnol. 6: 205–224.

Topp W, Hall JD, Massen N, AK. Teresky, D Rifkin, AJ. Levine, R. Pollack (1976). *In vitro* differentiation of teratomas and the distribution of creatine phosphokinase and plasminogen activator in terato-carcinoma-derived cells. Cancer Res. 36: 4217-4223.

Troll W, Sherry S, Wachman J (1954). The action of plasmin on synthetic substrates. J. Biol. Chem. 209: 80.

Turner BG, Avgerinos GC, Melnick LM, Moir DT (1991). Optimization of pro-urokinase secretion from recombinant *Saccharomyces cerevisiae*. Biotechnol. Bioengg. 37: 869-875.

Unkeless J, Gardon S, Reich E (1974). Secretion of plasminogen activator by stimulated macrophages. J. Expt. Med. 139: 834-850.

Valinskey E, Reich E, Le Douarin NM (1981). Plasminogen activator in the bursa of fabricius: correlations with morphogenetic remodeling and cell migrations. Cell 25: 471-476.

Van Wezel AL (1967). Growth of cell strains and primary cells on microcarriers in homogeneous cultures. Nat. 216: 64-65.

Wagner A, Marc A, Engasser JM (1990). Continuous production of prourokinase in feed harvest and perfusion cultures. Biotechnol. Bioengg. 36: 623-629.

Wang P, Sun Z, Chen H, Liu JN (2000). Glycosylation of pro-urokinase by *Pichia pastoris* impairs enzymatic activity but not secretion. Protein Expr. Purif. 20: 179-185.

Warner TG (1999). Enhancing therapeutic glycoprotein production in Chinese hamster ovary cells by metabolic engineering endogenous gene control with antisense DNA and gene targeting. Glycobiol. 9: 841-850.

Wigler M, Weinstein IB (1976). Tumor promoter induces plasminogen activator. Nat. 259: 232–233.

Williams JRB (1951). The fibrinolytic activity of urine. Br. J. Exp. Pathol. 32: 530-537.

Wojta J, Hoover RL, Daniel TO (1989). Vascular origin determines plasinogen activator expression in human endothelial cells: renal endothelial cells produce large amounts of scu-PA?. J. Biol. Chem. 264: 2846-2852.

Wu KK, Thaigarajan P (1996). Role of endothelium in thrombosis and hemostasis. Annu. Rev. Med. 47: 315-331.

Zhong J, Liang X, Huan L (2007). Method for expressing nattokinase with specific expression vector and engineering bacteria. Chinese Patent CN 1900289 A.

Ziche M, Parenti A, Ledda F, Dell'Era P, Granger HJ, Maggi CA, Presta M (1997). Nitric oxide promotes proliferation and plasminogen activator production by coronary venular endothelium through endogenous bFGF. Circ. Res. 80: 845–852.

Zimmerman M, Quigley JP, Ashe B, Dron C, Goldfarb R, Troll W (1978). Direct fluorescent assay of urokinase and plasminogen activators of normal and malignant cells: Kinetics and inhibitor profiles. Proc. Nalt. Acad. Sci. 75: 750-753.

17

Statistical analysis of the application of Wilcoxon and Mann-Whitney U test in medical research studies

U. M. Okeh

Department of Industrial Mathematics and Applied Statistics, Ebonyi State University, Abakaliki Nigeria. E-mail: umokeh1@yahoo.com.

Although non-normal data are widespread in biomedical research, parametric tests unnecessarily, predominate in statistical analyses. Five biomedical journals were surveyed and for all studies which contain at least the unpaired t-test or the non-parametric Wilcoxon and Mann-Whitney U test - investigated the relationship between the choice of a statistical test and other variables such as type of journal, sample size, randomization, sponsoring etc. The non-parametric Wilcoxon and Mann-Whitney U were used in 30% of the studies. In a multivariable logistic regression the type of journal, the test object, the scale of measurement and the statistical software were significant. The non-parametric test was more common in case of non-continuous data, in high-impact journals, in studies in humans, and when the statistical software is specified, in particular when SPSS was used.

Key words: Wilcoxon and Mann-Whitney U test, univariate analyses, non-parametric test, logistic regression.

INTRODUCTION

When looking into the medical literature one gets the impression that parametric statistical methods such as Student's t-test are common standard, although the underlying normal assumption is often not tenable, especially for small or moderate sample sizes. On the one hand, empirical work has shown that deviations from a normal distribution are frequent even for continuous data (Micceri, 1989). According to Nanna and Sawilowsky (1998), normality is the exception rather than the norm in applied research. However, for large sample sizes one may rely on the central limit theorem and apply a test designed for normally distributed data. On the other hand, ordinal data are widespread in biomedical research (Rabbee et al., 2003). For such data non-parametric tests based on ranks are appropriate, but the statistical analysis is often not performed properly, as shown e.g. by Jakobsson (2004) for the analysis of ordinal data in nursing research. Sometimes a transformation is applied in order to normalize continuous, but non-normal data. However, in case of non-normal data it is preferable to perform a nonparametric test. Transformations can often not be applied since the transformation "must be motivated from previous experimental or scientific evidence. Unless determined a priori, transforms can be misused to inflate or mitigate observed significance in a spurious fashion" (Piegorsch and Bailer, 1997 p. 130). Further-more, the hypotheses before and after the transformation

may differ (Games, 1984). Hence, the use of transformations for the sole purpose of complying with the assumptions of parametric tests is dangerous (Wilson, 2007). Investigation was made on how frequent the t-test and its nonparametric competitor, the Wilcoxon and Mann-Whitney (WMW) U test, are used in medical research. It is enquired which factors and variables are important for the choice between the non-parametric WMW test and the parametric t-test for research that compare two independent groups, published in medical journals with different scopes and impact. It will be discussed whether the decision for one of the methods is appropriate or not.

METHODS

All original work related to medical studies published in 2004 in five biomedical journals was surveyed. The three journals American Journal of Physiology (Heart Circ. Physiol.), Annals of Surgery, and Circulation Research were considered because they were also included in a previous study (Ludbrook and Dudley, 1998). In addition, The Lancet and The New England Journal of Medicine were included in my study. These journals were categorized into two groups with different topics and impact factors (Table 1). Each paper was thoroughly checked, on whether it included original material on not yet published data, irrespective of medical subject, study design or size/format of the paper. For the analyses presented here all research, which contain at least the unpaired t-

Table 1. Included journals and number of studies.

Journal	Surveyed studies	Included studies	Impact factor (2004)	Type of journal
American Journal of Physiology (Heart Circ. Physiology.)	645	251(38.9%)	3.5	Primarily specialized in subject
Annals of Surgery	227	83(36.6%)	5.9	,,
Circulation Research	343	143(41.7%)	10.0	,,
The Lancet	391		21.7	Articles of diverse topics
The New England Journal of Medicine			38.6	,,

test or the WMW test, were included. In addition to the test statistic the following factors and variables were also inspected: type of journal, sample size, kind of test objects, scale of measurements, information about randomization, sponsoring by pharmaceutical companies, and the used statistical software. Analyses were performed with logistic regressions. When the software used for analysis cannot perform both the t-test and the WMW test the respective study was excluded from the logistic regression analysis. The total sample size was categorized into three categories with an approx. equal number of research studies (<15, $15 - <50$, ≥ 50). Odds ratios (OR) and their 95% confidence intervals (95% -CI) were estimated by logistic regressions. A p-value ≤ 0.05 was considered as significant. Because of the exploratory nature of my research no multiplicity adjustment was applied (Neuhäuser, 2006).

RESULTS

In total, 1879 publications were surveyed, and 630 research studies could be included in the analyses (Table 1). Altogether the use of the unpaired t-test predominates in studies where two groups were compared. In 112 studies (18%) only the WMW and in 444 studies (70%) only the unpaired t-test is used; 74 times (12%) both tests are applied within one study. Please note that the two tests may be used to analyse different variables, however, it was also found that identical variables were analysed with both tests. In the logistic regressions presented below the studies without the WMW test are compared with the research studies with the WMW test. Two of the 630 studies were excluded from the logistic regression analyses because the specified software cannot perform the WMW test. The univariate analyses show significant relationships between the use of the WMW test and the journal type. The WMW test is more common in the diverse and high-impact journals The New England Journal of Medicine and The Lancet ($p \leq 0.001$, OR=5.21, 95% -CI: 3.53 - 7.69). Moreover, the WMW test is more common in studies in humans ($p \leq 0.001$, OR = 6.44, 95% -CI: 4.42 - 9.38), and, not surprisingly, in research studies with non-continuous variables ($p \leq 0.001$, OR = 8.49, 95% -CI: 4.73 - 15.27). In addition, the statistical software used is significantly related to the choice between the two statistical tests ($p \leq 0.001$). In particular, the WMW test is more common when one of the two common software packages SPSS ($p = 0.004$,

OR = 4.64, 95% -CI: 2.48 - 8.69) and SAS ($p = 0.030$, OR = 4.34, 95% -CI 1.96 - 9.61) is used. Another significant relationship was found regarding information about randomization ($p \leq 0.001$, Odds Ratio (OR) = 2.44, 95% -Confidence Interval (CI): 1.70 - 3.50). The WMW test seems to be more common when the study is sponsored by a pharmaceutical company ($p=0.028$, OR=2.32, 95% -CI: 1.10 - 4.90). The sample size was also significant in the univariate logistic regression ($p \leq 0.001$). In particular, the WMW test was applied more often in case of large samples (that is, $n \geq 50$) than in case of small samples (that is $n < 15$) ($p = 0.001$, OR = 5.88, 95% -CI: 3.68 - 9.39). Obviously, the different factors are not independent. Therefore, a multivariable logistic regression was applied in order to confirm the univariate results. The type of journal, the test object (research studies in humans or in other subjects), the scale of measurement (continuous or not) and the statistical software used remained significant (Table 2). The factors randomization, sponsoring and the categorized sample size are no longer significant. With regard to the software, SAS is no longer significant, either. The multivariate regression gives a significantly larger probability for performing the WMW test for SPSS, only. Sometimes, to be precise, in 57 studies, a reason is specified for using the WMW test. The most common reasons are "non-normal data" and "categorical data". Further correct reasons are "requirements for t-test not fulfilled" and "small sample sizes". However, the latter reason is correct only when applying the exact (permutation) version of the WMW test. There are also reasons that are problematic from a statistical point of view: In four research studies the WMW test was applied before or after the t-test, at least partly because the t-test was not significant. In one further study the WMW test was used because an observed heterogeneity in variances. However, the WMW test cannot guarantee the significance level in case of unequal variances (Kasuya, 2001). Moreover, the specified reason "in order to compare medians" is correct only if a pure location shift between the two distributions can be assumed. As mentioned above, one may rely on the central limit theorem when sample sizes are large and, consequently, one may apply a parametric test such as the t-test. However, in 395 out of the considered 630 research stu-

Table 2. Results of the univariate and multivariable logistic regressions.

Factor	Reference category	Univariate analysis			Multivariate analysis (n = 590)		
		n	p-value	OR	95%-CI	p-value	OR	95%-CI
Total sample size	<15	594	≤0.001			0.731		
15-<50			0.058	1.63	0.98-2.71	0.705	1.03	0.59-1.80
≤50			≤0.001	5.88	3.68-9.39	0.429	1.27	0.66-2.45
Randomization	No random or not specified	628	≤0.001	2.44	1.70-3.50	0.313	1.25	0.81-1.95
Sponsoring	No sponsoring[1]	628	0.028	2.32	1.10-4.90	0.631	0.80	0.33-1.98
Type of subject	Other than humans	624	≤0.001	6.44	4.42-9.38	0.012	2.08	1.18-3.67
Software used	Not specified	628	≤0.001			0.008		
SAS			0.030	4.34	1.96-9.61	0.528	2.08	0.86-5.06
SPSS			0.004	4.64	2.48-8.69	0.027	3.18	1.55-6.53
Other			0.007	1.19	0.70-2.04	0.175	1.18	0.65-2.17
Scale of measurement	Not only continuous variable	628	≤0.001	0.12	0.07-0.21	≤0.001	0.26	0.13-0.50
Journal Type	Primary specialized	628	≤0.001	5.21	3.53-7.69	0.004	2.25	1.30-3.87

[1]No sponsoring by a pharmaceutical company.

Table 3. Frequencies of study subject by scale of measurement.

		Study subject	
		Human	Other test objects
Scale of measurements	Only continuous variables	162	401
	Not only continuous variables	60	3

dies the (total) sample size is less than 50. In 89% (353) of these research studies with low sample size the t-test was applied, sometimes in addition to the WMW test (34 studies). In the remaining 319 studies with low sample size the t-test, but not the WMW test, was used. However, in 317 out of these 319 studies (99%) there are continuous variables. Hence, given the relatively high robustness of the t-test to skew continuous distributions (Posten, 1978), the basic assumptions seem to be fulfilled in the vast majority of studies when applying the t-test. In case of more than two groups the Kruskal-Wallis test can be applied as a non-parametric test instead of the WMW test. When considering the 1879 surveyed publications the Kruskal-Wallis test was applied in 53 research studies. Many of these studies have a low sample size smaller than 50 (23 studies) and/or non-continuous data (18 studies). The parametric analogue, an analysis of variance (ANOVA), was found in 658 studies. However, these 658 studies cannot be compared with the 53 studies with a Kruskal-Wallis test because an ANOVA is much more flexible than the Kruskal-Wallis test and can also be applied in studies with more complex designs.

DISCUSSION

The assertions some authors made about their decisions

for the WMW and the attributes of the published data indicate that the scale of measurement is the primary factor for a decision in favour of a non-parametric test. However, there are three further factors that remained significant in the multivariable logistic regression.

The study subject is one of these significant factors. The WMW test is more often used in studies in humans. However, in these studies non-continuous variables are more common as well (Table 3). Furthermore, the software has a significant influence. A further significant factor is the type of journal. A possible explanation is that the high-impact journals have a more detailed statistical review and that they may reject a paper because of an inappropriate statistical analysis. In line with this, studies published in journals with high impact factors often contain a more detailed methodical description compared to studies published in other journals. Please note in this context that The New England Journal of Medicine says in its instructions for authors that "nonparametric methods should be used to compare groups when the distribution of the dependent variable is not normal" (http://authors.nejm.org/help/newms.asp). In addition to The Lancet and The New England Journal of Medicine we included the three journals American Journal of Physiology (Heart Circ. Physiol.), Annals of Surgery, and Circulation Research in our study. These three latter journals were also included in a previous study (Ludbrook J, Dudley H (1998). This sample of five journals is not necessarily representative for the multitude of biomedical journals. However, we are able to compare our results towards the work of Ludbrook and Dudley (1998). This comparison indicates that the behaviour of medical scientists with parametric and non-parametric tests did not change considerably. Ludbrook and Dudley's [8] findings about the handling with statistical methods can be approved even ten years later. Given the higher efficiency of non-parametric tests for non-normal data (Lehmann, 1975), non-parametric tests such as the WMW test should be applied more often, especially when the sample size is not very large. In other areas of life sciences the WMW test seems to be more common. Ruxton (2006) surveyed one volume of the journal Behavioral Ecology. The WMW test was applied in 21/33 = 64% of the papers that used the two-sample t-test and/or the WMW test.

REFERENCES

Games PA (1984). Data transformation, power, and skew: a rebuttal to Levine and Dunlap. Psychol. Bull. 95: 345-7.

Jakobsson U (2004). Statistical presentation and analysis of ordinal data in nursing research. Scand. J. Caring. Sci.18(4): 437- 40.

Kasuya E (2001). Mann-Whitney U test when variances are unequal. Anim. Behav. 61(6): 1247-1249.

Lehmann EL (1975). Non-parametrics: Statistical methods based on ranks. San Francisco, CA: Holden-Day.

Ludbrook J, Dudley H (1998). Why permutation tests are superior to T and F tests in biomedical research. Am. Stat. 52(2): 127-32.

Micceri T (1989). The unicorn, the normal curve, and other improbable creatures. Psychol. Bull.105: 156-66.

Nanna MJ, Sawilowsky SS (1998). Analysis of Likert scale data in disability and medical rehabilitation research. Psychol. Methods. 3: 55-67.

Neuhäuser M (2006). How to deal with multiple endpoints in clinical trials. Fundam Clin. Pharmacol. 20(6): 515-23.

Piegorsch WW, Bailer AJ (1997). Statistics for environmental biology and toxicology. London, England: Chapman and Hall.

Posten HO (1978). The robustness of the two-sample t-test over the Pearson system. J. Stat .Comput. Simul. 6: 295-311.

Rabbee N, Coull BA, Mehta C, Patel N, Senchaudhuri P (2003). Power and sample size for ordered categorical data. Stat. Methods Med. Res. 12(1): 73-84.

Ruxton GD (2006). The unequal variance t-test is an underused alternative to Student's t-test and the Mann-Whitney U test. Behav. Ecol. 17(4): 688-90.

Wilson JB (2007). Priorities in statistics, the sensitive feet of elephants and don't transform data. Folia Geobot. 42: 161-7.

18

Probing yeast for insights into neurodegenerative disease: ORFeome-wide screens for genetic modifiers of α-synuclein cytotoxicity

Richard A. Manfready[1,2]

[1] Department of Biology, Massachusetts Institute of Technology, Cambridge, MA, USA.
[2]Whitehead Institute for Biomedical Research, Cambridge, MA, USA. E-mail: rman@mit.edu.

Several of the most devastating neurodegenerative disorders, including Parkinson's disease and dementia with Lewy bodies, belong to the synucleinopathy class of common neural disorders. A synucleinopathy is characterized by brain tissue plaques formed by the aggregation of misfolded protein—mainly misfolded α-synuclein. α-Synuclein has been extensively studied as the primary protein aggregate in brains afflicted by Parkinson's disease, but the toxic mechanism in which it is involved remains largely enigmatic. Fortunately, a simple but innovative yeast model of synucleinopathy has made possible high-throughput screens for genetic modifiers of α-synuclein toxicity. Deftly interpreted through the use of computational algorithms, these screens could reveal the genetic regulatory networks that underlie synuclein toxicity *in vivo*, and may enable therapeutic strategies to target the genetic root of neurodegeneration.

Key words: Parkinson's disease, α-synuclein, high-throughput screen, synucleinopathy.

INTRODUCTION

Protein misfolding and aggregation, problems typically associated with neurodegenerative diseases such as Creutzfeldt-Jakob disease and bovine spongiform encephalopathy, may also be responsible for the synucleinopathy class of neurodegenerative disorders in humans. One of the most clinically pervasive synucleino-pathies is Parkinson's disease (PD). About 1 - 2% of the general population over the age of 65 is thought to be affected by PD (Goedert, 2001). Since its first description by James Parkinson in 1817, PD and related disorders identified as Parkinsonisms have been diagnosed as movement disorders typified by muscle rigidity, resting tremor, and bradykinesia (Goedert, 2001). More recently, PD has been discovered to result from the selective degradation of dopaminergic neurons in the substantia nigra pars compacta, part of the midbrain basal ganglia that is associated with motor control. Importantly, pathological samples from the substantia nigra are studded with dense protein aggregates called Lewy bodies, which upon further analysis were found to be chiefly composed of natively-unfolded α-synuclein protein

(Ross and Poirier, 2004)—the hallmark of synucleino-pathy. It is believed that α-synuclein misfolding and aggregation are responsible for familial and idiopathic PD early pathological features, which include mitochondrial defects and vesicle mistrafficking.

In a rare case of familial PD, families originating from the village of Contursi Terme in Southwestern Italy contained heritable point mutations in SNCA, the α-synuclein gene locus, with over 90% penetrance. This rare case strengthened the association between α-synuclein and PD, but the most common idiopathic forms of PD are neither familial nor caused by point mutation; rather, unknown genetic and environmental factors are thought to play a role in PD etiology (Goedert, 2001). The neuro-pathology community is continually eluded by how and why α-synuclein accumulates in diseased brain tissue, and whether it does so inheritably or sporadically. To begin to understand the genetic mechanisms by which α-synuclein confers toxicity, investigators seek to identify genes responsible for producing proteins that enhance or suppress—modify—toxicity *in vivo*.

SCREENING FOR MODIFIERS OF α-SYNUCLEIN CYTOTOXICITY

Conducting a screen for α-synuclein toxicity modifiers is nearly impossible in mammals; mammalian neurons are cumbersome specimens to genetically manipulate, culture, and screen in large quantities. To the surprise of many, the solution to this screening problem came in the form of a tiny organism that lacks neurons. Outeiro and Lindquist (2003) first described a screenable model for α-synuclein toxicity in *Saccharomyces cerevisiae*, a single-celled budding yeast. Despite the enormous evolutionary gap separating yeast from humans, expressing α-synuclein, a distinctly mammalian protein, in *S. cerevisiae* recapitulates many mammalian toxic phenotypes. Outeiro and Lindquist (2003) revealed that toxicity is dependent upon the number of copies, or "dose," of α-synuclein expressed.

Following this discovery, Lindquist and colleagues developed a technique to use the yeast model in screens for genetic modifiers of α-synuclein toxicity. Rapid screening was made possible by expressing α-synuclein in combination with the overexpression of a gene or genes from a yeast open reading frame (ORF) library, then assessing resultant toxicity by a cell survival assay. Genes that increased cell survival when over expressed along with α-synuclein expression were deemed toxicity suppressors, while genes that decreased cell survival when over expressed along with α-synuclein expression were deemed toxicity enhancers (Cooper, 2006). Modifier gene "hits" uncovered by these screens were then verified by over expression in higher organisms such as nematodes, fruit flies, and rats to verify findings from the yeast model in organisms with neurons.

In early screens, Cooper et al. (2006) demonstrated that overexpression of the yeast gene *Ypt1* can rescue yeast from α-synuclein toxicity. When *Rab1*, the mammalian homolog of *Ypt1*, was then over expressed *in vivo* in rat brains, dopaminergic neurons were rescued from degeneration normally induced by α-synuclein expression. Since forward ER-Golgi trafficking genes such as *Rab1* suppressed toxicity and negative traffic regulating genes enhanced toxicity in the yeast model, Cooper et al. (2006) noted that vesicle misregulation may be an early step in the molecular etiology of PD and other synucleinopathies. The early screening experiments have definitively provided evidence that α-synuclein toxicity in yeast depends upon the expression levels of vesicle trafficking genes. It was hypothesized that these results may have implications for the neuron: If dopamine is improperly trafficked to synaptic vesicles, it could remain in the cytosol and oxidize to reactive oxygen species that impair mitochondrial function and precipitate cell death. Dopamine in the cytosol is prone to autooxidation or metabolism by monoamine oxidase, the products of which impair intracellular protein function; in contrast,

vesicular dopamine is protected from oxidation (Watabe and Nakaki, 2007). The results of Mosharov et al. (2006) suggest that overexpression of α-synuclein, particularly known pathological mutants of α-synuclein, causes dopamine to leak from its vesicles into the cytosol. Thus, at least when considering the role of α-synuclein, ER-Golgi trafficking in yeast may be analogous to synaptic vesicle trafficking in mammals.

Three years after the screens identified *Ypt1* as a suppressor, Gitler et al. (2009) discovered a strong genetic interaction between α-synuclein and *YPK9*, a yeast ortholog of the PD-linked gene *PARK9*. *YPK9* and *PARK9* rescued yeast cells and dopaminergic neurons, respectively, from α-synuclein toxicity. Furthermore, Gitler et al, (2009) demonstrated that *PARK9* can also protect cells from manganese toxicity. Epidemiological studies have documented an increase in Parkinsonism among miners and welders occupationally exposed to manganese (Jankovic, 2005). Exposure to the manganese-containing pesticide maneb has also been implicated as a causal event in the etiology of PD (Thiruchelvam et al., 2000). The *PARK9* discovery has therefore uncovered what could be the first direct connection between a genetic basis for neurotoxicity and environmental exposure to manganese.

THE SEARCH FOR MECHANISM

Despite the uncertainties surrounding the genetics of synucleinopathy, several explanations for α-synuclein toxicity have been proposed. Many agree that unfolded synuclein monomers slowly polymerize to form β-sheet-like oligomeric fibrils. Eventually, these fibrils become large enough to be regarded as Lewy bodies, which are accompanied by a host of toxic phenotypes including vesicle transport blockage and mitochondrial dysfunction (Cookson, 2009). These toxic insults may be coupled with oxidative stress (Dawson and Dawson, 2003), dispersed pigmentation (Halliday et al., 2005), and cell membrane injury (Takeda et al., 2006), finally resulting in the apoptosis of dopamine-producing neurons (Cookson, 2009). Yeast models have also established a potential role for serine phosphorylation in α-synuclein aggregation (Fiske and DebBurman, 2010). Although this cascade of neuronal insults may not follow directly from the formation of Lewy bodies, it is believed that neuronal cell death in synucleinopathies may critically depend upon α-synuclein aggregation (Cookson, 2009) and spreading (Desplats et al., 2009), providing evidence for prion-like behavior of α-synuclein. Like the infectious protein particles known as prions, altered α-synuclein proteins may bind together to form aggregates that are pathogenic under certain intracellular conditions.

Despite the enormous attention α-synuclein aggregation has received, it remains unknown whether aggregated fibrils constitute a direct cause of neurodegeneration (Lansbury and Lashuel, 2006; Cookson, 2009). Instead of drawing a causal link between α-synuclein fibrils and neuronal cell death, it is possible that misfolding and aggregation prevent α-synuclein from performing some essential wild-type role. Chandra et al, (2005) suggested that wild-type α-synuclein compliments the function of cysteine-string protein-α (CSPα), a chaperone protein responsible for the proper folding of SNARE proteins, which are needed for functional release of dopamine. Accordingly, transgenic expression of α-synuclein had been shown to abolish a lethal phenotype produced by deletion of CSPα in mice (Chandra et al., 2005). Scott et al. (2010) reported an absence of processed presynaptic SNARE proteins in synapses overexpressing α-synuclein.

These absences, potentially caused by the loss of WT α-synuclein function due to oligomerization, could be responsible for the observed disruption of endocytosis and vesicle recycling in affected neurons. In contrast, the familial A53T α-synuclein mutant binds to ER-Golgi SNAREs and directly inhibits their functions in vesicle trafficking (Thayanidhi et al., 2010). It had also been suggested that under non-pathological conditions, wild-type α-synuclein negatively regulates mitochondrial complex I activity (Loeb et al., 2010).

In assuming these wild-type functions, it is possible that sequestration of α-synuclein prevents it from properly assisting SNARE protein folding, producing vesicle trafficking impairments, toxic buildup of dopamine in the cytosol, and mitochondrial dysfunction. Thus, it seems likely that loss of normal α-synuclein function and not the proteinaceous fibrils themselves may have a causal role in the precipitation of toxic phenotypes. However, since α-synuclein is not native to yeast, and since the yeast model has nonetheless recapitulated mammalian toxic phenotypes, it is equally possible that α-synuclein itself, and not loss of its normal function, is responsible for cytotoxicity. If loss of wild-type α-synuclein function were required for toxicity, then introducing the mammalian protein into yeast would not produce the expression-dependent cytotoxicity effect observed. This evidence favors an intrinsic role of α-synuclein in conferring cytotoxicity. Yet, what causes α-synuclein to be expressed at a toxic dosage in idiopathic PD remains unknown. It has been hypothesized that loss of methylation at the *SNCA* locus, as observed in the substantia nigra of PD patients, may produce excessive expression of α-synuclein that leads to neuropathology (Jowaed et al., 2010). If the mechanism of α-synuclein toxicity, in part or whole, is indeed the same in yeast and mammals, then comprehensive screens of the yeast ORFeome are expected to supply valuable information on genetic processes that underlie α-synuclein toxicity in both organisms.

Yeast is thus a rather fortuitous model organism with which to study synucleinopathy, according to those engaged in the early toxicity screens. Not only does α-synuclein exert toxic phenotypes in yeast that are similar to those produced in mammals—the yeast genome is well-characterized and readily-manipulated, and many yeast gene interactions have already been documented (Cooper, 2006; Gitler, 2008; Yeger-Lotem et al., 2009). As investigators continue to screen for genes that modify α-synuclein toxicity, a more complete picture of neurodegenerative pathways is emerging. Yet, several challenges remain.

SCREENING CHALLENGES AHEAD

To improve screen methodology and analysis, yeast screening protocols must be optimized with suitable definitions for toxicity enhancement and suppression. Next, to gauge interactions between genes, a new type of screen is needed. This new screen will most likely take advantage of α-synuclein's dose dependence (Cooper et al., 2006). Overexpression of one suppressor gene alone is not enough to rescue a yeast cell from a highly toxic dose of α-synuclein; instead, two suppressor genes that rely on one another must be co-overexpressed in order to overcome high toxicity. Such combinatorial screens with two library genes may reveal protein-protein or protein-gene interactions with net suppressor effects. Conversely, toxicity enhancers can be identified by over-expressing libraries of yeast ORFs along with a low dose of α-synuclein, as a true enhancer is needed for the low dose to confer any toxicity. The Lindquist Laboratory has developed a methodology to screen the yeast ORFeome for genetic enhancers of α-synuclein cytotoxicity. This methodology relies on high-throughput matings between yeast containing integrated human α-synuclein and yeast containing a yeast ORF. Enhancer candidates identified by the high-throughput screen are then validated by further screens that rely on transformation instead of mating. Importantly, orthologs of all validated enhancers (originally yeast ORFs) must be confirmed in higher organisms with neurons.

The subsequent challenge will be piecing together the massive amounts information that these high-throughput screens will uncover. To help ease this data bottleneck, bioinformatics initiatives at the Whitehead Institute have developed a computational algorithm to elucidate genetic pathways that may be involved in the α-synuclein toxicity mechanism. The algorithm, ResponseNet, integrates screen data with data obtained from mRNA profiling to diagram the most likely molecular pathways given sets of screen results (Yeger-Lotem et al., 2009). ResponseNet output may reveal whether α-synuclein truly is a "nodal point" that integrates genetic and environmental information, as suggested by Yeger-Lotem et al. (2009).

As high-throughput screens are slowly unraveling the

fibers in our understanding of the α-synuclein toxicity mechanism, our broader perception of neurodegenerative disease is beginning to shift. In the clinic, PD is often seen as an idiopathic disorder that manifests itself spontaneously with age. As a result of the knowledge garnered from preliminary screens, incidence of PD and related synucleinopathies may be deemed less sporadic than previously thought. Synucleinopathies may result from the interplay of a variety of genetic and environmental nodes built into a network that affects the toxicity of α-synuclein or its aggregates. Luckily, evolution has conserved much of this network from yeast to human, rendering some findings from rapid genetic screens in yeast applicable to human disease. Although these screens may not capture the full range of genetic interactions present in human neural tissue, they may uncover useful clues for understanding, treating, and preventing a large class of neurodegenerative diseases.

ACKNOWLEDGEMENTS

The author would like to thank M. Geddie, B. Bevis, and S.L. Lindquist for excellent research collaboration. The author would also like to acknowledge support from the Howard Hughes Medical Institute and the National Parkinson Foundation, and is grateful to H. Han, S. Treusch, and N. Azubuine for laboratory assistance. N. Hopkins, A. Amsterdam, and K. Pepper assisted with manuscript communication.

REFERENCES

Chandra S, Gallardo G, Fernández-Chacón R, Schüter OM, Südhof TC (2005). α-Synuclein cooperates with CSPα in preventing neurodegeneration. Cell., 123: 383-396.

Cookson MR (2009). α-Synuclein and neuronal cell death. Mol. Neurodegener. 4.

Cooper AA, Gitler AD, Cashikar A, Haynes CM, Hill KJ, Bhullar B, Liu K, Xu, K, Strathearn KE, Liu F, Cao S, Caldwell, KA, Caldwell GA, Marsischky, G, Kolodner RD, LaBaer J, Rochet JC, Bonini NM, Lindquist S (2006). α-Synuclein blocks ER-Golgi traffic and Rab1 rescues neuron loss in Parkinson's models. Science, 313: 324-328.

Dawson TM, Dawson VL (2003). Molecular Pathways of Neurodegeneration in Parkinson's Disease. Sci., 302: 819-822.

Desplats P, Lee, HJ, Bae EJ, Patrick, C, Rockenstein E, Crews, L, Spencer B, Masliah E, Lee SJ (2009). Inclusion formation and neuronal cell death through neuron-to-neuron transmission of α-synuclein. Proc. Natl. Acad. Sci. USA., 106: 13010-13015.

Fiske M, DebBurman S (2010). Contributions of familial mutation E46K, phosphorylation, protein domains, and alanine-76 to α-synuclein toxicity in yeasts. FASEB J., 708.

Gitler AD (2008). Beer and bread to brains and beyond: Can yeast cells teach us about neurodegenerative disease? Neurosignals, 16: 52-62.

Gitler AD, Chesi A, Geddie ML, Strathearn KE, Hamamichi S, Hill KJ, Caldwell KA, Caldwell GA, Cooper AA, Rochet JC, Lindquist S (2009). α-Synuclein is part of a diverse and highly conserved interaction network that includes PARK9 and manganese toxicity. Nat. Genet., 41: 308-315.

Goedert M (2001). Alpha-synuclein and neurodegenerative diseases. Nat. Rev. Neurosci., 2: 492-501.

Halliday GM, Ophof A, Broe M, Jensen PH, Kettle E, Fedorow H, Cartwright MI, Griffiths FM, Shepherd CE, Double KL (2005). α-Synuclein redistributes to neuromelanin lipid in the substantia nigra early in Parkinson's disease. Brain, 128: 2654-2664.

Jankovic J (2005). Searching for a relationship between manganese and welding in Parkinson's disease. Neurology, 64: 2021-2028.

Jowaed A (2010). Methylation Regulates Alpha-Synuclein Expression and Is Decreased in Parkinson's Disease Patients' Brains. J. Neeurosci., 30: 6355-6359.

Lansbury PT, Lashuel HA (2006). A century-old debate on protein aggregation and neurodegeneration enters the clinic. Nature, 443: 774-779.

Mosharov EV, Staal RGW, Bové J, Prou D, Hananiya A, Markov D, Poulsen N, Larsen KE, Moore CMH, Troyer MD, Edwards RH, Przedborski S, Sulzer D (2006). α-Synuclein Overexpression Increases Cytosolic Catecholamine Concentration. J. Neurosci., 26: 9304-9311.

Outeiro TF, Lindquist S (2003). Yeast cells provide insight into alpha-synuclein biology and pathobiology. Science, 302: 1772-1775.

Ross CA, Poirier MA (2004). Protein aggregation and neurodegenerative disease. Nat. Med., 10: S10-S17.

Scott DA, Tabarean I, Tang Y, Cartier A, Masliah E, Roy S (2010). A Pathologic Cascade Leading to Synaptic Dysfunction in -Synuclein-Induced Neurodegeneration. J. Neurosci., 30: 8083-8095.

Takeda A, Hasegawa T, Matsuzaki-Kobayashi M, Sugeno N, Kikuchi A, Itoyama Y, Furukawa K (2006). Mechanisms of neuronal death in synucleinopathy. J. Biomed. Biotechnol., pp. 1-4.

Thayanidhi N, Helm JR, Nycz DC, Bentley M, Liang Y, Hay JC (2010). α-Synuclein Delays Endoplasmic Reticulum ER-to-Golgi Transport in mammalian Cells by Antagonizing ER/Golgi SNAREs. Mol. Biol. Cell., 21: 1850-1863.

Thiruchelvam M, Richfield EK, Baggs RB, Tank AW, Cory-Slechta DA (2000). The Nigrostriatal Dopaminergic System as a Preferential Target of Repeated Exposures to Combined Paraquat and Maneb: Implications for Parkinson's Disease. J. Neurosci., 20: 9207-9214.

Watabe M, Nakaki T (2007). Mitochondrial Complex I Inhibitor Rotenone-Elicited Dopamine Redistribution from Vesicles to Cytosol in Human Dopaminergic SH-SY5Y Cells. J. Pharmacol. Exp. Ther., 323: 499-507.

Yeger-Lotem E, Riva L, Su LJ, Gitler AD, Cashikar AG, King OD, Auluck PK, Geddie ML, Valastyan JS, Karger DR, Lindquist S, Fraenkel E (2009). Bridging high-throughput genetic and transcriptional data reveals cellular responses to alpha-synuclein toxicity. Nat. Genet., 41: 316-323.

19

Millet improvement through regeneration and transformation

Sonia Plaza-Wüthrich and Zerihun Tadele*

Institute of Plant Sciences, University of Bern, Altenbergrain 21, 3013 Bern, Switzerland.

Millets, comprising the small-seeded group of the Poaceae family, represent one of the major food- and feed-crops in the semi-arid tropical regions of Africa and Asia. Compared to major crops of the world, these indigenous crops possess a number of beneficial characteristics including tolerance to extreme climatic and soil conditions; hence, adapts to poor soil fertility and moisture deficient areas. Moreover, millets are also nutritionally rich especially in vitamins and minerals, and most of them are gluten-free. Despite all these benefits, millets are encountered with several production constraints. The major bottleneck affecting millets are their extremely low yield since they are mostly cultivated in marginal areas with poor moisture and fertility conditions. Inherent characteristics, such as susceptibility to lodging, also significantly affect the productivity of millets. Millets are also commonly known as orphan- or neglected-crops due to too little attention given to them by the world scientific community. Genetic improvement in millets could be achieved not only by conventional approaches but also through modern techniques such as genetic modification or transgenics. The main benefits of regeneration and transformation in millet improvement are: i) the multiplication of identical copies of plants that are free of diseases and pests, and ii) the regeneration of the whole plant from transformed tissues with desirable traits. Success in plant transformation is largely dependent on the efficiency of regeneration. Establishing optimum regeneration method for each plant species and ecotype is therefore, a pre-requisite before embarking on plant transformation. In this review, we present various studies made to identify optimum regeneration and transformation methods for major millets. The prospects of applying advanced regeneration and transformation techniques to these vital but under-studied crops of the developing world are also discussed.

Key words: Millets, under-researched crops, orphan crops, *in vitro* regeneration, transformation.

INTRODUCTION

Millets represent the small-seeded group of the Poaceae family. The similarities of millets are that they are grown under extreme environmental conditions and therefore, especially suited to areas with inadequate moisture or short-growing cycle and poor soil fertility (Baker, 2003). Although millets are many in number, the most widely-cultivated ones are pearl millet [*Pennisetum glaucum* (L.) R. Br.], finger millet [*Eleusine coracana* (L.) Gaertn], tef [*Eragrostis tef* (Zucc.) Trotter], fonio or acha [*Digitaria exilis* (Kippist) Stapf and *D. iburua* Stapf], foxtail millet [*Setaria italica* (L.) P. Beauvois], proso millet [*Panicum miliaceum* (L.)], barnyard millet [*Echinochloa crusgalli* (L.)P. Beauvois] and kodo millet [*Paspalum scrobiculatum* (L.)].

Millets play key role in the maintenance of food security in the developing world since they are the major food and feed sources. Together with sorghum, millets account for about half of the total cereal production in Africa (Belton and Taylor, 2004). However, the average yield for millets is only 0.8 ton ha^{-1} as compared to 3.5 ton ha^{-1} for other cereals (FAOSTAT: http://faostat.fao.org/ accessed 21.12.2011). Millets are rich sources of human and livestock nutrition in developing countries (NAS, 1996).

*Corresponding author. E-mail: zerihun.tadele@ips.unibe.ch.

They contain high amount of vitamin, calcium, iron, potassium, magnesium, and zinc (Leder, 2004). In addition to being nutritious, millets are also considered as healthy food. The grains of most millets do not contain gluten (Leder, 2004), a substance that causes coeliac disease or other forms of allergies. Six millet species (namely, kodo-, finger-, proso-, foxtail-, little- and pearl-millet) were recently shown to have an anti-proliferative property and might have a potential in the prevention of cancer initiation (Chandrasekara and Shahidi, 2011), due to the presence and amount of phenolic extracts (Rao et al., 2011). Similar to maize and sorghum, millets follow the C4 photosynthesis system (Brutnell et al., 2010; Warner and Edwards, 1988); hence they prevent photorespiration and as a consequence efficiently utilize scarce moisture in the semi-arid regions.

To meet the strong increase in cereal demand worldwide, new approaches and technologies for generating new varieties are necessary. One of these methods is the creation of transgenic plants with desirable traits. Although millets are economically important, especially in the developing world, little genetic improvement has been done so far specifically using wide- or cross- hybridization among closely related species. The incompatibilities due to interspecific hybridization are alleviated by directly transferring the desirable traits to millets using optimum or efficient transformation method. Hence, crossing barriers could be overcome, and genes from unrelated sources would be introduced asexually into crop plants. Monocots in general and cereals in specific were initially difficult to genetically engineer, mainly due to their recalcitrance to *in vitro* regeneration and their resistance to *Agrobacterium*-mediated infection. However, efficient transformation protocols have been later established for the major cereals including rice and maize. Gene transfer to millets would be facilitated once efficient or optimum regeneration and transformation techniques are established.

The optimization of regeneration method is, therefore, necessary for different millet types in order to increase the efficiency of transformation. In this review, we present various regeneration and transformation techniques studied for major millets. We also discuss the prospects of applying advanced techniques developed for major cereals to millets, vital but understudied crops of the developing world.

REGENERATION STUDIES IN MILLETS

The first regeneration studies in millets were performed in the 1970s for proso-, finger-, pearl- and kodo- millets (Rangan, 1973, 1976). Subsequent investigations were also made for other millet species. In the following sections, key parameters affecting millet regenerations are reviewed (Table 1). These important factors include

explants, plant growth regulators (PGRs) and media. Although environmental factors such as temperature, pH and light also affect the regeneration processes, they are not discussed here.

Regeneration processes

Plant regeneration is achieved by the process of either somatic embryogenesis or organogenesis. Somatic embryogenesis relies on plant regeneration through a process similar to zygotic embryo germination. Somatic embryos are developed either directly or indirectly through an intermediate step of callus formation. Direct embryogenesis occurs in plants rarely, compared to the indirect somatic embryogenesis.

The organogenesis process relies on the production of organs either directly from an explant or from a callus culture. It is a rare event in millets; to date, only finger millet and pearl millet were regenerated through organogenesis (George and Eapen, 1990; Jha et al., 2009).

Explants for regeneration

Explants refer to sterile pieces of the plant from which regeneration is initiated. The suitability of explants for regeneration depends on the type of the genotype and the culture media used. The maximum callus inductions obtained from different explants of pearl- and finger-millet are shown in Figure 1. Identifying the best explant is critical for increasing the competence of plant regeneration.

Roots were used as an explant in tef and finger millet. While the callus induction was more than 90% in finger millet (Mohanty et al., 1985) (Figure 1B), in tef, a maximum of 25% callus formation was obtained (Bekele et al., 1995). The difficulty of using root as an explant was also reported for major cereal crops such as wheat and barley (Bhojwani and Hayward, 1977; Chin and Scott, 1977). Another easily available explant is the shoot apical meristem (SAM) which contains the zone of actively dividing cells. The suitability of SAM as an explant was demonstrated in finger millet and pearl millet (Eapen and George, 1990; Lambe et al., 1999). Mesocotyl, the plant part between the cotyledon and the coleoptile, was also used as an explant for finger- (Rangan, 1976; Mohanty et al., 1985; Eapen and George, 1990), proso- (Rangan, 1973; Heyser and Nabors, 1982), kodo- (Rangan, 1973), and pearl-millet (Rangan, 1976).

Mature seeds and embryos were also studied in most millet types although mature embryos generated lower percentage of somatic embryos than immature embryos in kodo- and pearl- millet (Vikrant and Rashid, 2002b; Goldman et al., 2003; Campos et al., 2009). Immature inflorescences were also evaluated for their regenerative response especially in pearl millet where they gave the

Table 1. Summary of *in vitro* regeneration studies for important millets regarding explants, regeneration processes and growth regulators.

Millet type (species)	Explant	Processes	Growth regulators	Reference
Pearl millet (*Pennisetum glaucum*)	Mesocotyl	Somatic embryogenesis	2,4-D	Rangan (1976).
		Plant regeneration	IAA	
	Immature inflorescence; immature embryo; mature seed; leaf segment and shoot tip	Somatic embryogenesis	2,4-D or pCPA alone or with KIN or BA	Vasil and Vasil (1981, 1982), Pius et al. (1993), Lambe et al. (1995, 1999, 2000), Mythili et al. (1997), Oldach et al. (2001), Girgi et al. (2002, 2006), Srivastav and Kothari (2002), Goldman et al. (2003), O'Kennedy et al. (2004, 2011a, 2011b), Satyavathi et al. (2006), Muthuramu et al. (2008) and Jha et al. (2009).
		Plant regeneration	GA$_3$; BA or ABA alone; IAA with KIN or BA with IAA or KIN or TDZ or 2,4-D	
		Root formation	NAA alone or IBAor IAA; KIN	
		Organogenesis	BA	
	Mature embryo	Somatic embryogenesis	2,4-D	Campos et al. (2009).
	Root, mesocotyl and leaf base	Somatic embryogenesis	2,4-D	Rangan (1976) and Mohanty et al. (1985).
		Plant regeneration	None or NAA	
	Shoot tip, immature inflorescence and mesocotyl	Somatic embryogenesis	2,4-D or picloram; KIN or BA[2]	Eapen and George (1990), George and Eapen (1990), Latha et al. (2005) and Ceasar and Ignacimuthu (2008).
		Plant regeneration	KIN with TDZ or IAA	
		Organogenesis	2,4-D; zeatin[2]	
Finger millet (*Eleucine coracana*)	Mature seed	Somatic embryogenesis	2,4-D alone or with KIN	Sivadas et al. (1990), Poddar et al. (1997), Gupta et al. (2001), Kothari et al. (2004), Kothari-Chajer et al. (2008), Nethra et al. (2009) and Sharma et al. (2011).
		Plant regeneration	GA$_3$, BA or NAA alone or KIN; IAA	
	Mature embryo and epicotyl	Somatic embryogenesis	2,4-D	Patil et al. (2009).
		Plant regeneration	BA or KIN	
		Root formation	IBA; BA	
	Leaf and root explant and mature seed	Callus induction	2,4-D or 3,6-D or dicamba	Bekele et al. (1995) and Mekbib et al. (1997).
		Somatic embryogenesis	2,4-D or 3,6-D or dicamba; ABA, BA; KIN	
Tef (*Eragrostis tef*)	Mature seed	Somatic embryogenesis	2,4-D followed by TIBA	Assefa et al. (1998).
		Embryo promotion	2,4-D; KIN followed by IAA; BA	
		Plant regeneration	GA$_3$	
	Immature spikelet and panicle segment	Gynogenic tissue induction	2,4-D; BA	Gugsa et al. (2006)

Table 1. Contd.

Species	Explant	Type	Growth regulator	References
	Immature anther and embryo	Somatic embryogenesis Plant regeneration	2,4-D BA alone or with NAA	Tadesse et al. (2009) and Gugsa and Kumlehn (2011).
Fonio (*Digitaria exilis*)	Stem segment	Somatic embryogenesis Shoot development	2,4-D BA; GA$_3$[1]	Ntui et al. (2010).
Barnyard millet (*Echinochloa crusgalli*)	Mature seed	Somatic embryogenesis	2,4-D[2]	Gupta et al. (2001).
Proso millet (*Panicum miliaceum*)	Immature and mature embryo, mature seed, immature inflorescence, mesocotyl, shoot tip and leaf and stem segment	Somatic embryogenesis Plant regeneration	2,4-D alone or with KIN[2] 2,4-D or NAA	Rangan (1973), Bajaj et al. (1981), Heyser and Nabors (1982), Rangan and Vasil (1983) and Heyser (1984).
	Immature inflorescence, immature and mature embryo; mature seed; young leaf base and mesocotyl	Somatic embryogenesis Shoot regeneration Root formation	Picloram or 2,4-D alone or with TDZ[1]; KIN NAA alone or with BA PAA	Rangan (1976), Nayak and Sen (1989, 1991), Vikrant and Rashid (2001, 2002a, 2002b, 2003), Kaur and Kothari (2004) and Kothari-Chajer et al. (2008).
Kodo millet (*Paspalum scrobiculatum*)	Shoot tip	Somatic embryogenesis Plant regeneration Root formation	2,4-D alone or with KIN TDZ alone[1] or BA; NAA IBA	Arockiasamy et al. (2001) and Ceasar and Ignacimutu (2010).
Foxtail millet (*Setaria italica*)	Immature inflorescence; mature embryo; mature seed and shoot tip	Somatic embryogenesis Plant regeneration	2,4-D alone or with KIN or BA NAA with BA or KIN or 2,4-D with KIN	Xu et al. (1984), Rao et al. (1988), Reddy and Vaidyanath (1990), Osuna-Avila et al. (1995), Qin et al. (2008) and Wang et al. (2011).

[1] Transfer to medium without any growth regulator for root formation.
[2] Transfer to medium without any growth regulator for plant regeneration.

highest percentage of somatic embryos and shoot regeneration compared to shoot tips and seeds (Jha et al., 2009). Moreover, anthers were successfully used as an explant in tef (Tadesse et al., 2009). In general, immature embryos are the main source of an explant not only in the major cereal crops but also in millets. In pearl millet

alone, about 50% of the studies on regeneration used explants from immature embryo. In our laboratory, we routinely use immature embryos as an explant in order to regenerate tef (Figure 2A). We found that about 45% of immature embryos induced somatic embryos and about 55% of these somatic embryos formed plantlets (Figure 2B).

Plant genotypes

The types of genotypes or crop cultivars also determine the efficiency of regeneration. The investigation made on eight tef ecotypes indicated that although variations in calli weight were negligible, differences in the percentage of

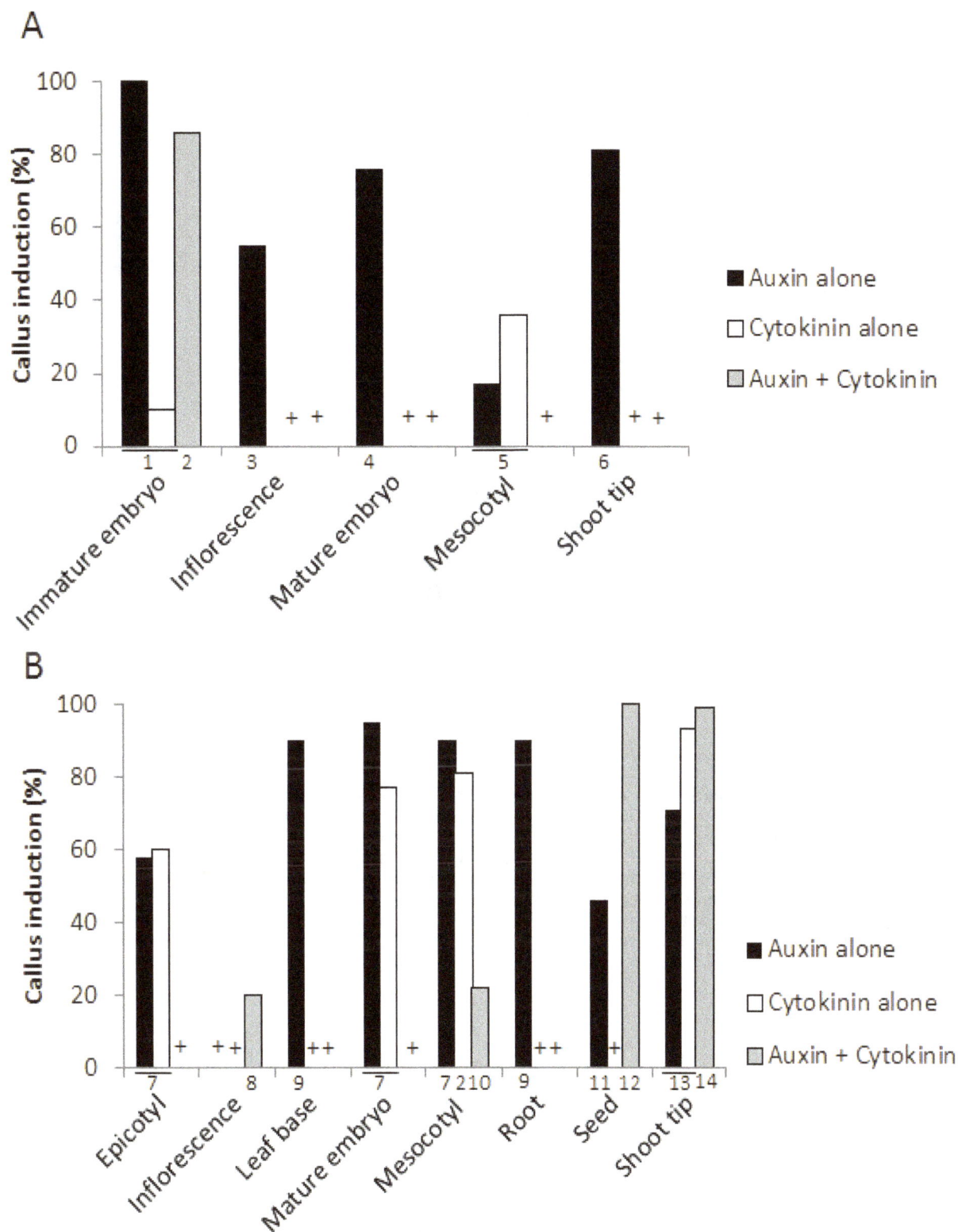

Figure 1. The maximum efficiency of callus formation in pearl millet (A) and finger millet (B). The callus induction for different explants depends on the type of hormone (auxin or cytokinin alone or by adding the two hormones together) applied. Data non-available or non-quantified were indicated as "+". The figure was made based on the results of the following authors: (1) Goldman et al. (2003), (2) Oldach et al. (2001), (3) O'Kennedy et al. (2004), (4) Campos et al. (2009), (5) Rangan (1976), (6) Lambe et al. (1999), (7) Patil et al. (2009), (8) George and Eapen (1990), (9) Mohanty et al. (1985), (10) Eapen and George (1990), (11) Gupta et al. (2001), (12) Kothari et al. (2004), (13) Ceasar and Ignacimuthu (2008) and (14) Latha et al. (2005). Numbers in the figure correspond to the references indicated.

A

B

Figure 2. *In vitro* regeneration of tef. (A) Tef variety *Tsedey* (also known as DZ-Cr-37) was used for *in vitro* regeneration based on Gugsa and Kumlehn (2011). Immature embryos were placed on K99 medium (Deutsch et al., 2004) facing the scutellum side up and allowed to grow in the dark for 3 weeks in the presence of 90 g/l maltose, 1 g/l glutamine and 2 mg/l 2,4-D. Somatic embryos formed were transferred to K4NB medium (Kumlehn et al., 2006) in the light with 36 g/l maltose, 0.15 g/l glutamine and 0.22 mg/l BAP. After four weeks, plantlets were transferred to soil and grown first for three weeks in the long-day (16 h light: 8 h dark) followed by short-day (8 h light: 16 h dark) until harvesting the seeds. (B) Percentage (+/- standard error) of immature embryos transformed to somatic embryos and somatic embryos transformed to plantlets. One hundred fifty initial explants were used for the experiment.

regenerants were considerable among the ecotypes tested (Bekele et al., 1995). Despite huge expected variability among different ecotypes or cultivars, most regeneration experiments in millets use a single line or cultivar without testing its performance. Hence, obtaining a genotype with high regenerative capacity is a widespread problem in millet improvement.

Plant growth regulators (PGRs)

PGRs are critical in determining the deve-lopmental pathway of the plant cells. Their roles in regeneration have been studied since the initial observations by Skoog

and Miller (1957) half a century ago. Auxins and cytokinins are the most widely employed PGRs in plant regeneration. They are usually applied together in the medium as the ratio of auxin to cytokinin determines the type of organ or tissue to be regenerated. While high auxin to cytokinin ratio promotes root development, low ratio stimulates shoot development. The intermediate ratio, on the other hand, facilitates the formation of undifferentiated organ called callus. This paradigm was also confirmed in somatic embryogenesis of millets (Kaur and Kothari, 2004).

On the other hand, somatic embryogenesis was also obtained by the application of auxin or cytokinin alone (Figure 1). Earlier studies showed that for the long-term maintenance of the callus, the concentration of auxin in the form of 2,4-D (2,4-Dicholorophenoxyacetic acid) and pCPA (p- chlorophenoxyacetic acid) need to decrease in finger millet and pearl millet, respectively (Sivadas et al., 1990; Kumar et al., 2001; Srivastav and Kothari, 2002) as prolonged exposure of cell cultures to high concentrations of auxin resulted in poor regeneration and caused chromosomal abnormalities (Deambrogio and Dale, 1980; Nabors et al., 1983). While in the majority of the somatic embryogenesis, 2,4-D was used as an auxin supplement, other types of auxin also showed good performance. Among these, picloram, a very potent growth regulator that induces somatic embryogenesis, was found to be superior to 2,4-D in kodo millet regeneration (Kaur and Kothari, 2004). A study in tef showed that the efficacy of auxin was dependent on the type of explant in which 2,4-D was best suited for leaf and root segments while 3,6-D for mature seeds (Bekele et al., 1995).

Shoot development was in the majority of cases formed once embryogenic calli were transferred to medium with a low auxin to cytokinin ratio (Girgi et al., 2002). However, many other reports indicated that successful regeneration were obtained by applying cytokinin alone in finger millet (Sankhla et al., 1992; Latha et al., 2005; Ceasar and Ignacimuthu, 2008; Yemets et al., 2003; Nethra et al., 2009), kodo millet (Ceasar and Ignacimuthu, 2010) and pearl millet (Mythili et al., 1997; Goldman et al., 2003; Satyavathi et al., 2006). Other workers also indicated that regeneration was promoted in diverse types of millets using either gibberellic acid alone (Sivadas et al., 1990; Nayak and Sen, 1991; Assefa et al., 1998; Kumar et al., 2001; Sharma et al., 2011) or together with cytokinin (Ntui et al., 2010). On the other hand, several other studies indicated that none of the known PGRs were necessary to regenerate shoots in finger millet (Eapen and George, 1990), proso millet (Heyser and Nabors, 1982), kodo millet (Vikrant and Rashid, 2001, 2002b) and pearl millet (Campos et al., 2009).

In general, in about half of regeneration studies on millets, PGRs were applied together in order to initiate shoots and roots simultaneously while in the remaining studies shoots were allowed to develop first followed by roots.

Culture media

The composition of the culture medium is another important parameter that determines the efficacy of regeneration independent of the explant. The medium has to supply all essential nutrients necessary for the growth and development of the plant.

Most in vitro culture studies use Murashige and Skoog (or commonly known as MS) medium (Murashige and Skoog, 1962). However, the N6 medium (Chu et al., 1975)

became popular in pearl millet since increased amount of embryogenic callus was obtained during long-term culturing (Lambe et al., 1999). In addition, compared to the MS medium, lower amount of auxin was required for the N6 medium (Vikrant and Rashid, 2001, 2002b).

Since the majority of plant cells are not photosynthetic, it is essential to add to the culture medium, a fixed carbon source. The type and concentration of carbon source also determine the competence of embryogenic calli to be formed. Carbon does not only serve as an energy source but also influences the osmolarity of the medium. Although sucrose is commonly applied in most tissue culture studies involving millets, maltose is preferentially used in pearl millet and tef (O'Kennedy et al., 2004; Tadesse et al., 2009; Gugsa and Kumlehn, 2011).

The concentration of micro-nutrients added to MS medium also affects the regeneration processes. The addition of higher concentration of cupric sulphate improved somatic embryogenesis, maintenance and regeneration in finger millet (Kothari et al., 2004). Another important component of the media is the ratio between nitrate and ammonia. Successful regeneration was reported in finger millet using high nitrate to ammonium ratio replacing PGRs (Poddar et al., 1997).

Organic compounds such as casein hydrolysate, glutamine and L-tryptophan were also proved to improve the initiation of embryogenic cultures in finger millet (Yemets et al., 2003). Although, the aforementioned report indicated the beneficial effects of amino acids on regeneration, another study in finger millet showed an adverse effect of certain amino acids on the initiation of shoots (Eapen and George, 1990).

Charcoal, which absorbs inhibitory compounds (Thomas, 2008), was also shown to increase the regeneration capacity in kodo millet and pearl millet (Vikrant et al., 2001; Lambe et al., 1999). Furthermore, ethylene inhibitors such as silver nitrate improved the regeneration process in pearl millet mainly by promoting the shoot formation (Pius et al., 1993; Oldach et al., 2001). Other ethylene inhibitors such as cefotaxime, carbenicillin and streptomycin similarly enhanced plant differentiation from somatic embryos in finger millet (Eapen and George, 1990). Cefotaxime and ASA (O-acetyl salicylic acid), another ethylene inhibitor, also enhanced regeneration efficiency in pearl millet (Pius et al., 1993).

TRANSFORMATION STUDIES IN MILLETS

Genetic engineering or transformation refers to the delivery of DNA, encoding a desirable trait to the plant cell. In order to deliver pieces of DNA to the plant of choice, two methods, namely physical and biological, are used. The physical method includes particle or microprojectile bombardment and electroporation while the only successfully applied biological technique is

Agrobacterium-mediated transformation. Since both physical and biological methods facilitate the transfer of the traits of importance to the plants of interest, a number of crop improvement studies benefited from the technique. Traits commonly employed in the transformation are those which increase resistance against biotic and abiotic stresses or those which improve the quality of food.

In cereal crops, *in vitro* regeneration is an essential component of the transformation because optimum transformation could not be achieved without having a reliable regeneration protocol. Cereals were until recently difficult to genetically engineer, mainly due to their recalcitrance to regeneration and their resistance to *Agrobacterium* infection. In developing optimum transformation techniques for millets, the following points need to be considered: suitable explants, appropriate transformation method, and appropriate promoters and selectable markers (Repellin et al., 2001; Kothari et al., 2005).

Optimum transformation methods have been studied for some millets (Table 2). Several of these transformations targeted agronomically important traits including resistance to pathogens (Latha et al., 2005, 2006; Girgi et al., 2006; O'Kennedy et al., 2011a).

Explants for transformation

Transformations of millets were largely dependent on embryogenic callus derived from seedlings, shoot tips, immature inflorescences and embryos, and mature seeds. However, initial explants such as leaf segments, pollen grains and immature embryos were also directly used in the transformation (Dong et al., 1999; Gupta et al., 2001; Girgi et al., 2002, 2006; Schreiber and Dresselhaus, 2003; O'Kennedy et al., 2004, 2011a, 2011b). The use of immature embryos instead of somatic embryos was found to be an ideal target for transformation of recalcitrant crop species especially cereals (Bartlett et al., 2008).

Transformation methods

Irrespective of the type of explant, most millet transformations applied either the microprojectile bombardment or the *Agrobacterium*-mediated method of transformation. These methods require specific conditions to boost the efficiency of transformation. For instance, osmotic treatment of the explant with sucrose was found to improve the gene delivery system in pearl millet transformed by microprojectile bombardment (Goldman et al., 2003). The *Agrobacterium*-mediated transformation is dependent on the choice of appropriate strain. The most widely used *Agrobacterium* strain for millet transformations are *LBA4404*, *EHA101* and derivatives of *EHA101* (namely *EHA105*, *AGL0* and *AGL1*). In foxtail millet transformation, *LBA4404*

performed significantly better than *EHA105* (Wang et al., 2011).

Although, *Agrobacterium*-mediated trans-formation is widely applied in cereals (Schrawat and Lörz, 2006), microprojectile bombardment is still the dominant method of transformation in millets despite its drawbacks, which includes multiple integration of the transgene into the target genome.

Promoters and selectable markers

The type of promoter used for driving the gene of interest has significant impact on the efficiency of transformation. While the CaMV 35S (commonly known as 35S) promoter works perfectly in dicots, it has a low activity in monocots (McElroy and Brettell, 1994). Among five promoters tested for finger millet transformation, the Actin 1 promoter isolated from rice and the ubiquitin 1 promoter from maize gave the highest transformation efficiency (Gupta et al., 2001). In barnyard millet, however, only ubiquitin 1 was effective (Gupta et al., 2001).

Plant transformation also requires the proper choice of the selectable marker(s). Commonly used selectable markers are antibiotic- and herbicide- resistance. These selectable markers also enabled millet researchers to identify the right transformants. The applicability of hygromycin and kanamycin markers were also tested on protoplast cultures derived from pearl millet (Hauptmann et al., 1988). Another selectable marker recently developed from the *phosphomannose isomerase* (*manA*) gene showed promising performance in pearl millet (O'Kennedy et al., 2004). Transgenic *manA* expressing cells acquired the ability to convert mannose 6-phosphate to fructose 6-phosphate while the non-transgenic cells lose the ability to convert this product, and eventually die due to excessive accumulation of mannose 6-phosphate which is toxic if present in high amount. A recently developed technique in which a modified alpha-tubulin gene was used as a selectable marker in the form of herbicide resistance, gave good performance in finger millet transformation (Yemets et al., 2008).

NEED FOR EFFICIENT REGENERATION AND TRANSFORMATION OF MILLETS

Regeneration has been studied since long time in diverse millet species. Rangan was a pioneer to investigate and successfully regenerate viable plants at least from three economically important millets, namely proso-, finger- and kodo- millets (Rangan, 1973, 1976). Later, optimum regeneration methods were also studied for other millets. Establishing efficient regeneration system requires optimization of various factors including the right type of explant and the proper composition of the medium. Compared to major cereals such as wheat and rice, little advancement was made in millet regeneration.

Table 2. Summary of transformation studies for economically important millets regarding explants, and method and purpose of transformation.

Millet type (species)	Initial explant	Transformed explant	Method[1]	Promoter[2]	Purpose[3]	Reference
	Immature embryo; pollen grain and shoot tip	Immature embryos; embryogenic cell suspension; embryogenic callus; pollen grain and shoot-tip clump	MB	Enhanced CaMV 35S, ZmAdh1, ZmUbi, CaMV 35S, OsAct, ZmMADS2	T	Taylor and Vasil (1991), Taylor et al. (1993), Dong et al. (1999), Devi and Stricklen (2002) and Schreiber and Dresselhaus (2003)
Pearl millet (Pennisetum glaucum)	Shoot tip; immature embryo; mature embryo and immature inflorescence	Embryogenic cell suspension, embryogenic callus, mature embryo	MB	CaMV 35S, ZmAdh1, Emu, ZmUbi, OsAct, double CaMV 35S, pin2	S	Lambe et al. (1995, 2000), Girgi et al. (2002, 2006), Goldman et al. (2003), O'Kennedy et al. (2004, 2011a, 2011b) and Latha et al. (2006)
Finger millet (Eleucine coracana)	Mature seed and shoot tip	Embryogenic callus	MB	ZmUbi, CaMV 35S, OsAct, RbcS, ppcA-L-Ft	S	Gupta et al. (2001), Latha et al. (2005) and Yemets et al. (2008)
	Mature seed	Green nodular callus	A (EHA105)	CaMV 35S	S	Sharma et al (2011)
Barnyard millet (Echinochloa crusgalli)	Mature seed and leaf segment	Embryogenic callus and leaf segment	MB	ZmUbi, CaMV 35S, OsAct, RbcS, ppcA-L-Ft	S	Gupta et al. (2001)
	Cell line	Protoplasts	E	CaMV 35S	T	Hauptmann et al. (1987, 1988)
Guinea grass (Panicum maximum)	Immature embryo	Embryogenic cell suspension and embryogenic callus	MB	ZmAdh1, ZmUbi	T	Taylor et al. (1993)
Foxtail millet (Setaria italica)	Immature inflorescence	Embryogenic callus	A (LBA4404; EHA105)	Zm13, PF128	S	Liu et al. (2005), Qin et al. (2008) and Wang et al. (2011)

[1] A: *Agrobacterium* transformation; E: Electroporation; MB: Microprojectile bombardment

[2] CaMV 35S: Cauliflower Mosaic Virus 35S; OsAct: rice actin ; ZmAdh1: maize alcohol dehydrogenase 1; Emu: engineered based on truncated Adh1; pin2: potato proteinase inhibitor IIk (wound inducible); ppcA-L-Ft: Flaveria trinervia phosphenolpyruvate carboxylase ; RbcS: rice small subunit of ribulose 1,5-biphosphate carboxylase; ZmMADS2: maize MADS-box gene 2 (pollen specific); ZmUbi: maize ubiquitin.

[3] S: Stable transformation of plants; T: Transient expression.

This was mainly because millets are crops of developing world that are limited by resources; hence investment towards improving these crops using tissue culture or regeneration, and transformation techniques is little advanced. As a result, these vital crops of resource-poor people in developing world did not benefit from agricultural revolutions such as Green Revolution that boosted the productivity of major crops. Once optimum transformation methods are established for millets, valuable agronomic and nutritional traits could be routinely transferred. Traits that needed to be incur-porated to millets include resistance to biotic (for example, pathogens) and abiotic stresses (for example, drought), biofor- tification of useful nutritional elements, and altered architecture of the plant (for example, semi-dwarfism).

Lessons from major cereals or model millets

Advances made for major cereals in the area of regeneration and transformation could be applied to millets either directly or after some optimization. Some tissue culture techniques developed for model millets such as large crabgrass millet (*Digitaria sanguinalis*) could also be transferred to less researched millets.

Regeneration method that uses immature embryo as an explant is dominantly applied in monocots; hence it has also prospects in millets. The main problem associated to using immature embryos or inflorescences is the need for continuous growth of donor plants. Therefore, efforts should be made to investigate for alternative explants regarding accessibility, quantity, and cost. Leaves are the most common source of explant especially in callus initiation and subsequent plant regeneration in dicot plants. Unlike dicots, the vegetative parts of monocots do not readily proliferate; hence no successful regeneration was reported for cereals when leaves were used as an explant (Saalbach and Koblitz, 1978). However, due to its continuous growth similar to the meristematic region, the basal part of the leaf was successfully used in sorghum and wheat regeneration (Wernicke and Brettell, 1980; Wernicke and Milkovits, 1984). Explants such as the transverse thin cell layers (tCLPs) did not only boost the regeneration capacity in recalcitrant genotypes of rice, sorghum and maize (Nhut et al., 2003) but also in the non-food millet called large crabgrass millet (Le et al., 1997, 1998).

Moreover, the regenerative competence of the genotype should be considered while choosing the appropriate explant as different genotypes of same species show huge variability. Hence, appropriate regeneration techniques need to be established at least for economically important millets. A large scale screening methodology has to be developed in order to determine the regeneration capacity for diverse genotypes of millets as it was investigated for rice (Dabul et al., 2009). Another important point in regeneration is the prolongation of the viability period of the explant. TDZ (thidiazuron), a cotton defoliant with cytokinin-like activity, was found to increase the viability by shortening the somatic embryogenesis phase (Mok et al., 1982). TDZ has been used for the enhancement of morphogenic competence in Poaceae since mid-1990 (Wenzhong et al., 1994). The beneficial effect of TDZ on shoot bud development was observed in millets such as kodo millet, finger millet and switchgrass (Gupta and Conger, 1998; Vikrant and Rashid, 2002; Ceasar and Ignacimuthu, 2008).

Another important parameter affecting the efficiency of *in vitro* regeneration is the composition of the culture medium. Diverse types of media were shown to improve regeneration in major crops (Wang et al., 1993; Kumlehn et al., 2006) and millets (Heyser, 1984; Nayak and Sen, 1991; Latha et al., 2005; Gugsa and Kumlehn, 2011). The type and concentration of carbon source affect the efficacy of regeneration. For example, an increased osmolarity due to sucrose, sorbitol, mannitol and maltose showed to improve embryo formation and maintenance in maize (Lu et al., 1983). The replacement of sucrose by maltose increased the efficiency of embryogenesis and regeneration in wheat and tef (Mendoza and Kaeppler, 2002; Gugsa and Kumlehn, 2011).

Optimum regeneration techniques targeting the rescuing of the progenies of crosses between economically important millets and their wild relatives need to be investigated in order to introduce important agronomic traits to the cultivated species. These introgressions between divergent species require a special regeneration procedure known as embryo rescue, a technique which allows the hybrids to become fertile. Although embryo rescue techniques are widely applied in crop plants (Sharma and Ohm, 1990; Price et al., 2005), they are not yet developed for millets.

Significant developments have also been made in transformation of major cereals (Repellin et al., 2001). However, optimum transformation methods are not yet established for most millet species. Although *Agrobacterium*-mediated transformation is becoming the main mode of transformation for major cereals (Komari and Kubo, 1999; Koichi et al., 2002) especially due to its simple integration in the plant genome, it is not widely practiced in millets. On the contrary, the microprojectile bombardment method is the dominant transformation technique in millets despite its pitfalls especially related to complex integration pattern of the transgene in the plant genome. Improvement in *Agrobacterium*-mediated transformation was achieved by applying acetosyringone in both the transformation and co-cultivation media. The addition of acetosyringone and cell extracts from dicot plants during the co-cultivation process increased the transformation efficiency of rice (Hiei et al., 1994) and recently also in millets (Liu et al., 2005; Sharma et al., 2011).

Another important point to be considered in monocot transformation is the selection of the right promoter. Ubiquitin and actin promoters are widely used in cereals transformation. Recently, two ubiquitin promoters, namely Ubi 1 and Ubi 2, which were isolated from switchgrass (*Panicum virgatum* L.), resulted in strong expression of reporter gene (Mann et al., 2011). In addition, except in few cases, transformation studies in millets did not address important agronomic problems or traits. Based on the available literature, the only two food-security important millets in which transformation was focused on transferring agronomically valuable traits were pearl millet and finger millet (Latha et al., 2005, 2006; Girgi et al.,

2006; O'Kennedy et al., 2011a).

CONCLUSION

In general, millets play huge role in the livelihood of the population of developing world especially due to their enormous contribution to food security. However, since these crops are not sufficiently studied, for which the name orphan crops is given to these groups of crops, they remain largely unimproved. Both conventional and modern improvement techniques were not adequately implemented. The regenerative competence of the explant should be considered while choosing the appropriate explant as different genotypes of same species show enormous variability in regeneration. Efforts need to be made to investigate appropriate regeneration techniques at least for economically important millets. A large scale screening methodology has to be developed in order to determine the regeneration capacity for diverse genotypes of millets as it was investigated for rice (Dabul et al., 2009). A broad range screening made for rice set a threshold of 85% of somatic embryogenesis as an earlier indicator for efficient regeneration. Although, extensive regeneration studies were made for different millets, only limited transformation experiments were conducted to date. Hence, future research needs to develop a robust transformation protocols for each type and ecotype of millet using *Agrobacterium* method.

ACKNOWLEDGEMENTS

We would like to thank Syngenta Foundation for Sustainable Agriculture and University of Bern for financial and technical support provided to our Tef Improvement Project.

Abbreviations: 2,4-D, (2,4-dichlorophenoxy) acetic acid; **3,6-D,** dichloromethoxybenzoic acid; **ABA,** abscisic acid; **BA,** 6-benzylaminopurine; **dicamba,** 3,6-dichloro-2-methoxybenzoic acid; **GA₃,** gibberellic acid; **IAA,** indole-3-acetic acid; **KIN,** kinetin; **NAA,** α-naphthaleneacetic acid; **PAA,** 2-phenylacetic acid; **pCPA,** 4-chlorophenoxyacetic acid; **picloram,** 4-Amino-3,5,6-trichloropicolinic acid; **TDZ,** thidiazuron; **TIBA,** 2,3, 5-triiodobenzoic acid.

REFERENCES

Arockiasamy S, Prakash S, Ignacimuthu S (2001). High regenerative nature of *Paspalum scrobiculatum* L, an important millet crop, Curr. Sci., 80: 496-498.

Assefa K, Gaj MD, Maluszynski M (1998). Somatic embryogenesis and plant regeneration in callus culture of tef, *Eragrostis tef* (Zucc.) Trotter, Plant Cell Rep., 18: 154-158.

Bajaj YPS, Sidhu BS, Dubey VK (1981). Regeneration of genetically of genetically diverse plants from tissue cultures of forage grass - *Panicum* sps, Euphytica, 30: 135-140.

Baker RD (2003). Millet production. Guide A-414. In: Cooperative Extension Service, College of Agriculture and Home Economics, New Mexico University, Las Cruces, p. 6.

Bartlett JG, Alves SC, Smedley M, Snape JW, Harwood WA (2008). High-throughput *Agrobacterium*-mediated barley transformation, Plant Methods, 4: 22.

Bekele E, Klock G, Zimmermann U (1995) Somatic embryogenesis and plant regeneration from leaf and root explants and from seeds of *Eragrostis tef* (Gramineae), Hereditas, 123: 183-189.

Belton PS, Taylor JRN (2004) Sorghum and millets: protein sources for Africa, Trends. Food Sci. Technol., 15: 94-98.

Bhojwani SS, Hayward C (1977). Some observations and comments on tissue culture of wheat, Z. Pflanzenphysiol., 85: 341-347.

Brutnell TP, Wang L, Swartwood K, Goldschmidt A, Jackson D, Zhu XG, Kellogg E, Van Eck J (2010). *Setaria viridis*: A Model for C4 Photosynthesis, Plant Cell, 22: 2537-2544.

Campos JMS, Calderano CA, Pereira AV, Davide LC, Viccini LF, Santos MO (2009). Embriogênese somática em híbridos de *Pennisetum* sp. e avaliação de estabilidade genômica por citometría, Pesq. Agropec. Bras, 44: 38-44.

Ceasar SA, Ignacimuthu S (2008). Efficient somatic embryogenesis and plant regeneration from shoot apex explants of different Indian genotypes of finger millet (*Eleusine coracana* (L.) Gaertn.), In Vitro Cell. Dev. Biol-Plant, 44: 427-435.

Ceasar SA, Ignacimuthu S (2010). Effects of cytokinins, carbohydrates and amino acids on induction and maturation of somatic embryos in kodo millet (*Paspalum scorbiculatum* Linn.), Plant Cell. Tiss. Organ. Cult., 102: 153-162.

Chandrasekara A, Shahidi F (2011). Antiproliferative potential and DNA scission inhibitory activity of phenolics from whole millet grains, J. Funct. Foods, 3: 159-170.

Chin JC, Scott KJ (1977). The isolation of a high rooting cereal callus line by recurrent selection with 2,4-D, Z. Pflanzenphysiol., 85: 117-124.

Chu CC, Wang CC, Sun CS, Hsu C, Yin KC, BI CV (1975). Establishment of an efficient medium for anther culture of rice through comparative experiments on the nitrogen source, Sci. Sin., 18: 659-668.

Dabul ANG, Belefant-Miller H, RoyChowdhury M, Hubstenberger JF, Lorence A, Philips GC (2009). Screening of a broad range of rice (*Oryza sativa* L.) germplasm for *in vitro* rapid plant regeneration and development of an early prediction system, In vitro Cell. Dev. Biol-Plant, 45: 414-420.

Deambrogio E, Dale PJ (1980). Effect of 2,4-D on the frequency of regenerated plants in barley and on genetic variability between them, Cereal. Res. Comm., 8: 417-423.

Deutsch F, Kumlehn J, Ziegenhagen B, Fladung M (2004). Stable haploid poplar callus lines from immature pollen culture, Physiol. Plant, 120: 613-622.

Devi P, Sticklen MB (2002). Culturing shoot-tip clumps of pearl millet (*Pennisetum glaucum* (L.) R. Br.) and optimal microprojectile bombardment parameters for transient expression, Euphytica, 125: 45-50.

Dong YZ, Duan SJ, Zhao LY, Yang XH, Jia SR (1999). Production of transgenic millet and maize plants by particle bombardment, Sci. Agric. Sinica, 32: 9-13.

Eapen S, George L (1990). Influence of phytohormones, carbohydrates, aminoacids, growth supplements and antibiotics on somatic embryogenesis and plant differentiation in finger millet, Plant Cell. Tiss.Organ. Cult., 22: 87-93.

FAOSTAT. http://faostat.fao.org.

George L, Eapen S (1990). High frequency plant-regeneration through direct shoot development and somatic embryogenesis from immature inflorescence cultures of finger millet (*Eleusine coracana* Gaertn). Euphytica, 48: 269-274.

Girgi M, O'Kennedy MM, Morgenstern A, Mayer G, Lörz H, Oldach KH (2002). Transgenic and herbicide resistant pearl millet (*Pennisetum glaucum* L.) R.Br. via microprojectile bombardment of scutellar tissue, Mol. Breed, 10: 243-252.

Girgi M, Breese WA, Lörz H, Oldach KH (2006). Rust and downy mildew resistance in pearl millet (*Pennisetum glaucum*) mediated by heterologous expression of the *afp* gene from Aspergillus giganteus,

Transgenic Res., 15: 313-324.

Goldman JJ, Hanna HH, Fleming G, Ozias-Akins P (2003). Fertile transgenic pearl millet (*Pennisetum glaucum* (L.) R. Br.) plants recovered through microprojectile bombardment and phosphinothricin selection of apical meristem-, inflorescence-, and immature embryo-derived embryogenic tissues, Plant Cell Rep., 21: 999-1009.

Gugsa L, Sarial AK, Lörz H, Kumlehn J (2006). Gynogenic plant regeneration from unpollinated flower explants of *Eragrostis tef* (Zuccagni) Trotter, Plant Cell Rep., 25: 1287-1293.

Gugsa L, Kumlehn J (2011). Somatic embryogenesis and massive shoot regeneration from immature embryo explants of tef. Biotechnol/ Res. Int., 2011: 30973.

Gupta SD, Conger BV (1998). *In vitro* differentiation of multiple shoot clumps from intact seedlings of switchgrass, *In vitro* Cell. Dev. Biol. Plant, 34: 196-202.

Gupta P, Raghuvanshi S, Tyagi AK (2001). Assessment of the efficiency of various gene promoters via biolistics in leaf and regenerating seed callus of millets, *Eleucine coracana* and *Echinochloa crusgalli*, Plant Biotechnol., 18: 275-282.

Hauptmann RM, Ozias-Akins P, Vasil V, Tabaeizadeh Z, Rogers SG, Horsch RB, Vasil IK, Fraley RT (1987). Transient expression of electroporated DNA in monocotyledonous and dicotyledonous species, Plant Cell Rep., 6: 265-270.

Hauptmann RM, Vasil V, Ozias-Akins P, Tabaeizadeh Z, Rogers SG, Fraley RT, Horsch RB, Vasil IK (1988). Evaluation of selectable markers for obtaining stable transformants in the Gramineae, Plant Physiol., 86: 602-606.

Heyser JW, Nabors MW (1982). Regeneration of proso millet from embryogenic calli derived from various plant parts, Crop Sci., 22: 1070-1074.

Heyser JW (1984). Callus and shoot regeneration from protoplasts of proso millet (*Panicum miliaceum* L.), Z. Pflanzenphysiol. Bd., 113: 293-299.

Hiei Y, Ohta S, Komari T, Kumashiro T (1994). Efficient transformation of rice (*Oryza sativa* L.) mediated by *Agrobacterium* and sequence analysis of the boundaries of the T-DNA, Plant J., 6: 271-282.

Jha P, Yadav CB, Anjaiah, V, Bhat, V (2009). *In vitro* plant regeneration through somatic embryogenesis and direct shoot organogenesis in *Pennisetum glaucum* (L.) R, Br. *In vitro* Cell. Dev. Biol. Plant, 45: 145-154.

Kaur P, Kothari SL (2004). *In vitro* culture of kodo millet: influence of 2,4-D and picloram in combination with kinetin on callus initiation and regeneration, Plant Cell. Tiss. Organ. Cult., 77: 73-79.

Koichi T, Bae CH, Seo MS, Song IJ, Lim YP, Song PS, Lee HY (2002). Overcoming of barriers to transformation in monocot plants, J. Plant. Biotechnol., 4: 135-141.

Komari T, Kubo T (1999). Methods of genetic transformation: *Agrobacterium* tumefasciens. In: Vasil IK (ed) Molecular improvement of Cereal Crops. Kluwer, pp. 43-82.

Kothari SL, Agarwal K, Kumar, S (2004) Inorganic nutrient manipulation for highly improved *in vitro* plant regeneration in finger millet - *Eleucine coracana* (L.) Gaertn. *In vitro* Cell Dev Biol. Plant, 40: 515-519.

Kothari SL, Kumar S, Vishnoi RK, Kothari A, Watanabe KN (2005). Applications of biotechnology for improvement of millet crops: Review of progress and future prospects, Plant Biotechnol., 22: 81-88.

Kothari-Chajer A, Sharma M, Kachhwaha S, Kothari SL (2008). Micronutrient optimization results into highly improved *in vitro* plant regeneration in kodo (*Paspalum scrobiculatum* L.) and finger (*Eleucine coracana* (L.) Gaertn.) millets, Plant Cell. Tiss. Organ. Cult., 94: 105-112.

Kumar S, Agarwal K, Kothari SL (2001). *In vitro* induction and enlargement of apical domes and formation of multiple shoots in finger millet, *Eleusine coracana* (L.) Gaertn and crowfoot grass, *Eleusine indica* (L.) Gaertn, Curr. Sci., 81: 1482-1485.

Kumlehn J, Serazetdinova L, Hensel G, Becker D, Loerz H (2006). Genetic transformation of barley (*Hordeum vulgare* L.) via infection of and rogenetic pollen cultures with *Agrobacterium tumefaciens*, Plant Biotechnol. J., 4: 251-261.

Lambe P, Dinant M, Matagne RF (1995). Differential long-term expression and methylation of the hygromycin phosphotransferase(hph) and B-glucuronidase (GUS) genes in transgenic pearl millet (*Pennisetum glaucum*), Plant Sci., 108: 51-62.

Lambe P, Mutambel HSN, Deltour R, Dinant M (1999). Somatic embryogenesis in pearl millet (*Pennisetum glaucum*): Strategies to reduce genotype limitation and to maintain long-term totipotency, Plant Cell. Tiss. Organ. Cult., 55: 23-29.

Lambe P, Dinant M, Deltour R (2000). Transgenic pearl millet (*Pennisetum glaucum*). In: Bajaj YPS (ed) Transgenic Crops I, Biotechnol. Agric. Forest., 46: 84- 108.

Latha AM, Rao KV, Reddy VD (2005). Production of transgenic plants resistant to leaf blast disease in finger millet (*Eleusine coracana* (L.) Gaertn.), Plant Sci .,169: 657-667.

Latha AM, Rao KV, Reddy TP, Reddy VD (2006). Development of transgenic pearl millet (*Pennisetum glaucum* (L.) R. Br.) plants resistant to downy mildew, Plant Cell. Rep., 25: 927-935.

Le BV, Thao DMN, Vidal GJ, Van TT (1997). Somatic embryogenesis on thin cell layers of a C4 species, *Digitaria sanguinalis* (L.) Scop, Plant Cell. Tiss. Organ. Cult., 49: 201-208.

Le BV, Janneau M, Do My NT, Vidal J, Thanh Vân KT (1998). Rapid regeneration of whole plants in large crabgrass (*Digitaria sanguinalis* L.) using thin-cell-layer culture. Plant Cell .Rep., 18: 166-172.

Leder I (2004). Sorghum and millets. In: Füleky G (ed.) Cultivated plants, primarily as food sources, in Encyclopedia of Life Support Systems (EOLSS), Developed under the auspices of the UNESCO, Eolss Publishers, Oxford ,UK..

Liu YH, Yu JJ, Zhao Q, Ao GM (2005). Genetic transformation of millet (*Setaria italica*) by *Agrobacterium*-mediated, Agric. Biotechnol. J., 13: 32-37.

Lu C, Vasil V, Vasil, JK (1983). Improved efficiency of somatic embryogenesis and plant regeneration in tissue cultures of maize (*Zea mays* L.) Theor. Appl. Genet., 66: 285-289.

Mann DGJ, King ZR, Liu W, Joyce BL, Percifield RJ, Hawkins JS, LaFayette PR, Artelt BJ, Burris JN, Mazarei M, Bennetzen JL, Parrott WA, Stewart CN (2011). Switchgrass (*Panicum virgatum* L.) ubiquitin gene (PvUbi1 and PvUbi2) promoters for use in plant transformation, BMC. Biotechnol., 11: 74.

McElroy D, Brettell RIS (1994). Foreign gene expression in transgenic cereals. Trends Biotechnol 12: 62-68.

Mekbib F, Mantell SH, Buchanan-Wollaston V (1997). Callus induction and *in vitro* regeneration of tef [*Eragrostis tef* (Zucc.) Trotter] from leaf. J Plant Physiol., 151: 368-372.

Mendoza MG, Kaeppler HF (2002). Auxin and sugar effects on callus induction and plant regeneration frequencies from mature embryos of wheat (*Triticum aestivum* L.) *In vitro* Cell . Dev. Biol-Plant, 38: 39-45.

Mohanty BD, Gupta SD, Ghosh PD (1985). Callus initiation and plant regeneration in ragi (*Eleusine coracana* Gaertn), Plant Cell. Tiss. Organ. Cult., 5: 147-150.

Mok MC, Mok DW, Amstrong DJ, Shudo K, Isogai Y, Okamanto T (1982). Cytokinin activity of N-phenyl-N'-1,2,3-thiadiazol-5-urea(thidiazuron). Phytochem., 21: 1509-1511.

Murashige T, Skoog F (1962). A revised medium for rapid growth and bioassays with tobacco tissue cultures, Physiol. Plant, 15: 473-497.

Muthuramu S, Ibrahim SM, Gunasekaran M, Balu PA, Gnanasekaran M (2008). *In vitro* response of CMS lines and their maintainers in pearl millet [*Pennisetum glaucum* (L.) R. Br.], Plant. Arch, 8: 229-232.

Mythili PK, Satyavathi V, Pavankumar G, Rao MVS, Manga V (1997). Genetic analysis of short term callus culture and morphogenesis in pearl millet, *Pennisetum glaucum*, Plant Cell. Tiss. Organ. Cult., 50: 171-178.

Nabors MW, Heyser JW, Dykes TA, De Mott KJ (1983). Long duration high frequency plant regeneration from cereal tissue cultures, Planta, 157: 385-391.

NAS (National Academy of Science) (1996). Lost crops of Africa. I. Grains. National Academy Press, Washington, DC, USA.

Nayak P, Sen SK (1989). Plant regeneration through somatic embryogenesis from suspension cultures of a minor millet, *Paspalum scrobiculatum*, Plant Cell. Rep., 8: 296-299.

Nayak P, Sen SK (1991). Plant regeneration through somatic embryogenesis from suspension culture-derived protoplasts of *Paspalum scrobiculatum* L, Plant Cell. Rep., 10: 362-365.

Nethra N, Gowda R, Gowda PHR (2009). Influence of culture medium on callus proliferation and morphogenesis in finger millet. In: Tadele Z. (ed) New approaches to plant breeding of orphan crops in Africa.

Proceedings of an International Conference, September 2007, 19-21. Bern, Switzerland. Univ. Bern., pp. 167-178.

Nhut DT, Silva JAT, Bui VL, Tran Thanh Van K (2003). Organogenesis of cereals and grasses by using thin cell layer technique. In: Nhut DT, Van Le B, Tran Thanh Van K, Thorpe T (eds.) Thin cell layer culture system: regeneration and transformation applications. Kluwer Academic Publishers, Dordrecht, The Netherlands, pp. 427-449.

Ntui VO, Azadi P, Supaporn H, Mii M (2010). Plant regeneration from stem segment-derived friable callus of "Fonio" (Digitaria exilis (L.) Stapf.). Sci. Hortic., 125: 494-499.

O'Kennedy MM, Burger JT, Botha FC (2004). Pearl millet transformation system using the positive selectable marker gene phosphomannose isomerise, Plant Cell Rep., 22: 684-690.

O'Kennedy MM, Crampton BG, Lorito M, Chakauya E, Breese WA, Burger JT, Botha FC (2011a). Expression of a beta-1,3-glucanase form a biocontrol fungus in transgenic pearl millet, South Afri. J. Bot., 77: 335-345.

O'Kennedy MM, Martha M, Stark HC, Dube N (2011b) Biolistic-mediated transformation protocols for maize and pearl millet using pre-cultured immature zygotic embryos and embryogenic tissue. Plant Embryo Cult: Methods Protoc., 343-354.

Oldach KH, Morgenstern A,·Rother S, Girgi M, O,Kennedy M, Lörz H (2001). Efficient in vitro plant regeneration from immature zygotic embryos of pearl millet (Pennisetum glaucum (L.) R. Br.) and Sorghum bicolor (L.) Moench Plant Cell Rep., 20: 416-421.

Osuna-Avila P, Nava-Cedillo A, Jofre-Garfias AE, Cabrera-Ponce JL (1995). Plant regeneration from shoot apex explants of foxtail millet, Plant Cell Tiss. Organ. Cult., 40: 33-35.

Patil SM, Sawardekar SV, Bhave SG, Sawant SS, Jambhale ND, Gokhale NB (2009). Development of somaclones and their genetic diversity analysis through RAPD in finger millet (Eleusine coracana L. Gaertn.), Indian J. Genet., 69: 132-139.

Pius J, George L, Eapen S, Rao PS (1993). Enhanced plant regeneration in pearl millet (Pennisetum americanum) by ethylene inhibitors and cefotaxime, Plant Cell Tiss. Organ. Cult., 32: 91-96.

Poddar K, Vishnoi RK, Kothari SL (1997). Plant regeneration from embryogenic callus of finger millet [Eleucine coracana (L.) Gaertn.] on higher concentrations of NH_4NO_3 as a replacement of NAA in the medium, Plant Sci., 129: 101-106.

Price HJ, Hodnett GL, Burson BL, Dillon SL, Rooney WL (2005). A Sorghum bicolor x S. macrospermum hybrid recovered by embryo rescue and culture, Austr. J. Bot., 53: 579-582.

Qin FF, Zhao Q, Ming Ao G (2008). Co-suppression of Si401, a maize pollen specific Zm401 homologous gene, results in aberrant anther development in foxtail millet, Euphytica, 163: 103-111.

Rangan TS (1973). Morphogenic investigations on tissue cultures of Panicum miliaceum, Z .Pflanzenpyhsiol. Bd., 72: 456-459.

Rangan TS (1976). Growth and plantlet regeneration in tissue cultures of some Indian millets: Paspalum scrobiculatum L., Eleusine coracana GAERTN. and Pennisetum typhoideum PERS, Z. Pflanzenphysiol. Bd., 78: 208-216.

Rangan TS, Vasil LK (1983). Somatic embryogenesis and plant regeneration in tissue cultures of Panicum miliaceum L. and Panicum miliare Lamk. Z. Pflanzenphysiol. Bd., 109: 41-48.

Rao AM, Kishor PBK, Reddy LA, Vaidyanath K (1988). Callus induction and high frequency plant regeneration in Italian millet (Setaria italica), Plant Cell Rep., 7: 557-559.

Rao BR, Nagasampige MH, Ravikiran M (2011). Evaluation of nutraceutical properties of selected small millets, J. Pharm. Bioallied. Sci., 3:277-279.

Reddy LA, Vaidyanath K (1990). Callus formation and regeneration in two induced mutants of foxtail millet (Setaria italica), J. Genet. Breed, 44: 133-138.

Repellin A, Baga M, Jauhar PP, Chibbar RN (2001). Genetic enrichment of cereal crops via alien gene transfer: New challenges, Plant Cell Tiss. Organ. Cult., 64: 159-183.

Saalbach G, Koblitz H (1978). Attempts to initiate callus formation from barley leaves, Plant Sci. Lett., 13: 165-169.

Sankhla A, Davis TD, Sankhla D, Sankhla N, Upadhyaya A, Joshi S (1992). Influence of growth regulators on somatic embryogenesis, plantlet regereneration, and post-transplant survival of Echinochloa frumentacea, Plant Cell Rep., 11: 368-371.

Satyavathi V, Rao MVS, Manga V, Chittibabu M (2006). Genetics of some in vitro characters in pearl millet, Euphytica, 148: 243-249.

Schreiber DN, Dresselhaus T (2003). In vitro pollen germination and transient transformation of Zea mays and other plant species, Plant. Mol. Biol. Rep., 21: 31-41.

Sharma HC, Ohm HW (1990). Crossability and embryo rescue enhancement in wide crosses between wheat and three Agropyron species. Euphytica, 49: 209-214.

Sharma M, Kothari-Chajer A, Jagga-Chugh S, Kothari SL (2011). Factors influencing Agrobacterium tumefasciens-mediated genetic transformation of Eleucine coracana (L.) Gaertn, Plant Cell. Tiss. Organ. Cult, 105: 93-104.

Schrawat AK, Lörz H (2006). Agrobacterium-mediated transformation of cereals: A promising approach crossing barriers, Plant. Biotechnol., J 4: 575-603.

Skoog F, Miller CO (1957). Chemical regulation of growth and organ formation in plant tissue cultures in vitro, Symp. Soc. Exp. Biol., 11: 118-131.

Sivadas P, Kothari SL, Chandra N (1990). High frequency embryoid and plantlet formation from tissue cultures of the finger millet - Eleusine coracana (L.) Gaertn, Plant. Cell. Rep., 9: 93-96.

Srivastav S, Kothari SL (2002). Embryogenic callus induction and efficient plant regeneration in pearl millet, Cer. Res. Comm., 30: 69-74.

Tadesse A, Tefera H, Guzmann M, Zapata FJ, Afza R, Mba C (2009). Androgenesis. In: Tournaev A (eds). Advances in Haploid Production in Higher Plants, pp. 274- 283.

Taylor MG, Vasil IK (1991). Histology of, and physical factors affecting, transient GUS expression in pearl millet (Pennisetum glaucum (L.) R. Br.) embryos following microprojectile bombardment, Plant Cell. Rep., 10: 120-125.

Taylor MG, Vasil V, Vasil IK (1993). Enhanced GUS gene expression in cereal/grass cell suspensions and immature embryos using the maize ubiquitin-based plasmid pAHC25, Plant Cell. Rep., 12: 491-495.

Thomas TD (2008). The role of activated charcoal in plant tissue culture, Biotechnol. Advan., 26: 618-631.

Vasil V, Vasil IK (1981). Somatic embryogenesis and plant regeneration from tissue cultures of Pennisetum americanum, and P. americanum x P. purpureum hybrid. Am. J. Bot., 68: 864-872.

Vasil V, Vasil IK (1982). Characterization of an embryogenic cell suspension culture derived from cultured inflorescences of Pennisetum americanum (pearl millet, Gramineae), Am. J. Bot., 69: 1441-1449.

Vikrant, Rashid A (2001) Direct as well as indirect somatic embryogenesis from immature (unemerged) inflorescence of a minor millet Paspalum scrobiculatum L. Euphytica, 120: 167-172.

Vikrant, Rashid A (2002a). Induction of multiple shoots by thidiazuron from caryopsis cultures of minor millet (Paspalum scrobiculatum L.) and its effect on the regeneration of embryogenic callus cultures, Plant Cell. Rep., 21: 9-13.

Vikrant, Rashid A (2002b). Somatic embryogenesis from immature and mature embryos of a minor millet Paspalum scrobiculatum L, Plant Cell. Tiss. Organ. Cult., 69: 71–77.

Vikrant, Rashid A (2003). Somatic embryogenesis or shoot formation following high 2,4-D pulse-treatment of mature embryos of Paspalum scrobiculatum, Biol. Plant, 46: 297-300.

Wang XH, Lazzeri PA, Lörz H (1993). Regeneration of haploid, dihaploid and diploid plants from anther- and embryo-derived cell suspensions of wild barley (Hordeum murinum L.), J. Plant. Physiol., 141: 726-732.

Wang MZ, Pan YL, Li C (2011). Culturing of immature inflorescences and Agrobacterium-mediated transformation of foxtail millet (Setaria italica), Afr. J. Biotechnol., 10: 16466-16479.

Warner DA, Edwards GE (1988). C_4 photosynthesis and leaf anatomy in diploid and autotetraploid Pennisetum americanum (pearl millet), Plant Sci., 56: 85-92.

Wenzhong T, Rance I, Sivamani E, Fauquet C, Beachy RN (1994). Improvements of plant regeneration frequency in vitro in Indica rice, Chinese. J. Genet., 21: 105-112.

Wernicke W, Brettell R (1980). Somatic embryogenesis from Sorghum bicolor leaves. Nature, 287: 138-139.

Wernicke W, Milkovits L (1984). Developmental gradients in wheat leaves - Response of leaf segments in different genotypes culture *in vitro*. J. Plant Physiol., 115: 49-58.

Xu ZH, Wang DY, Yang LJ, Wei ZM (1984). Somatic embryogenesis and plant regeneration in cultured immature inflorescences of *Setaria italic,* Plant Cell Rep., 3: 149-150.

Yemets AI, Klimkina LA, Tarassenko LV, Blume YB (2003). Efficient callus formation and plant regeneration of goosegrass (*Eleusine indica* (L.) Gaertn.), Plant Cell. Rep., 21: 503-510.

Yemets AI, Radchuk V, Bayer O, Bayer G, Pakhomov A, Vance Baird W, Blume Y (2008). Development of transformation vectors based upon a modified plant α-tubulin gene as the selectable marker, Cell. Biol. Int., 32: 566-570.

20

Emerging trends in nanobiotechnology

K. Sobha[1]*, K. Surendranath[2], V. Meena[3], T. Keerthi Jwala[1], N. Swetha[1] and K. S. M. Latha[1]

[1]Department of Biotechnology, RVR and JC College of Engineering, Chowdavaram, Guntur – 522 019, Andhra Pradesh, India.
[2]Department of Physics, RVR and JC College of Engineering, Chowdavaram, Guntur – 522 019, Andhra Pradesh, India.
[3]Department of Chemical Engineering, Andhra University, Visakhapatnam – 530 003, Andhra Pradesh, India.

Nanobiotechnology, an exciting interdisciplinary field of science, is making rapid progress in recent years with the development of new kinds of materials with all the desired physico-chemical properties needed for their successful application in various fields, in particular, medicine. Nanomaterials find applications in different thrust areas of medicine like therapeutics, diagnostics, surgical devices/implants, novel drug delivery systems etc. Recent advancements in this field include the development of semiconductor nanocrystals called "Quantum Dots" (QDs) and their very recent modifications called "Cornell Dots" (CU). Both QDs and CUs have extra-ordinary physico-chemical properties and have either low or no toxicity at all depending on the type of shell coated around the heavy metal. Of late, the toxic heavy metal core is also being replaced suitably for avoiding any potential risk during the long accumulation periods of these particles in biological tissues. This review focuses on the emerging trends in the development of wide array of nanomaterials for biological applications. The areas of emphasis include mainly the QDs - their properties, toxicity studies and some of their biological applications like labeling of cellular structures/molecules, cell uptake, biocompatibility, bioconjugation etc. Also, a short note is added on Cornell dots.

Key words: Nanobiotechnology, nanomaterials, quantum dots, Cornell dots, biological applications, biocompatibility, bioconjugation etc.

INTRODUCTION

"Nanobiotechnology", an extended term, can be defined as the Science and Engineering involved in the design, synthesis and characterization of non-toxic bioactive nanomaterials and devices which interact with cells and tissues at a molecular level with a high degree of specificity. These engineered materials and devices at the nanometer scale are constituted by molecules and atoms that were manipulated for specific and controlled physico-chemical properties.

The different synthetic methods of nanoengineered materials and devices, employing precursors from any of the three states of matter viz., solid, liquid and gas, have been broadly classified under "Top down" or "Bottom up" approaches. "Top down" technique encompasses the methodology of incorporating smaller-scale details into macroscopic material and is best exemplified by the

photolithography technique used in the manufacture of integrated circuits by the semiconductor industry (Hu and Shaw, 1999).

The classic biological example of the lithographic technique is provided by the neuron-astrocyte communication studies (Takano et al., 2002). In this study, neuron and astrocyte cell cultures were placed in adjacent wells in agar that were connected by a channel made of poly dimethyl siloxane, allowing the diffusion of soluble factors. "Bottom-up" approaches, on the contrary, begin by designing and synthesizing custom-made molecules that have the inherent ability to self-assemble into structures of higher order. The critical part of this approach lies in the design of molecules which when subjected to physical/chemical trigger (change in pH, specific solute concentration, non-covalent interactive forces, application of electric field etc.) undergo self-assembly into macroscopic structures displaying desirable and unique physico-chemical properties that are not manifested by the constituents (Silva, 2004). The best examples for this

*Corresponding author. E-mail: sobha_kota@yahoo.co.in.

approach come from the field of orthopedics. Bone is one type of connective tissue subjected to enormous use and abuse and consequent stress leading to frequent wear and tear. The repair process of this tissue is attempted in several studies using various types of artificial bone and biomaterials (Hartgerink et al., 2001; Stupp and Ciegler, 1992; Stupp et al., 1993). These materials are constantly being improved for better mechanical and cell signaling properties as well (Abbot and Cyranoski, 2003). Gradually nanomaterials are finding sustainable applications in almost every branch of medicine and diagnostics.

The major complications in building up nanomaterials come from their extremely minute size and time scales at the level of atomic bond oscillations. Therefore these studies warrant highly sophisticated theoretical and experimental tools. For visualization, characterization and manipulation of nanomaterials, novel physical characterization and sophisticated imaging techniques are under constant developpment. The ultimate aim of this technological development is to understand and evaluate the subtilities of intended interactions between the cells and the nanomaterials. Several significant studies are being carried out across the globe to harness the potential of the emerging field of nanotechnology for wide applications in Medicine such as development of novel drug delivery systems (Hanes et al., 1997; La Van Da et al., 2002; Lockman et al., 2002), porous self-assembling bilayer tubule systems (Schnur, 1993; Schnur et al., 1994), dendrimers that could be used as gene therapy agents or as contrast agents in diagnostic imaging (Karak and Maiti, 1997; Laus et al., 2003). Biological applications are innumerable ranging from the study of molecular motors such as flagella of bacteria (Imae and Atsumi, 1989; Noji et al., 1997) to the development of molecular computers (Adleman, 1994; Birge, 1995).

Visualization of structures and compartments within cells and molecules involved in biochemistry is not possible by routine microscopy as they are transparent to visible light and therefore molecules of interest have to be labeled with a marker. One of the common labeling techniques is Fluorescence labeling (Stephens and Allan, 2003; Weijer, 2003; Miyawaki et al., 2003) in which fluorophores can either be directly attached to target or can be attached to a molecule that binds to the target (Geiger and Volberg, 1994) by molecular recognition. Traditional fluorophores have limitations because of their conformational sensitivity to the local environment and consequent loss of fluorescence, inherent low resolution to optical microscopy as well as photobleaching (Pawley and Centonze, 1994). These limitations are to a considerable extent overcome by the use of colloidal metal particles particularly gold nanoparticles that give adequate contrast for imaging with electron microscopy. Yet again the disadvantage is that only fixed (dead) samples could be visualized. Therefore, in recent years, semiconductor nanocrystals, also called Quantum Dots (QDs) are introduced as a novel type of colloids for biolabeling,

imaging and targeting. These QDs are nm size luminescent semiconductor crystals and have unique physical and chemical properties due to their size and highly compact structure. They emit different wavelengths over a broad range of the light spectrum from visible to infrared, depending on their size and chemical composition. Since these are fluorescent, inorganic nanosolids, they can resist photobleaching and could be observed with high resolution by electron microscopy.

CHARACTERISTIC FEATURES OF NANOSYSTEMS

Nanosystems, mostly inorganic, are defined as nanosized chemical objects with special features because of their quantum size and geometric effects. There are also inorganic-organic hybrid nanosystems and systems with specific applications in biology. Nanosystems, which are via media of isolated molecules with properties that follow quantum mechanical rules and the bulk materials that obey laws of classical mechanics, exhibit unique electronic, photochemical, electrochemical, optical, magnetic, mechanical or catalytic properties differing from those of molecular units and the macroscopic systems. The nanosystems are classified hierarchically as zero-, one-, two- and three dimensional nanosystems. Zero-dimensional systems include pseudo-spherical objects such as nanoclusters, nanoparticles or ceramic nanopowders. One dimensional systems account for carbon-based, metal-based or oxide-based systems in which the extension over one dimension is predominant over the other two. Ex: Solid nanowires, nanofibers or nanorods and hollow nanotubes. Two dimensional nanosystems are exemplified by the crystalline flat nanometric materials such as nanodiscs or nanoprisms and the amorphous nanofilms and nanomembranes. Finally, the three dimensional nanosystems, which can also be generated from simpler nanocomponents, consist of both crystalline and amorphous nanostructures such as nanocrystals and a wide array of ordered nanoarranged porous materials. The physical behaviour and optical responses of nanoparticles are determined by the wavelengths of light absorbed/scattered which in turn are dependent on the particle size and shape. Metallic nanoparticles (Au and Ag) are used as sensors to detect analytes through surface enhanced Raman scattering and other optical effects characteristic to the size range of 10 to 100 nm. There are several synthetic methods available in literature for the production of nanoparticles and these methods could be manipulated to control the particle size, morphology, crystallinity, shape and properties depending upon the intended application. Also, several kinds of materials like glass, metal etc. could be used for the production of nanoparticles.

Three dimensional super structures such as super lattices containing nanoparticles in predictable and periodic lattice points could be formed easily from 0-D nanosystems by chemical interparticle interactions. The symmetry of the close-packed super lattice could be modulated by a careful control of the assembling parameters like particle size, shape and interparticle distances. Nanoporous materials, crystalline or non-crystalline display voids/pores ranging from 1 - 100 nm. Depending on pore size, they are categorized according to IUPAC as microporous (<2 nm), mesoporous (between 2 and 50 nm) and macromaterials (>50 nm). Inorganic-organic hybrid nanocomposites are a class of materials defined as inorganic nanostructures included in an organic matrix. Interestingly, the constituents of the hybrid system show special characteristics different from the ones they would have in the absence of the other. The organic matrices are mainly the polymers of varied compositions as they undergo phase transformation easily from fluid to solid and therefore could be moulded into desired shapes.

MAGNETIC NANOPARTICLES

These offer controlled size (a few nm to tens of nm) and are comparable to the dimensions of a cell (10 - 100 μm), a virus (20 - 450 nm), a protein (5 - 50 nm) or a gene (2 nm wide and 10 - 100 nm long). For this reason, the magnetic nanoparticles can be coated with biological molecules and make them interact with or bind to a biological entity. These particles could also be manipulated by an external magnetic field gradient. Since custom designed and fabricated multifunctional nanoparticles have wide applications in biology and medicine, researchers are currently modifying the magnetic nanoparticles by attaching them to antibodies, proteins, dyes etc. and / or integrate the magnetic nanoparticles by sequential growth / coating with quantum dots or metallic nanoparticles to attain multifunctionality like specific targeting of a drug to the region of tumour/cancer. The magnetic nanoparticles can also be made to heat up such that they could be used as hyperthermia agents, delivering toxic amounts of thermal energy to targeted bodies such as tumours or as chemotherapy and radiotherapy enhancement agents, where a moderate degree of tissue warming results in more effective malignant cell destruction. Conjugated magnetic nanoparticles with ligands, antibodies or proteins exhibit highly selective binding and show special properties like paramagnetism, fluorescence or enhanced optical contrast. Therefore these could be applied to biological/ medical problems such as protein purification, bacterial detection, toxin decorporation, enhanced medical imaging and controlled drug delivery. Heterodimer nanostructures with both magnetic and fluorescent properties would definitely serve as good candidates for dual-functional molecular imaging like a combination of MRI (Magnetic Resonance Imaging) and fluorescence imaging. Similarly, the integration of magnetic and metallic nanoparticles into heterodimers allows attachment of different kinds of functional molecules onto the specific parts of heterodimers, which then could bind to multiple receptors or act as agents for multimodality imaging. For example, yolk-shell nanostructures developed through encapsulation of a potential anti-cancer drug by iron oxide nanoshells serve as ideal nanodevices for controlled drug delivery.

Early detection of pathogens at their ultra low concentrations enables efficient clinical diagnosis and environmental monitoring. This is exemplified by the use of vancomycin-conjugated FePt nanoparticles to capture and detect pathogens such as vancomycin-resistant enterococci (VRE) and other gram +ve bacteria as well as Gram −ve bacteria at ultra low concentrations (Gu et al., 2003a, 2003b). It is reported that the detection limit achieved using FePt@Van magnetic nanoparticles matches with that of PCR (Polymerase Chain Reaction) based assays and in addition this protocol is faster and useful when PCR is inapplicable. Combination of FePt@Van biofunctional magnetic nanoparticles with fluorescence dyes provides quick, sensitive and low cost detection of bacteria in blood (Gao et al., 2006).

Protein purification is one critical step in down stream processing and greatly influences the economy of production of valuable biological products used in various industries, more particularly pharmaceutics. Research studies indicate that, of the existing protocols, magnetic separation and purification is a convenient method for selective and reliable capture of specific proteins, genetic materials, organelles and cells (Saiyeed et al., 2003; Safarik and Safarikova, 2004; Xu et al., 2004a, 2004b). Magnetic nanoparticles are also found to be of use in toxin decorporation as exemplified by the removal of UO_2^{2+} to the extent of 99% and 69% from water and blood respectively (Leroux, 2007). Conjugation of magnetic nanoparticles with organic dyes such as Cy5.5, FITC etc. will offer the multifunctional nanoprobes combining MRI and optical imaging (Cheon and Lee, 2008). A good example for this kind of bimodal imaging nanoprobe is obtained by the conjugation of Fe_3O_4 nanoparticles and porphyrin. Since porphyrin has low systemic toxicity and well understood pharmacokinetics, porphyrin derivatives are put to use in photodynamic therapy (PDT). Thus porphyrin modified Fe_3O_4 nanoparticles can act as a multifunctional nanomedicine that combine PDT anti-cancer treatment and non-invasive MRI imaging.

METALLIC NANOPARTICLES

Nanoparticles of metals like Au, Ag and Pt have excellent optical properties and hence could be used as optical contrast agents. They could also be used as multimodal sensors (a combination of optical and scattering imaging) and in photothermal therapy (Skrabalak et al., 2008).

Heterodimers of two distinct nanospheres like that of Fe_3O_4-Ag could be produced by sequential growth of metallic components (nucleation) on the exposed surface of the magnetic nanoparticles. Thus the general method for the formation of heterodimer nanoparticles is liquid-liquid interface heterogeneous growth (Gao et al., 2009). Another method to fabricate Fe_3O_4-Au heterodimers in a homogeneous organic solvent was reported in which $Fe(CO)_5$ is subjected to thermal decomposition onto the surface of Au nanoparticles and the following oxidation of intermediates produces uniform Fe_3O_4-Au heterodimers (Yu et al., 2005). This kind of multifunctional heterodimers (Fe_3O_4-Ag or Fe_3O_4-Au), along with the retention of their individual distinctive functionalities, can respond to magnetic forces, demonstrate enhanced resonance absorption and scattering and bind with specific receptors (Jiang et al., 2008; Xu et al., 2008).

YOLK-SHELL NANOSTRUCTURES

In conventional use of magnetic nanoparticles for drug delivery, they are coated with polymers and drugs are encapsulated to form nanocapsules or micelles (Gupta and Gupta, 2005). However, $FePt@CoS_2$ yolk-shell nanoparticles were developed as novel and potential nanodevices for controlled drug release in the treatment of cancer (Gao et al., 2007). Since FePt nanoparticles without any surface coating could act as potential anticancer drug like that of Cisplatin, the sequential growth of CoS_2 porous nanoshells by Kirkendall effect (Yin et al., 2004; Gao et al., 2006a) produced $FePt@CoS_2$ nanoparticles with an IC_{50} value lower than that of Cisplatin. Electron microscopic studies of these particles in vivo indicated that they are well taken up by cells and cellular organelles and that the FePt yolks disintegrate after the cellular uptake (Gao et al., 2009). Further developments include the production of $FePt@Fe_2O_3$ yolk-shell nanoparticles with dual functions viz., high cytotoxicity and strong MR contrast enhancement (Peng and Sun, 2007; Gao et al., 2008a). The essential characteristics in the above two kinds of nanoparticles that is $FePt@CoS_2$ and $FePt@Fe_2O_3$ for serving as effective cancer agents were their ability to release Platinum (II) species from the yolk and good permeability of shells to allow the metal ions to get released from the shells.

DENDRIMERS AND DIAMONDOIDS

Dendrimers are cylindrical structures providing new and unique properties that could be applied within the emerging field of nanotechnology (Zhang et al., 2003). These unique macromolecules generally possess multiple branches which can be used to carrya variety of agents. Dendronized polymers (with thickness in the range of several nanometers) composed of a linear polymeric backbone and dendritic side chains, attributed with high transfective efficiency and very low toxicity, could be used to form complexes with biomolecules that are to be delivered to cell and its constituents. Polyamidoamine (PAMAM) dendrimers have well defined surface functionality, good water solubility, low polydispersity and lack immunogenicity. For these properties, they serve as good candidates for use as the backbone of multitasking therapeutics (Choi et al., 2001). Similarly, Diamondoids are cage hydrocarbons with better therapeutic actions and less adverse effects. The smallest diamondoid molecule named adamantine and its derivatives can readily be synthesized and used for cancer treatment (Mansoori et al., 2007).

Liposomes

These are nanoscale closed vesicles consisting of a single lipid bilayer and are biodegradable. Liposomes are manufactured to encapsulate drugs for drug delivery like chemotherapy. The enclosed drug is delivered to targeted site only when the liposome adheres to the outer membrane of target cancer cells thus preventing drug toxicity to healthy cells (Silva et al., 2001; Torchilin and Weissig, 2003; Duncan et al., 2005).

QUANTUM DOTS (QD)

These are heterogeneous spherical nanocrystals constituted by a colloidal core and one or more surface coatings. They could be made of wide range of semiconductor metals like Cd, CdSe, CdTe, ZnS, PbS etc. and also by alloys and metals like Au. The size range of QDs is 2-10 nm in diameter (10 to 50 atoms). The surface coatings play a significant role in determining the QD applications as they prevent agglomeration, encapsulate toxic metals, affect stability in aqueous buffers, absorption and transport, modulate immunological responses, determine toxicity and aid in tissue elimination. QD could be customized for specific applications by varying their surface coatings (Ballou et al., 2004; Chang et al., 2006; Ghasemi et al., 2009). QDs are reported to have potential value in medicine, particularly in drug discovery due to their bright fluorescence, narrow emission, broad UV excitation, tunable size and high photostability (Tan et al., 2006). At the same time, they do pose challenges like biocompatibility, toxicity, photo-oxidation etc. which need to be addressed in detail for improving the applications of QDs in drug discovery.

Research on biomedical applications of QDs reveal that QDs are sensitive, stable, non-toxic, versatile fluorescent probes (Ghasemi et al., 2009). Specific targeting of biomolecule- labeled QDs in vivo (Ex: Peptide GFE labeled QDs recognize the membrane dipeptidase on the endothelial cells in lung blood vessels; Peptide F3 labeled QDs bind to blood vessel and tumor cells) without toxic effects is achieved (Akerman et al., 2002). Another tool developed for the identification of target biomolecules is the QD-tagged microbead that gives a specific optical code for each molecule in fluorescent imaging (Han et al., 2001; Battersby et al., 2002). The microbeads are constituted by certain number of beads with prede-termined ratios of colors and emission intensities. Suppose m kinds of colored dots with n kinds of light intensities are used, then $n^m - 1$ microbeads with specific optical codes could be composed. As the target biomolecules could be identified from the specific optical codes of microbeads, it is also called the "barcode" of the target molecule. This novel technique is applied to multiplexed bioanalysis including gene expression (Gao and Nie, 2001; Kralj and Pavelic, 2003).

Synthesis

Early synthesis of QDs employed traditional methods like a combination of electron beam lithography and etching limiting the particle size to the scale of nm only in one dimension but not in other two. Later QDs were prepared in aqueous solutions with added stabilizing agents but the quality of QDs so obtained was poor with low fluore-scence efficiencies and large size variations. Then evolved the high temperature organometallic procedure for growing QDs with perfect crystal structures and narrow size variations. But as the fluorescence property did not improve deposition of surface-capping layer such as ZnS or CdS was employed that dramatically increased the fluorescence properties of CdSe nanocrystals. Alternative precursor materials such as CdO are also used to prepare high quality CdS, CdSe and CdTe nanocrystals. Further, the size of the QD could be controlled by high temperatures (>300^{0}C) and time duration ranging from minutes to hours depending on the desired particle size (Murray et al., 1993; Hines and Guyot-Sionnest, 1996; Peng and Peng, 2001).

Physico-chemical properties

Solubility in aqueous buffers

Synthesis of high quality nanocrystals of various semi-conductor materials could be carried out in organic solvents at high temperatures in the presence of surfactants yields monodisperse and stable particles. This procedure produces nanoparticles with the polar surfactant head group attached to the inorganic surface and the hydrophobic chain protruding into the organic solvent mediating colloidal stability. The problem with these surfactant coated particles is that they are insoluble in water and hence limit their biological applications. This in turn could be overcome either by replacing the surfactant layer or by coating with an additional layer

thereby introducing either electric charge or hydrophilic polymers for mediating solubility in water. In general, coulomb repulsion between nanocrystals of same surface charge prevents aggregation in water but in salt containing solutions such as cell culture media, charged particles tend to aggregate resulting in flocculation. On the other hand, use of hydrophilic polymers like PEG or dextrane involves steric stabilization such that vander-waal forces leading to flocculation, as in the above case, does not happen here. In practice, hydrophobic nanocry-stals can be made hydrophilic by exchanging surfactant coatings with ligand molecules which have reactive functional groups towards nanocrystal surface and hydrophilic groups on other end that ensure water stability. The anchoring groups used frequenty is thiol and carboxyl as hydrophilic head groups for pH 5. Mercaptohydrocarbonic acids like mercaptoacetic acid (Chan and Nie, 1998; Kloepfer et al., 2003), mercaptopropionic acid (Mitchell et al., 1999), synthetic peptides with multiple cysteines (Sukhanova et al., 2002; Sukhanova et al., 2004; Pinaud et al., 2004), Dithiothreitol (DTT) (Pathak et al., 2001), organic dendrons (Wang et al., 2002) and polyethylene glycol (PEG) (Skaff and Emrick, 2003) have been used for solubility in aqueous solution. In order to further stabilize the polymer shell around the nanocrystal, the individual polymer chains are cross-linked.

Photophysical properties

Energy difference between excited and ground state of a quantum dot strongly depends on its size. Upon optical excitation, electrons are excited from the valence shell to the conduction band which is analogous to exciting electrons from the highest occupied molecular orbital (HOMO) to the lowest unoccupied molecular orbital (LUMO). The recombination of electron-hole pairs results in emission of fluorescence light. Typical water soluble nanocrystal comprises a semiconductor core, a shell of semiconductor material with a higher band gap and a hydrophilic coating to warrant water solubility. Organic fluorophores (Ex: Rhodamine, Fluorescein) can be optically excited within a narrow window of wavelengths and consequently the fluorescence emission is also limited to a certain window of wavelength. The fluore-scence spectra of these organic fluorophores are not symmetric but exhibit a tail to longer wavelength called the "red tail". Contrastingly, colloidal quantum dots have a continuous broad absorption spectrum and the fluorescence can be excited with any wavelength shorter than the wavelength of fluorescence. Since the fluores-cence emission spectra of QD are relatively narrow, symmetric and do not exhibit a red tail, many different colours can be distinguished without spectral overlap. This property is highly advantageous with respect to biological samples as their several different compartments /structures/processes could simultaneously be uniquely labeled (Wu et al., 2003; Mattheakis et al., 2004). Thus several quantum dots with different colours of fluorescence can be excited with one single wavelength unlike the case with organic fluorophores. This flexible optical property of QD helps to reduce autofluorescence of biological samples by just selecting the appropriate excitation wavelength for which the autofluorescence is minimum. Further, the sensitivity of detection is greatly enhanced owing to the large separation between excitation and emission wavelengths leading to broad frequency windows in the emission spectra. The biggest advantage of QDs is their reduced tendency to photobleach because of their inorganic nature and this is useful for long term imaging such as fluorescence labeling of transport processes in cells or tracking the path of single membrane bound molecules (Dahan et al., 2003). Colloidal QDs are known to have long fluorescence time and interestingly, can also have fluorescence life times of a few tens of nanoseconds. If time-gated detection is employed that is, fluorescence is recorded after a few nanoseconds of optical excitation, then improved signal-to-noise ration could be achieved (Dahan et al., 2001). For this reason, CdSe/ZnS quantum dots have been used as contrast agents for imaging of blood vessels in living mice (Larson et al., 2003). Since colloidal QDs have a significant two photon cross-section, multiphoton excitation could be used for imaging of structures deep inside the biological systems. At this juncture, one should note that the quantum yield of hydrophilic QDs in aqueous solutions is lower than that of hydrophobic QDs.

Bioconjugation of QDs

Biomolecules can be fluorescence labeled by attaching quantum dots by one of the two strategies: First, biomolecules can be functionalized with a chemical group like mercapto group (-SH) that is reactive towards the semiconductor surface of materials like CdSe/CdS/CdTe /ZnS (Akerman et al., 2002; Rosenthal et al., 2002). The second strategy uses the covalent linking of biomolecules to the outer hydrophilic shell of the QD surface either nonspecifically or specifically by electrostatic interaction (Dubertret et al., 2002). Also, this could be done by means of cross-linker molecules which require the presence of hydrophilic surfactant shells with reactive groups such as $-COOH$, $-NH_2$ or $-SH$. Thus conjugation of QDs with biomolecules is well established for small molecules like biotin, folic acid, serotonin, avidin, albumin, transferring, trichosanthin, lectin, wheat germ agglutinin, antibodies and DNA. Streptavidin coated quantum dots are commercially avail-able and could be readily conjugated with biotinylated proteins and antibodies (Parak et al., 2005).

There are several alternatives in QDs with NIR emission for in vivo imaging compared with organic fluorophores. These special properties of QDs have inspired

the fabrication of hybrid nanostructures that exhibit both fluorescence and magnetism such as Co@CdSe coreshell nanocomposites (Kim et al., 2005) and FePt-ZnS nanosponges (Gu et al., 2005). The combination of super paramagnetism and fluorescence at nm scale should help the biological applications of multifunctional nanomaterials. As the fluorescence of QDs is partially quenched by metallic nanoparticles in FePt-CdX hybrid nanostructures, their replacement with metal oxide nanoparticles results in a good quantum yield. It is shown that the Fe_3O_4 – CdSe heterodimer nanoparticles exhibit the emission wavelength peak at 610 nm with quantum yield of about 38% and these resulting fluorescent magnetic nanoparticles have both superparamagnetism and fluorescence of high quality. Hence the intracellular movements of these particles could be controlled using magnetic force and monitored using a fluorescent microscope (Gao et al., 2008). The magnetic nanoparticles drift to the magnet (external/applied magnetic force) due to the magnetic field gradient (H) and their significant magnetic moment. Although the movement of the magnetic nanoparticles inside the cell is slowed down due to high viscosity of cytosol, yet when the particles approach each other, the magnetic dipolar-dipolar (F_D – $_D$) interactions become the dominant forces and cause aggregation of nanoparticles inside the cells.

Efficacy of conjugated biomolecules with QDs should be experimentally verified for their potential applications in biology since it has been shown that the hybridizing ability of the oligonucleotides, bound to gold (Au) nanoparticles, with their complementary molecules is reduced (Demers et al., 2000). One of the major applications of QD in cell biology is their use as fluorescent markers for labeling of cellular constituents. This is achieved by conjugating receptor-specific ligand molecules such as antibodies to QD for cell surface target recognition and binding. It means that the conjugation should not hamper the molecular recognition mechanism. Various experimental studies using QDs demonstrated that the molecular recognizing abilities of the bound ligands are not compromised (Parak et al., 2005). Studies on neurotransmission, ion transport, enzyme catalysis etc. revealed that the functionality of biomolecules on conjugation with QDs is not significantly reduced and hence could be of vital importance in biological studies (Zhang et al., 2000, Kloepfer et al., 2003). However, contradictory results have also been reported by some investigators as in the case of serotonin-QD conjugate showing drastic reduction in its binding affinity to serotonin-transporter proteins (Rosenthal et al., 2002). Further research studies could alone reveal the extent to which the QD-bioconjugates could be useful in biological applications.

Biocompatibility of QDs

Colloidal QDs constituted by elements like lead, selenium,

cadmium etc. become toxic when they lose their hydrophilic shell coating. However, some cells have developed mechanisms to assemble these toxic ions as a part of biomineralizaton process. Toxicity studies on colloidal QDs which include the study of response of biological material to the presence of toxic ions and the stability of hydrophilic shell coating on the surface of QDs are vital for their enhanced bioapplications. A few investigations have already been carried out on the early embryonic stages of *Xenopus* and the results suggest that the dosage of the introduced particles / cell decide the extent of cellular and phenotypic abnormalities like variations in cell size, movement, axis elongation etc. There are also good number of reports suggesting no harmful effect of colloidal QDs tested on cells grown at various concentrations in culture media which is an encouraging sign for developing more applications for the colloidal QDs in cell proliferation and adhesion studies (Jaiswal et al., 2003; Winter et al., 2001). Colorimetric assays such as MTT (3-(4,5-dimethylthiazol-2-yl)-2,5-diphenyl tetrazolium bromide) were used by investigators to measure the survival rate of cells on incubation with coated toxin (CdSe/ZnS) containing QDs (Mattheakis et al., 2004). Similar other studies revealed no interference of QDs with cellular functions like ligand binding to cellular compartments, protein trafficking, signal transduction etc. It is quite evident from the results of various experiments that the harmful effects on cells are directly proportional to the extent of ion (Cd^{2+}) release (Aldama et al., 2001; Rikans and Yamano, 2000) and therefore appropriate stable encapsulation (like adsorption with bovine serum albumin, BSA) of the QDs would definitely suppress their cytotoxic effects (Derfus et al., 2004). Yet it is imperative that materials other than Cd, Se etc. should be explored for better utilization as no encapsulation mechanism would confer zero degradation on the particles which tend to remain for several years in biological tissues.

Labeling of cellular structures

Fluorescence microscopy is a widely used technique for visualization of molecules and cellular structures (Bruchez et al., 1998) present at the surface (Dahan et al., 2003; Lidke et al., 2004), and for labeling of intracellular constituents, the fluorescence labeled ligands will have to be artificially introduced by any one of the different methods like microinjection, electroporation etc. Researchers interested in these studies could use any of the standardized protocols available in literature (Herzog et al., 1994; Osborn, 1994). If antibodies are used as ligands, first a primary antibody is reacted with the target and then a biotinylated secondary antibody labeled by streptavidin-quantum dot conjugate via biotin-avidin interaction is an advantageous methodology since biotinylated antibodies are commercially available and a universal conjugation of QDs with streptavidin is

adequate enough. For signal amplification, TSA enzyme amplification technique in which the secondary antibody is conjugated with horse-radish peroxidase and quantum dots conjugated to tyramide could be employed (Ness et al., 2003). Although colloidal quantum dots have an edge over organic fluorophores, yet the former cannot completely replace the latter.

Uptake of quantum dots by living cells

After the first report by the Chan and Nie group (1998) on the uptake and incorporation of QDs by living cells, several investigators made similar observations. The detailed study of this property of living cells throws light on the cellular communication modes for functions like exo and endocytoses apart from other functions like the processing of nutrients, uptake of viruses, drugs, nucleic acids and gene delivery. It is not enough if one can make the cells ingest QDs and fluoresce but they should be guided through the membrane into the nucleus for interaction with the chemical molecules like DNA, mRNA etc. as in the case of drug/gene delivery. Release of these QDs from internalized vesicles could be mediated through various methods like use of membrane-disrupting peptides (Wagner et al., 1992; Plank et al., 1994, 1998) and osmotic destabilization (Sonawane et al., 2003). With the use of fluorescein labeled dextrane as marker, Hanaki et al. (2003) showed that endosomes/lysosomes colocalize with ingested quantum dots. Further Jaiswal et al. (2003) demonstrated that quantum dot uptake could be blocked by cooling the cells to 4°C. Interestingly, the ingested QDs were found to be distributed between daughter cells during cell division suggesting an analogy to the nucleic acid delivery. This property could be made use of for arresting the cellular proliferation in malignancies through appropriate drug coated QD delivery.

Cell tracking

QDs by virtue of their optical properties can highlight cells and subcellular structures and hence can be classified as "Labels" or "Contrast agents" for they can greatly enhance the signal-to-noise ratio in medical diagnostic imaging procedures. When used for this purpose, QDs should possess the properties of molecular recognition facilitating the display of miniature structures and/or physiological processes and meet the safety requirements laid down for any labeling substance. "Cell tracking" is an important area of study in developmental biology wherein the cellular predetermination, differenttiation, de-differentiation/re-differentiation, cell lineage (fate maps) and "Homing" of embryonic cells are studied at cellular and molecular level. Also, cell tracking is essential for understanding the molecular mechanisms underlying the development of dreaded cellular malfunctions such as cancer development and metastasis. In all these studies, "cell labeling" is done with microinjection of oil drops/organic fluorophores or transfecting cells with genes for fluorescent proteins like GFP (Tombolini and Jansson, 1998). In recent years, QDs are being tested for their suitability and efficacy as markers (Dubertret et al., 2002) in embryonic development.

"Cell migration" (morphogenetic movements) and "Cell adhesion" are phenomena that play essential role in designing biocompatible surfaces of medical implants which are relevant in orthopedics and in tissue engineering. First studies on cell migration came to light through a popular method called Albrecht-Buehler "Phagokinetic track" method to follow cellular migration on cell culture substrates (Albrecht-Buehler, 1977a; Albrecht-Buehler, 1977b; Albrecht-Buehler and Lancaster, 1976). Originally Au particles were used as markers and were visualized by dark field microscopy (Albrecht-Buehler and Lancaster, 1976) and by Transmission Electron Microscopy (TEM). Later colloidal gold polystyrene microbeads were employed (Obeso and Auerbach, 1984) and presently QDs are put to fruitful use for recording phagokinetic tracks. The advantages of QDs are that they allow tracking the migration behaviour of cells in three dimensional cultures as they could be observed in stacked layers with different colours of fluorescence.

QDs as contrast agents

In medical diagnostics, various imaging procedures such as magnetic resonance imaging (MRI) (a non-invasive technique), Ultrasound (non-invasive), computerized tomography scan (CT scan) (invasive technique) and radioactive imaging such as scintigraphy and positron emission tomography (PET) employ contrast agents like paramagnetic iron oxide nanoparticles (MRI), halogenated organic compounds for X-ray, fluoro-deoxy glucose for PET or fluorocarbon filled microbubbles for ultrasound. For in vivo imaging of organs/tissues, preferential accumulation of a contrast agent in sub-lethal doses in the target structure is the desired principle. This could be achieved either through direct injection into the target organ or administering a substance systemically that would "passively" accumulate in the intended tissue owing to its biophysical properties or combine "passive" targeting with molecular recognition and thereby make it "active targeting". In the latter two methods i.e. passive and active targeting, a long circulation time in blood evading extrusion from the body by defense systems like reticulo-endothelial system is an essential feature to be endowed upon the contrast agent. Relevant studies demonstrated long blood circulation times with colloidal stability for PEGylated particles (PEG of appropriate chain length) (Ballou et al., 2004). If the intended targeting is of "active" mode, receptor ligands should be

incorporated into the surface coating. Several studies in recent times have successfully employed QDs (with/without fluorescence), with optical detection in imaging (Larson et al., 2003; Ballou et al., 2004; Lim et al., 2003; Kim et al., 2003).

Kim et al. (2004) in their studies reported an exciting medical application which involved local administration of QDs as contrast agents intraepidermally (paw of mouse and thigh of pig) and their subsequent uptake by lymphatic vessels passing to the sentinel nodes (in the axilla of the mouse and in the groin of the pig) were imaged. These results reveal that QDs as contrast agents could be well used in surgical procedures (Uren, 2004) with "ease" for the surgeon in locating and delimiting the part of the tissue to be excised and thereby the extent of incision to be made. The only limitation, hitherto, is the paucity of information on the toxicity to tissues. Gao et al. (2004) described the use of multifunctional QD probes for cancer targeting and imaging in living animals. Their cell tracking-imaging study included passive targeting to experimental tumors through systemic administration and also performed active targeting by coupling a tumour specific antibody to the QD probes. They were able to specifically label "nuclei" of cells in culture before they were inoculated in the animals in order to grow tumors. Subsequently, the labeled cells could be tracked and imaged by virtue of their fluorescence.

Toxicity tests on QDs

Studies indicate that QDs cannot be considered as a uniform group of substances and their characteristics, more particularly toxicity depend on multiple factors which include their inherent physico-chemical properties and environmental conditions as well. Toxicity of QDs is reported to be influenced by their size, charge, concentrations, outer coating bioactivity and oxidative, photolytic and mechanical stability (Ron Hardman, 2005). Although QD nanoparticles have proved to be excellent materials for biomedical imaging, drug targeting and in the electronics industries, it is suggestive to investigate the toxic effects of these particles on various tissues like skin. Since the QD core consists of heavy metals, any alteration in their shell or surface coating will lead to the leaching of heavy metal and cause a potential health risk. With the increase in use of QD, the penetration, localization and toxicity of QDs in skin and skin cells became an issue of concern and Zhang et al. (2008) carried out a study on biological interactions of QD in skin and in human epidermal keratinocytes. Results of the study indicate that QD penetration of skin depends on the type, shape of the rigid core and/or size and more essentially on the surface charge. The study reports that of the several types of PEG coated QDs tested QD 621 remains on the surface of the skin and sometimes near hair follicles. All the three PEGylated QDs (QD 621, QD

565 and QD 655) had the same chemical composition with respect to their core and surface coating but showed varying penetration properties. Earlier studies suggest that elastic properties enable the particles to penetrate faster through the epidermis while rigid particles remain on the surface of the upper skin cells (Honeywell-Nguyen et al., 2004). The common route of penetration in skin is through the intercellular spaces between the corneocytes and studies show that the vertical and lateral gaps between corneocytes are 19nm (Van der Merwe et al., 2006). Since the characteristic nail shaped QD 621 employed by Zhang et al. (2008) had an overall size of 39-40 nm and as the surface PEG coating was soft, these particles alone could penetrate and remain lodged with the skin cells' lipid bilayers. On the other hand, spherical QD 565 and elliptical QD 655 which were smaller and "more" regular in shape could well penetrate the skin. The extent of penetration and the process depend on the intercellular lipid structure or hair follicle density (Monteiro-Riviere, 2008). There are some important reports on the penetration behaviour of nanomaterials like TiO_2 and ZnO which are used as key ingredients in sunscreen creams (Gamer et al., 2006; Cross et al., 2007). In this context, the findings of the study by Zhang et al. (2008) throw light on safe biomedical applications of the QD and suggest that their characteristics like core, shell, surface coatings, shape, size and charge should be tested on skin and human epidermal keratinocytes for minimizing dermal toxicity or irritation. Further, if QD are to be used in biological applications, the lowest concentration with lowest toxicity and high fluorescence intensity need to be identified for individual types rather than a general formula.

In recent years, non-heavy metal quantum dots for life science research have been introduced commercially by the Evident Technology (Troy, New York). These new QDs, called T2-MP EviTags™ possess a ternary core consisting of indium, gallium phosphide coated with a metallic plating shell and a natural coating on the outer layer that offers them low toxicity and a wide range of colours into the near infrared.

CORNELL DOTS

An exciting research study was reported in the year 2005 by the researchers of Cornell University. They have created fluorescent nanoparticles by encapsulating fluorescent dyes with a protective silica shell. These particles with possible applications in biological imaging, optical computing, sensors and microarrays such as DNA chips appear to gradually replace QDs because of their greater chemical inertness and reduced cost. Cornell dots, also known as CU dots, are nanoparticles consisting of a core (about 2.2 nm in diameter) constituted by several dye molecules and surrounded by a protective silica shell, making the entire particle about 25 nm in diameter. CU dots are 20-30 times brighter than single dye molecules in solution, resist photobleaching and produce a large assortment of colours. Therefore, Cornell dots could preferentially be used for biological tagging, imaging and optical computing (Wiesner et al., 2005). CU dots fluoresce so brightly that they can be seen through the skin of a mouse. Experimental studies revealed that they accumulate in organs like liver and bladder within hours after injection and could be harmlessly excreted after they perform their function. Since, CU dots are biologically safe, stable and small enough to be easily transported across the body's structures accompanied by harmless excretion, they are being put to use to "light up" cancerous tumors enabling the surgeons to excise them efficiently. PEGylation of CU dots protects them from being recognized as antigens by body's defense system and enables them to find targeted tumors in considerable time. This technology could be used to know the extent of angiogenesis in tumors, cell death, treatment response and metastatic spread to lymph nodes and other organs of the body. High sensitivity and specificity of CU dots as probes in molecular imaging pave the way for early diagnosis with precision by medical practitioners which is vital for effective treatment of patients.

CONCLUSION

A wide array of nanomaterials and nanodevices are being developed through extensive research in laboratories. These research findings should be evaluated in terms of safety, easy administration and detection, efficacy and affordability for the benefit of extensive practical applications in the field of biomedicine. Quantum dots and Cornell dots, which are semiconductor nanocrystal materials, are being synthesized with a variety of protocols depending on their application. Customized quantum dots are also made commercially available by Quantum Dot Corporation and others. Further, need - based clinical research including toxicological studies will unravel the potential applications of these magic materials in the fields of biology and medicine.

ACKNOWLEDGEMENTS

We sincerely thank the Management and the Principal of RVR and JC College of Engineering, Guntur, Andhra Pradesh, India for their encouragement and support. Also, we gratefully acknowledge the help received from Mr. N. Aravind and Mrs. Ch. V. Ramani in the final preparation of manuscript.

REFERENCES

Abbot A, Cyranoski D (2003). Biology's new dimension. Nature 424: 870-872.

Adleman LM (1994). Molecular computation of solutions of combinatorial problems. Science 266: 1021-1024.

Akerman ME, Chan WC, Laakkonen P, Bhatia SN, Ruoslahti E (2002). Nanocrystal targeting *in vivo*. Proc. Natl. Acad. Sci. USA 99: 12617-12621.

Albrecht-Buehler G (1977a). The phagokinetic tracks of 3T3 cells. Cell 11: 395-404.

Albrecht-Buehler G (1977b). Phagokinetic tracks of 3T3 cells: parallels between the orientation of track segments and of cellular structures which contain actin or tubulin. Cell 12: 333-339.

Albrecht-Buehler G, Lancaster RM (1976). A quantitative description of the extension and retraction of surface protrusions in spreading 3T3 mouse fibroblasts. J. Cell Biol. 71: 370-382.

Aldama J, Wang YA, Peng X (2001). Photochemical instability of CdSe nanocrystals coated by hydrophilic thiols. J. Am. Chem. Soc. 123: 8844-8850.

Ballou B, Lagerholm BC, Ernst LA, Bruchez MP, Waggoner AS (2004). Non invasive imaging of quantum dots in mice. Bioconjugate Chem. 15: 79-86.

Battersby BJ, Lawrie GA, Johnston APR, Trau M (2002). Optical barcoding of colloidal suspensions: applications in genomics, proteomics and drug discovery. Chem. Commun. 14: 1435-1441.

Birge RR (1995). Protein based Computers. Sci. Am. 3: 90-95.

Bruchez MJ, Moronne M, Gin P, Weiss S, Alivisatos AP (1998). Semiconductor nanocrystals as fluorescent biological labels. Science 281: 2013-2016.

Chan WCW, Nie SM (1998). Quantum dot bioconjugates for ultrasensitive nonisotropic detection. Science 281: 2016-2018.

Chang E, Thekkek N, Yu WW, Colvin VL, Drezek R (2006). Evaluation of quantum dot cytotoxicity based on intracellular uptake. Small 12: 1412-1417.

Cheon J, Lee JH (2008). Synergistically integrated nanoparticles as multimodal probes for nanobiotechnology. Acc. Chem. Res. 41: 1630-1640.

Choi Y, Thomas T, Kotlyan A, Islam MT, Baker Jr. JR (2001). Synthesis and functional evaluation of DNA – assembled polyamidoamine dendrimer clusters for cancer cell-specific targeting. Chem. Biol. 12: 3543-3552.

Cross SE, Innes B, Roberts MS, Tsuzuki T, Robertson TA, McCormick P (2007). Human skin penetration of sunscreen nanoparticles: *in-vitro* assessment of a novel micronized zinc oxide formulation. Skin Pharmacol. Physiol. 20: 148-154.

Dahan M, Laurence T, Pinaud F, Chemla DS, Alivisatos AP, Sauer M, Weiss S (2001). Time-gated biological imaging by use of colloidal quantum dots. Opt. Lett. 26: 825-827.

Dahan M, Levi S, Luccardini C, Rostaing P, Riveau B, Triller A (2003). Diffusion dynamics of glycine receptors revealed by single-quantum dot tracking. Science 302: 442-445.

Demers LM, Mirkin CA, Mucic RC, Reynolds RA, Letsinger RL, Elghanian R, Viswanadham G (2000). A fluorescence-based method for determining the surface coverage and hybridization efficiency of thiol-capped oligonucleotides bound to gold thin films and nanoparticles. Anal. Chem. 72: 5535-5541.

Derfus AM, Chan WCW and Bhatia SN (2004). Probing the cytotoxicity of semiconductor quantum dots. Nano. Lett. 4: 11-18.

Dubertret B, Skourides P, Norris DJ, Noireaux V, Brivanlou AH, Libchaber A (2002). *In vivo* imaging of quantum dots encapsulated in phospholipids micelles. Science 298: 1759-1762.

Duncan R, Vicent MJ, Greco F, Nicholson RI (2005). Polymer-drug conjugates: towards a novel approach for the treatment of endocrine-related cancer. Endocrine-Related Cancer 12: S189-S199.

Gamer AO, Leibold E, Van Ravenzwaay B (2006). The *in vitro* absorption of microfine zinc oxide and titanium dioxide through porcine skin. Toxicol. *In Vitro* 20: 301-307.

Gao J, Gu H, Xu B (2009). Multifunctional magnetic nanoparticles: Design, Synthesis and Biomedical applications. Acc. Chem. Res. 42(8): 1097-1107.

Gao JH, Li L, Ho PL, Mak GC, Gu HW, Xu B (2006). Combining fluorescent probes and biofunctional magnetic nanoparticles for rapid detection of bacteria in human blood. Adv. Mater. 18: 3145-3148.

Gao JH, Liang GL, Cheung JS, Pan Y, Kuang Y, Zhao F, Zhang B, Zhang XX, Wu EX, Xu B (2008a). Multifunctional yolk-shell nanoparticles: A potential MRI contrast and anticancer agent. J. Am. Chem. Soc. 130: 11828-11833.

Gao JH, Liang GL, Zhang B, Kuang Y, Zhang XX, Xu B (2007). FePt@CoS₂ yolk-shell nanocrystals as a potent agent to kill HeLa cells. J. Am. Chem. Soc. 129: 1428-1433.

Gao JH, Zhang B, Zhang XX, Xu B (2006a). Magnetic dipolar interaction induced self-assembly affords wires of hollow nanocrystals of cobalt selenide. Angew. Chem. Int. Ed. 45: 1220-1223.

Gao JH, Zhang W, Huang PB, Zhang B, Zhang XX, Xu B (2008). Intracellular spatial control of fluorescent magnetic nanoparticles. J. Am. Chem. Soc. 130: 3710-3711.

Gao X, Nie S (2001). Biologists join the dots. Nature 13: 450-452.

Gao X, Cui Y, Levenson RM, Chung LW, Nie S. (2004). *In vivo* cancer targeting and imaging with semiconductor quantum dots. Nat. Biotechnol. 22: 969-976.

Geiger B, Volberg T (1994). Conjugation of fluorescent dyes to antibodies . Cell Biology – A Laboratory Hand Book ed JE Celis (San Diego, CS: Academic) pp. 387-393.

Ghasemi Y, Peymani P, Afifi S (2009). Quantum dot: magic nanoparticle for imaging, detection and targeting. Acta Biomed. 80: 156-165.

Gu HW, Ho PL, Tsang KWT, Wang L, Xu B (2003a). Using biofunctional magnetic nanoparticles to capture vancomycin-resistant enterococci and other gram positive bacteria at ultralow concentration. J. Am. Chem. Soc. 125: 15702-15703.

Gu HW, Ho PL, Tsang KWT, Yu CW, Xu B (2003b). Using biofunctional magnetic nanoparticles to capture gram negative bacteria at ultra-low concentration. Chem. Commun. pp. 1966-1967.

Gu HW, Zheng RK, Liu H, Zhang XX, Xu B (2005). Direct synthesis of bimodal nanosponge based on FePt and ZnS. Small 1: 402-406.

Gupta AK, Gupta M (2005). Synthesis and surface engineering of iron oxide nanoparticles for biomedical applications. Biomaterials 26: 3995-4021.

Han MY, Gao XH, Su JZ, Nie S (2001). Quantum-dot-tagged microbeads for multiplexed optical coding of biomolecules. Nat. Biotechnol. 19: 631-635.

Hanaki K, Momo A, Oku T, Komoto A, Maenosono S, Yamaguchi Y, Yamamoto K (2003). Semiconductor quantum dot/albumin complex is a long-life and highly photostable endosome marker. Biochem. Biophys. Res. Commun. 302: 496-501.

Hanes J, Cleland JL, Langer R (1997). New advances in microsphere-based single dose vaccines. Adv. Drug Del. Rev. 28: 97-119.

Hartgerink JD, Beniash E, Stupp SI (2001). Self-assembly and mineralization of peptide- amphiphile nanofibers. Science 294: 1684-1688.

Herzog M, Draeger A, Ehler E, Small JV (1994). Immunofluorescence microscopy of the cytoskeleton: double and triple immunofluorescence. Cell Biology – A Laboratory Handbook ed JE Celis (San Diego, CA/Academic) pp. 355-360.

Hines MA, Guyot-Sionnest P (1996). Synthesis and characterization of strongly luminescing ZnS-capped CdSe nanocrystals. J. Phys. Chem. 100: 468-471.

Honeywell-Nguyen PL, Gooris GS, Bouwstra JA (2004). Quantitative assessment of the transport of elastic and rigid vesicle components and a model drug from these vesicle formulations into human skin *in vivo*. J. Invest. Dermatol. 123: 902-910.

Hu EL, Shaw DT (1999). Synthesis and assembly. In: Siegel RW, Hu EL, Roco MC, eds. Nanoscience Structure and Technology. National Science and Technology Council.

Imae Y, Atsumi T (1989). Na⁺ driven bacterial flagellar motors: a mini-review. J. Bioenerg Biomembr. 21: 705-716.

Jaiswal JK, Mattoussi H, Mauro JM, Simon SM (2003). Long-term multiple colour imaging of live cells using quantum dot bioconjugates. Nat. Biotechnol. 21: 47-51.

Jiang J, Gu HW, Shao HL, Devlin B, Papaetthymiou GC, Ying JY (2008). Bifunctional Fe₃O₄-Ag heterodimer nanoparticles for two-photon fluorescence imaging and magnetic manipulation. Adv. Mater. 20: 4403-4407.

Karak N, Maiti S (1997) Dendritic Polymers: a class of novel material. J. Polym. Mater. 14: 105.

Kim H, Achermann M, Balet LP, Hollingsworth JA, Klimov VI (2005).

Synthesis and characterization of Co/CdSe core/shell nanocomposites: Bifunctional magnetic-optical nanocrystals. J. Am. Chem. Soc. 127: 544-546.

Kim S, Fisher B, Eisler HJ, Bawendi M (2003). Type – II quantum dots: CdTe/CdSe (Core/Shell) and CdSe/ZnTe (Core/Shell) hetero-structures. J. Am. Chem. Soc. 125: 11466-11467.

Kim S, Lim YT, Soltesz EG, De Grand AM, Lee J, Nakayama A, Parker JA, Mihaljevic T, Laurence RG, Dor DM, Cohn LH, Bawendi MG, Frangioni JV (2004). Near-infrared fluorescence type II quantum dots for sentinel lymph node mapping. Nat. Biotechnol. 22: 93-97.

Kloepfer JA, Mielke RE, Wong MS, Nealson KH, Stucky G, Nadeau JL (2003). Quantum dots as strain and metabolism specific microbiological labels. Appl. Environ. Microbiol. 69: 4205-4213.

Kralj M, Pavelic K (2003). Medicine on a small scale. EMBO Reports 4: 1008-1012.

La VD, Lynn DM, Langer R (2002). Moving smaller in drug discovery and delivery. Nat. Rev. Drug Discov. 1: 77-84.

Larson DR, Zipfel WR, Williams RM, Clark SW, Bruchez MP, Wise FW, Webb WW (2003). Water-soluble quantum dots for multiphoton fluorescence imaging in vivo. Science 300: 1434-1436.

Laus S, Ruloff R, Toth E, Merbach AE (2003). GdIII Complexes with fast water exchange and high thermodynamic stability: potential building blocks for high-relaxivity MRI contrast agents. Chem. 9: 3555-3566.

Leroux JC (2007). Injectable nanocarriers for biodetoxification. Nat. Nanotechnol. 2: 679-684.

Lidke DS, Nagy P, Heintzmann R, Jovin DJ, Post JN, Grecco HE (2004). Quantum dot ligands provide new insights into erbB/HER receptor-mediated signal transduction. Nat. Biotechnol. 22: 198-203.

Lim YT, Kim S, Nakayama A, Stott NE, Bawendi MF, Frangioni JV (2003). Selection of quantum dot wavelengths for biomedical assays and imaging. Mol. Imaging 2: 50-64.

Lockman PR, Mumper RJ, Khan MA, Allen DD (2002). Nanoparticle technology for drug delivery across the blood-brain barrier. Drug Dev. Ind. Pharm. 28: 1-13.

Mansoori GA, Mohazzabi P, McCormack P, Jabbari S (2007). Nanotechnology in cancer prevention, detection and treatment: bright future lies ahead. World Review of Science, Technol. Sustain. Dev. 4: 2-3.

Mattheakis LC, Dias JM, Choi YJ, Gong J, Bruchez MP, Liu J, Wang E (2004). Optical coding of mammalian cells using semiconductor quantum dots. Anal. Biochem. 327(2): 200-208.

Mitchell GP, Mirkin CA, Letsinger RL (1999). Programmed assembly of DNA functionalized quantum dots. J. Am. Chem. Soc. 121: 8122-8123.

Miyawaki A, Sawano A, Kogure T (2003). Lighting up cells: labeling proteins with fluorophores. Nat. Cell Biol. 5S: 1-7.

Monteiro-Riviere NA (2008). In: Zhai H, Wilhelm KP, Maibach HI (Eds.).

Murray CB, Norris DJ, Bawendi MG (1993). Synthesis and Characterization of Nearly Monodisperse Cde (e = S, Se, Te) Semiconductor Nanocrystallites. J. Am. Chem. Soc. 115: 8706-8715.

Ness JM, Akhtar RS, Latham CB, Roth KA (2003). Combined tyramide signal amplification and quantum dots for sensitive and photostable immunofluorescence detection. J. Histochem. Cytochem. 51: 981-987.

Noji H, Yasuda R, Yoshida M, Kinosita K (1997). Direct observation of the rotation of F-1-ATPase. Nature 386: 299-302.

Obeso JL, Auerbach R (1984). A new microtechnique for quantitating cell movement in vitro using polystyrene bead monolayers. J. Immunol. Methods 70: 141-152.

Osborn M (1994). Immunofluorescence microscopy of cultured cells. Cell Biology – A Laboratory Handbook ed JE Celis (San Diego, CA/Academic) pp. 347-54.

Parak WJ, Pellegrino T, Plank C (2005). Labelling of cells with quantum dots. Nanotechnol. 16: R9-R25.

Pathak S, Choi SK, Amheim N, Thompson ME (2001). Hydroxylated quantum dots as luminescent probes for in situ hybridization. J. Am. Chem. Soc. 123: 4103-4104.

Pawley JB, Centonze VE (1994). Practical laser-scanning confocal light microscopy: obtaining optimal performance from your instrument. Cell Biology - A Laboratory Hand Book ed JE Celis (San Diego, CS: Academic) pp. 44-64.

Peng S, Sun SH (2007). Synthesis and Characterization of mono-disperse hollow Fe_3O_4 nanoparticles. Angew. Chem. Int. Ed. 46: 4155-4158.

Peng ZA, Peng XG (2001). Formation of high quality CdTe, CdSe, and CdS nanocrystals using CdO as precursor. J. Am. Chem. Soc. 123: 183-184.

Pinaud F, King D, Moore HP, Weiss S (2004). Bioactivation and cell targeting of semiconductor CsSe/ZnS nanocrystals with phytochelatin-related peptides. J. Am. Chem. Soc. 126: 6115-6123.

Plank C, Oberhauser B, Mechtler K, Koch C, Wagner E (1994). The influence of endosome-disruptive peptides on gene transfer using synthetic virus-like gene transfer systems. J. Biol. Chem. 269: 12918-12924.

Plank C, Zauner W, Wagner E (1998). Application of membrane-active peptides for drug and gene delivery across cellular membranes. Adv. Drug Deliv. Rev. 34: 21-35.

Rikans LE, Yamano T (2000). Mechanisms of cadmium-mediated acute hepatotoxicity. J. Biochem. Mol. Toxicol. 14: 110-117

Ron H (2005). A Toxicological review of Quantum Dots: Toxicity depends on physico-chemical and environmental factos. Environ. Health Perspect. 114(2): 165-172.

Rosenthal SJ, Tomlinson A, Adkins EM (2002). Targeting cell surface receptors with ligand-conjugated nanocrystals. J. Am. Chem. Soc. 124: 4586-4594.

Safarik I, Safarikova M (2004). Magnetic techniques for the isolation and purification of proteins and peptides. Biomagn. Res. Technol. 2: 7.

Saiyeed Z, Telang S, Ramchand C (2003). Application of magnetic techniques in the field of drug discovery and biomedicine. Biomagn. Res. Technol. 1: 2.

Schnur JM (1993). Lipid tubules: a paradigm for molecularly engineering structures. Sci. 262: 1669-1676.

Schnur JM, Price R, Rudolph AS (1994). Biologically engineered microstructures: controlled release applications. J. Control. Release 28: 3-13.

Silva A, Antonio T, Chung MC, Castro FF, Guido RVC, Ferreira EI (2001). Advances in Prodrug design. Mini Rev. Med. Chem. 5(10): 893-914.

Silva GA (2004). Introduction to Nanotechnology and its Applications to Medicine. Surg. Neuro. 61: 216-220.

Skaff H, Emrick T (2003). The use of 4-substituted pyridines to afford amphiphilic, pegylated cadmium selenide nanoparticles. Chem. Commun. 52: 3.

Skrabalak SE, Chen JY, Sun YG, Lu XM, Au L, Cobley CM, Xia YN (2008). Gold nanocages: Synthesis, properties and applications. Acc. Chem. Res. 41: 1587-1959.

Sonawane ND, Szoka FC, Verkman AS (2003). Chloride accumulation and swelling in endosomes enhances DNA transfer by polyamine-DNA polyplexes. J. Biol. Chem. 278: 44826-44831.

Stephens DJ, Allan VJ (2003). Light microscopy techniques for live cell imaging. Sci. 300: 82-86.

Stupp SI, Ciegler GW (1992). Organoapatites: materials for artificial bone. I. Synthesis and microstructure. J. Biomed. Mater. Res. 26: 169-183.

Stupp SI, Mejicano GC, Hanson JA (1993). Organoapatites: materials for artificial bone. II. Hardening reactions and properties. J. Biomed. Mater. Res. 27: 289-299.

Sukhanova A, Devy M, Venteo L, Kaplan H, Artemyev M, Oleinikov V, Klinov D, Pluot M, Cohen JH, Nabiev I (2004). Biocompatible fluorescent nanocrystals for immunolabeling of membrane proteins and cells. Anal. Biochem. 324: 60-67.

Sukhanova A, Venteo L, Devy J, Artemyev M, Oleinikov V, Pluot M, Nabiev I (2002). Highly stable fluorescent nanocrystals as a novel class of labels for immunohistochemical analysis of paraffin-embedded tissue sections. Lab. Invest. 82: 1259-1261.

Takano H, Sul JY, Mazzanti MI, Doyle RT, Haydon PG, Porter MD (2002). Micropatterned substrates: approach to probing intercellular communication pathways. Anal. Chem. 74: 4640-4646.

Tombolini R, Jansson JK (1998). Monitoring of GFP-tagged bacterial cells. Bioluminiscence Methods and Protocols ed RA La Rossa (Totowa, NJ: Humana Press) pp. 285-98.

Tan J, Wang B, Zhu L (2006). Quantum Dots: A novel tool to

discovery. Pak. J. Biol. Sci. 9(5): 917-922.

Torchilin V, Weissig V (2003). Liposomes: a practical approach. The Practical Approach Series #264, Oxford Univ. Press, Aug. 7, Oxford, GB.

Uren RF (2004). Cancer surgery joins the dots. Nat. Biotechnol. 22: 38-39.

Van der MD, Brooks JD, Gehring R, Baynes RE, Monteiro_Riviere NA, Riviere JE (2006). A physiologically based pharmacokinetic model of organophosphate dermal absorption. Toxicol. Sci. 89: 188-204.

Wagner E, Plank C, Zatloukal K, Cotton M, Birnstiel ML (1992). Influenza virus hemagglutinin HA-2 N-terminal fusogenic peptides augment gene transfer by transferring-polylysine-DNA complexes: toward a synthetic virus-like gene-transfer vehicle. Proc. Nat. Acad. Sci. USA 89: 7934-7938.

Wang YA, Li JJ, Chen HY, Peng XG (2002). Stabilization of inorganic nanocrystals by organic dendrons. J. Am. Chem. Soc. 124: 2293-2298.

Weijer CJ (2003). Visualizing signals moving in cells. Sci. 300: 96-100.

Weisner, Hooisweng Ow, Daniel R. Larson, Mamta Srivastava, Barbara A. Baird, Watt WW (2005). "Bright and stable Core-shell Fluorescent Silica Nanoparticles". Nano Lett. 5(1).

Winter JO, Liu TY, Korgel BA, Schmidt CE (2001). Recognition molecule directed interfacing between semiconductor quantum dots and nerve cells. Adv. Mater. 13: 1673-1677.

Wu X, Liu H, Liu J, Haley KN, Treadway, JA, Larson JP, Ge N, Peale F, Bruchez MP (2003). Immunofluorescent labeling of cancer marker Her2 and other cellular targets with semiconductor quantum dots. Nat. Biotechnol. 21: 41-46.

Xu CJ, Xie J, Ho D, Wang C, Kohler N, Walsh EG, Morgan JR, Chin YE, Sun SH (2008). Au-Fe$_3$O$_4$ dumbbell nanoparticles as dual-functional probes. Angew. Chem. Int. Ed. 47: 173-176.

Xu CJ, Xu KM, Gu HW, Zheng RK, Liu H, Zhang XX, Guo ZH, Xu B (2004). Dopamine as a robust anchor to immobilize functional molecules on the iron oxide shell of magnetic nanoparticles. J. Am. Chem. Soc. 126: 9938-9939.

Xu CJ, Xu KM, Gu HW, Zhong XF, Liu H, Guo ZH, Zheng RK, Zhang XX, Xu B (2004b). Nitrilotriacetic acid-modified magnetic nanoparticles as a general agent t bind histidine-tagged proteins. J. Am. Chem. Soc. 126: 3392-3393.

Yin YD, Rioux RM, Erdonmez CK, Hughes S, Somorjai GA, Alivisatos AP (2004). Formation of hollow nanocrystals through the nanoscale kirkendall effect. Sci. 304: 711-714.

Yu H, Chen M, Rice PM, Wang SX, White RL, Sun SH (2005). Dumbbell-like bifunctional Au- Fe$_3$O$_4$ nanoparticles. Nano Lett. 5: 379-382.

Zhang A, Zhang B, Wachtersbach E, Schmidt M, Schluter AD (2003). Efficient synthesis of high molar mass first to fourth generation distributed dendronized polymers by the macromonomer approach. Chem. Eur. J. 9(24): 6083-6092.

Zhang C, Ma H, Ding Y, Jin L, Chen D, Nie S (2000). Quantum dot-labeled trichosanthin. Analyst 125: 1029-1031.

Zhang LW, Yu WW, Colvin VL, Monteiro RNA (2008). Biological interactions of Quantum Dot nanoparticles in skin and in human epidermal keratinocytes. Toxicol. Appl. Pharmacol. 228: 200-211.

21

Microtubers in yam germplasm conservation and propagation: The status, the prospects and the constraints

Morufat Oloruntoyin Balogun

Institute of Agricultural Research and Training, Obafemi Awolowo University, P. M. B. 5029, Ibadan, Nigeria. E-mail: kemtoy2003@yahoo.com

The conservation of yam genetic resources using field genebanks, *in vitro* plantlets, pollen and seed storage are constrained by high losses and space requirements, maintenance cost and an irregular flowering, respectively. Microtubers produced from *in vitro* plantlets are proposed for conservation and propagation, as they have a longer shelf-life due to dormancy, and are also hardier and less bulky than plantlets. A lot of work has been done on microtuber production, especially the use of temporary immersion systems in production of larger, multiple microtubers. However, there have been different degrees of success, and, very few reports on microtuber dormancy. Also, research findings on post-sprout management and efficiency of microtubers relative to other systems in terms of cost, ease of handling and savings on time are sparse. These research gaps limit the practical use of microtubers in conservation and propagation. Future research should be on dormancy control and post-sprout management. A microtuber to microtuber cycle for the conservation and propagation of yam germplasm is proposed in this review, and the invaluable potentials of microtubers in these regards is emphasised.

Key words: Yams, *Dioscorea* species, germplasm conservation, propagation, microtuberization.

Table of contents

INTRODUCTION

The yams, family *Dioscoreaceae*, genus *Dioscorea*, are staples (Hahn 1995) in many countries of the tropics where it provides 200 dietary calories daily to 300 million people (Coursey, 1967; Nweke et al., 1991; FAO, 2000). Yam production is however constrained by abiotic factors, pests and diseases (Emehute et al., 1998) and scarcity of propagules (Nweke et al., 1991). Tuber dormancy also prevents year-round production and this hampers productivity. In addition, uncontrolled sprouting after dormancy break causes tremendous storage losses (Asiedu et al., 1998; FAO, 2000; Craufurd et al., 2001). Selection for desirable traits for breeding of improved varieties and development of an efficient, cost-effective propagation system is thus crucial for sustainable yam production (Asiedu et al., 1998; Quin, 1998). To provide a broad germplasm base for selection and breeding, conservation of yam genetic resources must continue (Quin, 1998).

Plant tissue culture techniques of meristem culture combined with heat therapy have been successfully used to produce high-yielding plantlets tested to be virus-free which are not only conserved in *in vitro* genebanks but also used in rapid multiplication of superior clones (Mantell et al., 1980; Ng, 1984, 1992). However, the stress of transportation causes low survival rates during transplanting and germplasm exchange despite the specialized handling (Ng, 1988). There is also the need for frequent subculturing when plantlets show signs of deterioration. Microtubers produced from pathogen-tested, *in vitro* plantlets have been proposed as an addi-tional means of germplasm conservation and propa-gation (Ammirato, 1984). This is because they are less vulnerable to transportation hazards during germplasm exchange, less bulky and can be kept for several months due to microtuber dormancy (Ng, 1988; Balogun et al., 2004). They can also be easily established in the soil, not requiring acclimatization and transplanting (Ng, 1988; Ng and Ng, 1997). A lot of work has been done on yam microtuberization in recent times. This review discusses the status and current trends in the production and utilization of microtubers for germplasm conservation and propagation.

Other options in yam conservation and propagation

Conservation

The rich diversity among yam collections (Ng and Ng, 1997) calls for germplasm conservation to prevent their genetic erosion and enhance the use of these genetic resources in breeding programmes (Ng and Ng, 1994; Acheampong, 1996). Field conservation requires considerable space, maintenance and time while annual losses from collections, as high as 10% have been reported in field genebanks due to pest and disease pro-

blems, poor sprouting, unsatisfactory storage of tubers, drought and poor handling (Okoli, 1991; Acheampong, 1996; Taylor, 1996a).

Other options for conserving the yam genepool have therefore been investigated. These include seed conservation for a considerable period at low temperature and low seed moisture content (Daniel et al., 1999), pollen storage at 0% relative humidity and $-5°C$ for about one year (Akoroda, 1983) and at $-80°C$ for more than two years (Daniel et al., 2002). However, seed conservation can only be applied to female plants while pollen conservation is only applicable to male plants while non-flowering genotypes can only be conserved vegetatively.

In vitro methods of plantlet storage at reduced temperature to slow down their growth rates, thereby extending the viable storage period, is also used for several species of yam. This serves as a duplication and backup to the field genebank (Zamura and Paet, 1996). At the International Institute of Tropical Agriculture (IITA) for example (Ng and Ng, 1997), axillary buds and nodal cultures are most frequently used for the rapid clonal propagation of meristem-derived, pathogen-tested plantlets in the establishment of an *in vitro* gene-bank under slow growth. This is however for short to medium term conservation, after which the plantlets are subcultured when signs of deterioration are visible. With this method, plantlet storage period of *Dioscorea alata* and *Dioscorea rotundata* could be extended to 1 - 2 years by reduction of the incubation temperature from $28 - 30°C$ to $18 - 22°C$ (Ng and Ng, 1997). In some cases however, storage period was only 9 - 12 months at $20°C$ (Taylor 1996b) and one year at $25 - 28°C$ (Zamora and Paet, 1996). In addition, the need for subculturing when deterioration signs are visible increases labour costs for the maintenance of an *in vitro* genebank.

The use of artificial (synthetic) seeds (Rao et al., 1998) in storage and exchange of yam germplasm has also been reported (Hasan and Takagi, 1995). Synthetic seeds are artificially encapsulated somatic embryos, shoots, or other tissues that are able to grow into plantlets after sowing under *in vitro* or *ex vitro* conditions (Standardi and Piccioni, 1998). The efficiency of using these artificially encapsulated propagules lies in their small size and relative ease of handling.

The alginate coat which protects the micropropagules makes them more tolerable than plantlets to the stress of transit during yam germplasm conservation and exchange (Hasan and Takagi, 1995). However, viability of the synthetic seeds decreased as the sucrose concentration in both the matrix and polymerization medium increased (Hasan and Takagi, 1995) while high levels of sucrose were toxic to the explants (Kitto and Janick, 1985; Uragami, 1993). An optimum sucrose concentration of 0.1 - 0.3 M was found suitable for keeping the micropropagules viable for up to 2 weeks during inter-national germplasm transfer before direct trans-planting to soil (Hasan and Takagi, 1995). To achieve long-term storage of desirable elite genotypes, the use of artificial seeds for cryopreservation via encapsulation dehydration or encapsulation vitrification (Pennycooke and Towil, 2001; Wang et al., 2002; Tessereau et al., 1995; Ng and Daniel 2000; Ng and Ng, 2000) was reported.

Cryopreservation is a process where micropropagules are preserved by cooling to low, sub-zero temperatures, such as 77K or –196 °C. However, yam genotypes differ in their response to cryopreservation protocols (Leunufna and Keller, 2003). *Dioscorea bulbifera, Dioscorea oppositifolia* and *Dioscorea cayenensis* seemed to be more able to withstand the stress imposed during the cryopreservation procedure than *D. alata*. In terms of survival, *D. oppositifolia* exhibited the highest rate, but only one-third of the surviving explants developed further. *D. cayenensis* showed a lower (not significant) survival rate than D. *oppositifolia*, but most of the surviving explants grew further. *D. bulbifera* also showed a similar survival rate to that of *D. cayenensis* and *D. oppositifolia*, but very low numbers of surviving explants regrew.

The report of Reed et al. (2001) also indicated an inconsistence with respect to the reproducibility of cryopreservation protocols from one laboratory to the other. Kyesmu (1998) reported a shoot recovery of 47, 85 and 91% for three different cultivars of *D. alata*, respectively, using the vitrification method. Using encapsulation-dehydration, Mandal et al. (1996) obtained survival as high as 64% and a recovery of 21.8% for *D. alata* and 26% survival with no recovery for *D. bulbifera*.

Other *in vitro* methods for conservation include embryo, callus and suspension cultures although their use is limited by lack of successful regeneration protocols.

Propagation

Traditionally, yams are propagated by planting whole tubers or large pieces weighing 200 g or more (Okoli et al., 1982). A sizable portion of otherwise consumable tubers are therefore reserved for planting yearly, and this leads to scarcity of planting materials. Multiplication ratio for seed yam production in the field is 1:10 compared to 1:300 in cereals. Planting materials alone constitute about 50% of production costs (Nweke et al. 1991, Akoroda and Hahn, 1995). Most farmers propagate yams by "milking". In this technique, tubers are harvested two-thirds into the growing season without destroying the root system. This provides early yam for home consumption and market. There is regeneration of fresh small tubers from the corm at the base of the vine and these are used as planting materials for the following season.

The major constraint of planting materials to yam production is being tackled by the development of more efficient propagation methods (Orkwor and Asadu 1998). These include partial sectioning technique (Nwosu, 1975) and minisett technique (Okoli et al., 1982). Although the latter has significantly increased propagation rates (Okoli

et al. 1982), it has been associated with less uniform and poor rate of sprouting when applied to white yam (George, 1990; Sreekantan et al., 1995; Craufurd et al., 2001). Although multiplication rates are doubled using the partial sectioning technique, it requires considerable manpower for the repeated examining and digging out of tubers to excise sprouted sections for field planting (Nwosu, 1975). In the vine rooting technique, either tubers did not develop due to early senescence of rooted vines (Acha et al., 2005), or small tubers are produced when applied to *D. rotundata* relative to other species.

Also, the layering technique is unsuitable for farm use due to rigorous procedures involved (Acha et al., 2004).

Research is needed in aerial tuber production, which was proposed as an alternative means of seed tuber production (H. Shiwachi, personal communication), although they are seldom produced in *D. rotundata*, unless induced by stem girdling (Okonkwo, 1985). In addition, plants raised from sexual seeds and the tubers produced are small relative to plants raised from tubers probably due to small amount of stored food reserves in the seed (Okonkwo, 1985).

The synthetic seed technology is also useful for the propagation of vegetatively propagated plants (Kumar, 1998) whose true seeds are not used or readily available for multiplication (e.g. yam and potato). There is also the possibility of using synthetic seeds to time production cycles in micropropagation laboratories if the development of the plant could be properly directed towards proliferation and rooting. However, conversion is the most important aspect of the synthetic seed technology, and one of the factors that have limited its practical use (Standardi and Piccioni, 1998). In contrast to somatic embryos which are bipolar structures, shoots and buds do not have root meristems and they must regenerate roots in order to be able to convert (Bapat, 1993; Piccioni, 1997).

The microtuber option in yam conservation and propagation

The process of *in vitro* tuberization in yams

In vitro tuberization has been reported in *Dioscorea abyssinica* (Jean and Cappadocia, 1991); *Dioscorea floribunda* (Sengupta et al. 1984.); *D. alata* (Ammirato, 1976; Alhassan and Mantell, 1991); *D. bulbifera* (Forsyth and Van Staden, 1984; Mantell, 1987); *D. rotundata* and *D. cayenensis* (Ng and Mantell, 1996) and *Dioscorea opposita* (Mantell and Hugo 1986) with various degrees of success.

The origin of *in vitro* tubers is essentially similar to that in the development of aerial tubers produced on greenhouse-grown plants (Mantell, 1987). Microtubers develop from primary nodal complexes (Mantel, 1987). In the leaf axils of old nodes, there are two axillary buds and one shoot primordium, which also later develop into an axillary bud. A primary nodal complex (PNC), preceded by a meristematic PNC-initial is developed at the base of the first-formed axillary bud. This PNC-initial has capacity for multiple bud production, roots and a tuberous storage organ (Wickham et al. 1982). The PNC-initial is the organ of renewed growth and the only true organ of vegetative propagation in *Dioscorea* species (Wickham et al., 1981, 1982). The morphogenetic expression of the PNC activity is also under hormonal control (Wickham et al., 1982).

Effects of cultural factors

Investigations on microtuberization have revealed the effect of a number of factors on the phenomenon. The Murashige and Skoog (1962) basal medium formulation (MS) inhibited microtuberization in *D. alata* and *D. bulbifera* (Mantell and Hugo, 1989) and *D. opposita* (Asahira and Yazawa, 1979). Microtubers were however produced on glycine-free MS medium in *D. alata* and *D. abyssinica* while half-strength MS medium enhanced microtuberization in *D. alata* (Chang and Hayashi 1995a; Chang et al. 1995b). Tuberization ('T') medium, specially designed for delivering reduced nitrogen (6% w/w total nitrogen present in full strength MS on a molar basis) to yam shoot cultures was optimum for microtuberization in *D. alata* (Mantell and Hugo, 1989). Microtuber formation was stimulated in *Dioscorea batatas* (Asahira and Nitsch, 1968) and *D. alata* (Mantell and Hugo, 1989; Jean and Cappadocia, 1991) in the absence of ammonium. It should be confirmed, however, whether this is due to the absence of NH_4^+ or reduction of total nitrogen content or high NO_3^- : NH_4^+ ratio, as observed in *D. opposita* (Asahira and Yazawa, 1979). Balogun et al. (2006) reported that optimum basal medium formulation for microtuberization varied with other factors like medium matrix, sucrose concentration, and light and temperature regimes.

Microtuberization is affected by plant growth regulators (Koda and Kikuta, 1991; Jean and Cappadocia, 1992; John et al., 1993; Kikuno et al., 2002a), the effect being greater when applied at culture initiation than at later stages (Balogun 2005). Naphthalene acetic acid (NAA) enhanced microtuber production in *D. rotundata* (Ng and Mantell, 1996). NAA was also superior to indole acetic acid in microtuber induction in *D. alata* (Chang and Hayashi, 1995). In *D. alata*, kinetin did not significantly affect MTZ (Ng and Mantell, 1996) but enhanced MTZ in *D. bulbifera* (Mantell and Hugo, 1989). In *D. rotundata*, kinetin induced the highest percentage microtuberization, followed by 2ip, while tuberization was poor in BAP. Also, supplementing the tuberization (T') medium with 1.0μM kinetin improved microtuber frequencies but not individual microtuber weights in *D. alata* (Alhassan and Mantell 1991). In *D. alata*, 50μM gibberrellic acid (GA_3) increased the tuber fresh weight (Ng and Mantell, 1996, Onjo et al., 2001), but was detrimental to shoot fresh weight and number of nodes per plantlet. GA_3 also induced fewer nodes at 5.0 μM in *D. rotundata* (Ng and Mantell 1996). In *D. alata* tuber weight increased with increasing con-

centration of abscisic acid at 8 h photoperiod (Jean and Cappadocia, 1992). Interaction of kinetin with abscisic acid reduced microtuber frequency but increased individual microtuber weights, suggesting a possible re-stimulation of tuberization by kinetin, earlier suppressed by abscisic acid (Alhassan and Mantell, 1991). GA_3 and JA-like substances might be related to initiation of tuber enlargement while ABA is not directly related (Kikuno et al., 2002b).

Tuberization in yam was reported to be controlled by jasmonic acid, a 12-carbon acid that has fragrant and plant growth-regulating properties, inhibiting growth and promotes senescence (Vick and Zimmerman, 1984). It was isolated from yam and potato leaves (Koda and Kikuta, 1991; Koda and Okazawa, 1988) and found to have strong tuber-inducing properties in both species. The threshold concentration of JA for induction of yam tuberization *in vitro* was found to be 10^{-7}M (Vick and Zimmerman, 1984). Kikuno et al. (2002b) reported that in *D. alata*, jasmonic acid synthesis is activated by short day length, while the peak of jasmonic acid content coincided with initiation of tuber enlargement and decreased significantly afterwards, although vigorous growth and enlargement of tuber continued. In addition, sensitivity to JA was greater in late maturing genotypes than earlier ones (Balogun, 2005). Uniconazole-p was inhibitory to MTZ (Balogun, 2005) regardless of the growth phase at which it was applied, probably due to its inhibitory effect on GA synthesis (Izumi et al., 1984). However, MTZ can be achieved without PGRs if dormancy control is not desired (Balogun et al., 2006).

Sucrose was found to be best for microtuberization in *D. rotundata* relative to fructose, although the reverse is the case for shoot multiplication. Sorbitol and galactose were however inhibitory to tuberization. Higher concentration (5%) of all carbon sources gave higher tuberization than lower (3%) concentration. 8% sucrose also enhanced MTZ in *D. rotundata* (Balogun, 2005; Balogun et al., 2006).

Short days stimulate yam tuberization depending on the growth stage at which it was administered and the earliness of varieties while long days inhibit it (Koda and Kikuta, 1991; Shiwachi et al., 2002). In *D. rotundata*, increase in day-length, (up to 16 h photoperiod) increased shoot and root weights, number of nodes and microtubers; but 24 hr photoperiod was detrimental. In *D. alata*, however, shorter day length was necessary to consistently produce more and larger microtubers (Ng and Mantell, 1996). At 8 h photoperiod, only basal nodes produced MTs in *D. alata*, but other nodes did at longer (16 and 24 h) photoperiods. The tubers were however larger at 8 h d^{-1} (Jean and Cappadocia 1991). Under 12 h photoperiod, MTZ was significantly higher than in complete darkness (Balogun, 2005).

Whether in liquid or agar-solidified medium, culture aeration was found to be beneficial to yam shoot development and microtuber production. This might be due to the stimulation of photosynthesis *in vitro* by aeration (Ng and Mantell, 1996). Yam shoot growth and microtuber production in terms of weight and size was significantly enhanced in temporary immersion system relative to solid medium (Jova et al., 2005). It offers automation in culture systems, wherein temporary immer-sion of explants in medium allows all buds to be in con-tact with culture medium at the same time. This effect is not given by static culture systems. Explant growth is favoured because contact with media has a short duration with time to renew atmosphere (aeration) in flask. So, disadvantages of solid culture media are decreased and a high photosynthetic activity is obtained (Etienne and Berthouly, 2002). The technique can also be used for shoot multiplication during the planting season, when *in vitro* plants can be immediately acclimatized and transplanted. A similar observation on potato crop was previously carried out (Akita and Takayama, 1994).

Effects due to the cultured explant

Previous reports have shown that genotypes differ in their response to the above conditions and ability to tuberize. For example, *D. alata* produced microtubers more readily than *D. rotundata* (Balogun 2005). This confirms endogenous control of tuberization (Shiwachi et al. 2002). For propagation purposes therefore, microtuber production and dormancy control protocols specific to economically important genotypes will have to be developed. For conservation purposes however, it should be expected that not all genotypes of *D. rotundata* might be conserved using microtubers.

The source of explant affects MTZ frequencies ((Balogun et al., 2005). Screen house explants produced more MTs than *in vitro* explants. This may be due to better aeration and hence shoot vigour in the former than the latter growth environment (Balogun et al. 2004).

Tuber dormancy in yam conservation and propagation

The efficiency of any conservation and propagation system lies in the ability of regeneration of propagules as may be desired. The single most important factor that limits regeneration of yam propagules as needed is tuber dormancy. It is a physiological rest period in which there is no visible physiological or biochemical activity, the inability of growth in plant meristems in spite of suitable environmental conditions (Lang, 1996). It allows propagules to survive prolonged dry seasons, and hence, is ecologically significant (Craufurd et al., 2001). Thus, all techniques aimed at yam germplasm conservation and propagation, including microtuberization systems, need to be improved in the area of dormancy breaks so that yam sprouting can be achieved as and when desired (Craufurd et al. 2001). Dormancy break will allow easy regeneration of plantlets while prolonging the dormancy period

will increase the viable shelf life during storage (Ng and Ng, 1997; Craufurd et al., 2001), as reported for potatoes (Keller and Schuler, 1996).

Ammirato (1982) reported the non-sprouting of yam microtubers, while Ng (1998) reported in studies on influence of carbon source on *in vitro* tuberization and growth of *D. rotundata*, that microtubers harvested from mannose, fructose and sucrose treatments did not sprout until eight months after harvest. Also, bigger MTs sprouted later than smaller ones although the frequency of sprouting was higher (Ammirato, 1982; Balogun, 2005).

Although GA_3 promoted the enlargement of microtubers in *D. alata* (Onjo et al., 2001), it also stimulated the thickening of tubers and hence extends the period of tuber dormancy (Onjo et al., 1999; Balogun, 2005; Giradin et al., 1998; Tschannen et al., 2003). Microtuber sprouting is enhanced by Jasmonic acid (Bazabakana et al., 1999, Balogun, 2005).

Yam tuber dormancy was reported to respond to plant growth regulators (Wickham et al., 1984).The production and dormancy of MTs vary with the growth phase of the plantlet at which specific Plant growth regulators are applied (Balogun, 2005). In a particular genotype, the phase of plant growth (that is, vine development, PNC formation, tuber initiation) whose length mostly affects the maturity period (e.g long versus short PNC formation phase) may determine the optimum PGR regime for dormancy control (Balogun 2005). Uniconazole-p inhibits MTZ but shortens the dormancy period of microtubers (Balogun 2005) as was reported *in vivo* (Park et al., 2003).

Microtubers versus synthetic seeds

Optimum protocol for cryopreservation and conversion of cryopreserved germplasm to plantlets differs among genotypes of yam ((Kyesmu, 1998; Kyesmu et al., 1997; Malaurie et al., 1998; Mandal, 2000; Mandal et al., 1996). In contrast, groups of genotypes can produce micro-tubers under similar protocols while all genotypes will eventually break dormancy under natural conditions. This will circumvent the rigorous procedures of conversion and recovery associated with cryopreservation.

The use of both microtubers (Ng, 1988) and encapsulated embryos (Hasan and Takagi, 1995) in germplasm exchange are not as vulnerable as plantlets to unfavourable conditions of transportation and this reduces germplasm losses. Both options produce virus-free, true-to-type materials while each is limited to those varieties amenable to them. This is because protocols for both procedures have not been optimized for all yam species (Jean and Cappadocia, 1991, Hasan and Takagi, 1995). Although conversion of synthetic seeds is comparable to dormancy break in microtubers, the former is artificial and while the latter is the natural physiology of the germplasm. The advantage of microtuber use in germplasm storage over synthetic seeds is that without any exoge-

Table 1. Previous reports on microtuberization in yams.

Author/Year	Species
Microtuber production	
Alhassan and Mantell, 1991	D. alata
Ammirato, 1976, 1982,1984	D. alata, D. bulbifera
Asahira and Yazawa, 1979	D. opposita
Asahira and Nitsch, 1968	D. batatas
Balogun et al., 2004, 2005	D. alata, D. rotundata
Bazabakana et al., 2003	D. alata
Chang et al., 1995	D. alata
Forsyth and Van Staden, 1984	D. bulbifera
John et al., 1993	D. alata
Jova et al., 2005	D. alata
Jean and Cappadocia, 1991, 1992.	D. alata, D. abyssinica
Kikuno, 2004, 2005	D. alata, D. rotundata
Mantell, 1987	D. alata, D. bulbifera, D. rotundata
Mantell and Hugo, 1986; 1989	D. bulbifera, D. alata, D. opposita
Ng and Mantell, 1996	D. alata, D. cayenensis, D. rotundata
Ng, 1988, 1998	D. rotundata
Onjo et al., 2001	D. alata
Sengupta et al., 1984	D. floribunda
Microtuber dormancy	
Ammirato, 1982	D. alata, D. bulbifera
Ng, 1998	D. rotundata
Bazabakana et al., 1999	D. alata
Balogun , 2005	D. alata, D. rotundata

nous influence, microtuber dormancy will break naturally after a minimum of 4 months without loss of viability (Balogun, 2005). The length of the dormancy period can also be controlled by exogenous application of plant growth regulators to extend or reduce the storage period. Cryopreservation of synthetic seeds will alsoex-tend their viability period although recovery and survival rates differ among genotypes and may be unpredictable

In contrast, high frequency of sprouting and survival were recorded in microtubers with exogenously extended dormancy period (Balogun, 2005). Use of microtubers could be more reliable than cryopreserved synthetic seeds if the dormancy period can be exogenously controlled for any desirable length of time. However, this is yet to be perfected and constitutes a research gap.

Implications of research findings and future trends

Although a lot of work has been done on microtuber production, there have been very few reports on yam microtuber dormancy (Table 1) while, only one report has proposed a protocol for yam conservation using MTs, and it is yet to be applied. In addition, more work has been done on *D. alata* than any other species (Figure 1). These reports may however not be exhaustive. The use of microtubers in germplasm conservation, propagation

and exchange will be impossible without adequate protocols for microtuber production and dormancy control, as only this will allow for storage and regeneration as may be desired (Craufurd et al., 2001). Based on the available reports, a scheme for utilization of MTs in conservation is proposed here (Figure 2), although this may be far from optimum. Most of the research gaps hindering the use of MTs are related to dormancy and post-sprout studies. These include:

1. The genetic variation among yam collections in terms of ability to form MTs should be determined so as to know the scope of its applicability.
2. What is the optimum condition for storing MTs? This will involve investigations into the effects of light, temperature and humidity on microtuber dormancy.
3. What is the condition of plantlets regenerated from nodes excised from plantlets relative to those from MTs in terms of vigour?
4. What are the optimum conditions or protocols for raising seedlings from MTs?
5. How many generations of planting seedlings from microtubers will give tuber yield comparable to or bigger than those from field tubers?
6. What is the survival rate of plants from MTs relative to transplanted *in vitro* plantlets?

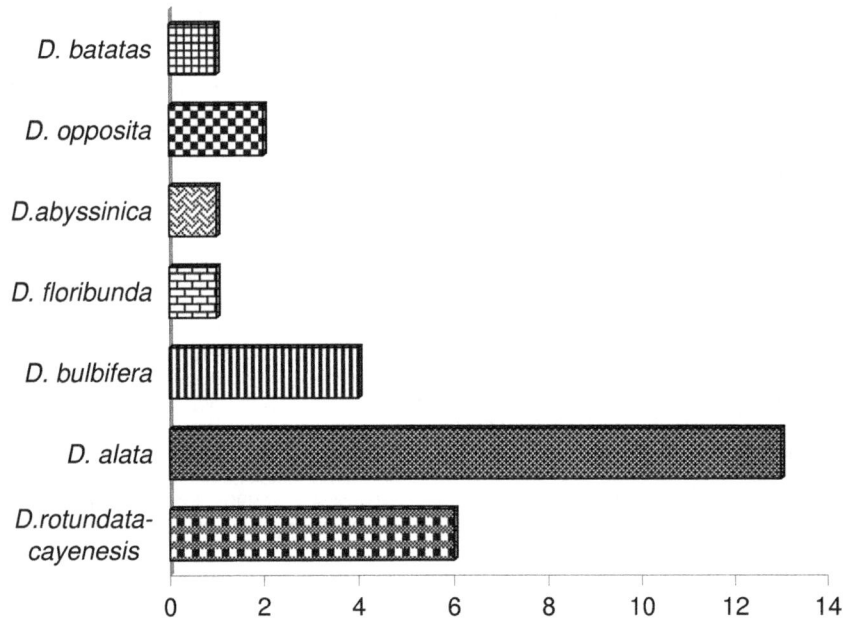

Figure 1. Number of authors who have worked on microtuberization in each species

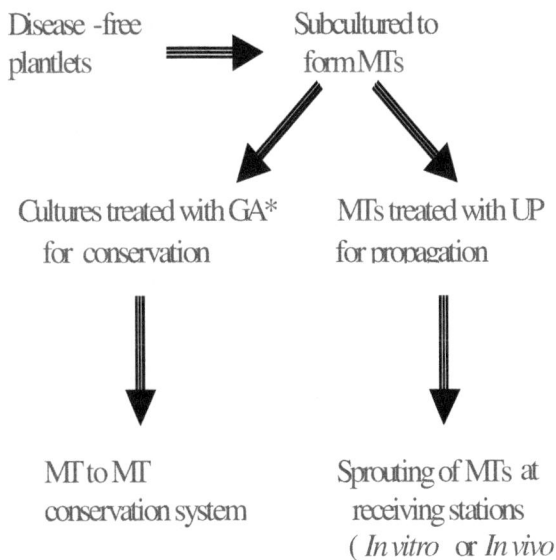

Figure 2. Proposed protocol for MT use in yam germplasm conservation and propagation.

7. What is the cost of conservation, propagation and exchange using MTs relative to other options? This will include investigations for possible economic benefits in the establishment of specialist producers who can produce healthy seed yams of desirable varieties to meet growers' needs.

Conclusion

The ability of yam plants to produce tubers *in vitro* has been established, and many factors which affect it are known. These range from genotype, micro-and macro-nutrients, light and temperature regime through plant growth regulators to sources of explants. However, investigations on the control of microtuber dormancy, post-sprout management and efficiency of microtubers relative to other systems in terms of cost, ease of handling and savings on time are sparse. Thus, for effective use of MTs in conservation and propagation of yam germplasm, future research should be on dormancy control and post-sprout management. Also, emphasis should be laid on the ultimate goal of a seed production system which is to produce good quality propagules that will yield optimally, good enough for reasonable econo-mic returns.

So far, research reports indicate a possibility of developing a MT to MT cycle for the conservation and propagation of yam germplasm. Specifically, a highly valuable alternative for the commercial production of microtubers as seed yam is offered by temporary immersion system which induces more tubers per plant and increases the size and weight of tubers. With the availability of reliable microtuber production and dormancy control systems, the germplasm propagation, conservation and exchange of *in vitro* propagated, pathogen-tested elite clones will be facilitated. However, different methods of conservation should still be combined for a better security of germplasm collections.

ACKNOWLEDGEMENT

I am grateful to the management and staff of the International Institute of Tropical Agriculture and the Institute of

Agricultural Research and Training, Ibadan, Nigeria for their respective supports toward the doctoral research conducted by the author at the IITA. I appreciate Dr. H. Kikuno of the IITA for providing useful literature and Dr. (Mrs.) S.R. Akande of the Institute of Agricultural Research and Training, Ibadan, for a critical review of the write-up.

REFERENCES

Acha IA, Shiwachi H, Asiedu R, Akoroda MO (2004). Effect of auxins on root development in yam (*Dioscorea rotundata*) vine. Trop. Sci. 44, in press.

Acheampong E (1996). Issues affecting the field genebank management in Ghana. Paper presented at the consultation meeting on the management of field and *in vitro* genebank, 15-20 January, 1996, CIAT, Cali, Colombia.

Akita M, Takayama S (1994). Stimulation of potato (*Solanum tuberosum* L.) tuberization by semicontinuos liquid medium surface level control. Plant Cell Reports 13: 184–187.

Akoroda MO (1983). Long-term storage of yam pollen. Sci. Hortic. 20: 225-230.

Akoroda MO, Hahn SK (1995). Yams in Nigeria: Status and trends. – Afr. J. Root and tuber crops 1(1): 38-41.

Alhassan AY, Mantell AH (1991). Manipulation of cultural factors to increase microtuber size and frequency in shoot cultures of food yam *Dioscorea alata* L. cv. Oriental Libson. In: Proceedings of the ninth Symposium of the International Society for Tropical Root crops. (Eds.) Ofori F, Hahn SK 20-26, Accra, Ghana, pp. 342-348.

Ammirato PV (1976). Hormonal control of tuber formation in cultured axillary buds of *D. bulbifera* and *D.alata*, Plant Physiol. (Supplement) 57: 66.

Ammirato PV (1982). Growth and Morphogenesis in Cultures of the Monocot yam, *Dioscorea*. In: Plant Tissue Culture. Ed. Fujiwara M.A, Tokyo. pp. 169-170.

Ammirato PV (1984). Yams. In: Handbook of Plant Cell Culture, Vol. 3. Eds. Ammirato PV Evans DA, Sharp WR, Yamada Y. Macmillan, New York. pp. 327-354.

Asahira T, Nitsch JP (1968). Tuberization *In vitro*: Ullucus tuberosus et *Dioscorea*. Bulletin de la Societe Botanique de France 115: 345-352.

Asahira T, Yazawa S (1979). Bulbil Formation of *Dioscorea opposita* Cultured *in vitro* Mem. Coll. Agric. Kyoto Univ. 113: 39-51.

Asiedu R, Ng SYC, Bai KV, Ekanayake IJ, Wanyera NMW (1998). Genetic Improvement. In: Food yams. Advances in research. Eds. Orkwor GC, Asiedu R, Ekanayake IJ. IITA / NRCRI. pp. 63-104.

Balogun MO, Fawole I, Ng SYC, Ng NQ, Shiwachi H, Kikuno H (2006). Interactions among cultural factors in microtuberization of white yam *Dioscorea rotundata*). (United Kingdom). Trop. Sci. 46(1): 55-59.

Balogun MO, Ng SYC, Shiwachi H, Ng NQ, Fawole I (2004). Comparative effects of explant source and genotype on microtuberization in *Dioscorea alata* and *D. rotundata*. (United Kingdom). Trop. Sci. 44: 196-200.

Balogun, MO (2005). Development of microtuber production and dormancy control protocols for yams (*Dioscorea spp*) germplasm conservation.Ph.D. Thesis. University of Ibadan, Nigeria.

Bapat VA (1993). Studies on synthetic seeds of sandalwood *(Santalum album* L.) and mulberry (*Morus indica* L.). *In:* Synseeds: applications of synthetic seeds to crop improvement Redenbaugh, K. Eds.. Boca Raton, CA, United States, CRC Press Inc. pp. 381-407.

Bazabakana R, Fauconnier B, Dialo JP, Dupont J Homes, Jaziri M (1999). Control of *Dioscorea alata* microtuber dormancy and sprouting by jasmonic acid. Plant Growth Regul. 27: 113-117.

Chang KJ, Hayashi M (1995). Ecophysiological studies on growth and enlargement of tubers in yams (*Dioscorea spp*). Development of bioassay method using microtubers of yams. Jpn. J. Agric. 39(1): 39-46.

Chang KJ, Shiwachi H, Hayashi M (1995). Ecophysiological studies on growth and enlargement of tubers in yams (*Dioscorea spp*) II. Detection of effects of plant growth regulators on growth and

enlargement of microtubers of yams. Jpn. J. Agric. 32(2): 69-75.

Coursey DG (1967). Yams. Longmans, Green and Co. Ltd. London. p. 230.

Craufurd PO, Summerfield RJ, Asiedu R, Prasad V (2001). Dormancy in Yams. Exp. Agric. 37: 75-109.

Daniel IO, Ng NQ, Tayo TO, Togun AO (1999). West African yam seeds stored under dessicated and cold storage conditions are orthodox. Seed Sci. Technol. 27: 969-975.

Daniel IO, Ng NQ, Tayo TO, Togun AO (2002). Wet-cold preservation of West African yam (*Dioscorea* spp.) pollen. J. Agric. Sci. 138: 57-62.

Emehute JKU, Ikotun T, Nwauzor EC, Nwokocha HN (1998). Crop Protection. In: Food yams. Advances in research. Eds. Orkwor GC, Asiedu R, Ekanayake IJ. IITA / NRCRI. pp 143-186.

Etienne H, Berthouly M (2002). Temporary immersion systems in plant micropropagation. Plant Cell Tiss. Org. Cult. 69: 221–231.

Food and Agricultural Organization (2000). FAO's position references paper, 2000. Food and agricultural organisation of the united nations. Rome. On the Internet.

Forsyth C, Van Staden J (1984). Tuberization of *Dioscorea bulbifera* stem nodes in culture. J. Plant Physiol. 115:

George J (1990). Effect of minisett size and nursery media on the sprouting of yams. Root Crops 16: 71-75.

Giradin OC, Nindjin Z, Farah F, Escher P Stamp, Otokore D (1998). Use of gibberellic acid to prolong dormancy and reduce losses during traditional storage of yams. J. Sci. Food Agric. 77: 172-178.

Hahn SK (1995). Yams: *Dioscorea spp*. (*Dioscoreaceae*). In: Evolution of crop plants. Eds. Smartt J, Simmonds NW. Longman scientific and Technical Essex. pp. 112-120.

Hasan SMZ, Takagi H (1995) Alginate-coated nodal segments of yam (*Dioscorea* spp) for germplasm exchange and distribution. Plant Genetic Resources Newsletter 103: 32–35.

Izumi K, Yamaguchi I, Wada A, Oshio H, Takanashi N (1984). Effects of a new growth retardant {(E)-1-(4-chlorophenyl)-4, 4-dimethyl-2-(1,2,4-triazol-1-yl)-1-penten-3-ol (S-330) }on the growth and gibberellin content of rice plant. Plant Cell Physiol. 25: 611-617.

Jean M, Cappadocia M (1991). *in vitro* tuberization in *Dioscorea alata* L. 'Brazo Fuerte' and Florido' and D. *abyssinica* Hoch. Plant Cell, Tissue and Organ Culture; 26:147-152.

Jean M, Cappadocia M (1992). Effects of some growth regulators on *In vitro* tuberization in *Dioscorea alata* L. 'Brazo fuerte' and D. *abyssinica* Hoch. Plant Cell Reports 11: 34-38.

John JL, Courtney WH, Decoteau DR (1993). The influence of Plant Growth Regulators and light on microtuber induction and formation in *Dioscorea alata* L. cultures. Plant Cell, Tissue and Organ Culture 34: 245-252.

Jova MC, Kosky RG, Pe´rez1 MB, Pino AS, Vega1 VM, Torres JL, Cabrera1 AR, Garci´ a1 MG, de Ventura JC (2005). Production of yam microtubers using a temporary immersion system. Plant Cell, Tissue Organ Cul., 83: 103–107.

Keller ERJ, Schuler K (1996). Present situation of the *in vitro* collections in the Gatersleben Genebank. Paper presented at the consultation meeting on the management of Field and *in vitro* Genebank,. CIAT, Cali, Colombia. pp. 15-20

Kikuno H, Onjo M, Kusigemati K, Hayashi M (2002a). A Relationship between the initiation of tuber enlargement and endogenous plant hormones in water yam (*Discorea alata* L.). Jpn. J. Trop. Agric. 46 (1): 39-46.

Kikuno H, Onjo M, Kusigemati K, Hayashi M (2002b). A Relationship between the initiation of tuber enlargement and changes in the content of endogenous jasmonic acid in water yam (*Discorea alata* L.). Jpn. J. Trop. Agric. 46 (2): 109-113.

Kitto SL, Janick J (1985). Hardening treatments increases survival of synthetically-coated asexual embryos of carrot. J. Am. Society Hortic. Sci. 110(2): 283-286.

Koda Y, Kikuta Y (1991). Possible Involvement of Jasmonic acid In Tuberization of yam plants. Plant Cell Physiol. 32(5): 629-633.

Koda Y, Okazawa Y (1988). Detection of potato tuber-inducing activity in potato leaves and old tubers. Plant Cell Physiol. 29(6): 969-974.

Kumar U (1998). Synthetic Seeds for Commercial Crop Production. p. 160.

Kyesmu PM (1998). Cryopreservation of shoot apices of *Dioscorea* species by vitrification: application of D. *rotundata*'s protocol to other species. ICRS (Int Collaboration Res Sect) progress report, final reports of the visiting research fellows, vol 5, 1996. ICRS, Okinawa Subtropical Station, JIRCAS, Japan.

Kyesmu PM, Takagi H (2000). Cryopreservations of shoot apices of yams (Dioscorea species) by vitrification. In: Cryopreservation of tropical plant germplasm. (Eds.) Engelmann F, Takagi H JIRCAS/PGRI, Rome, Italy. pp. 411-413.

Kyesmu PM, Takagi H, Yashima S (1997). Cryopreservation of white yam (Dioscorea rotundata) shoot apices by vitrification In: Proc. Annu. Meet Jpn. Mol. Biol. Kumamoto, Japan, p. 162.

Lang GA (1996). Plant Dormancy, Physiology, Biochemisry and Molecular Biology. Wallingford: CABI Publishing.

Leunufna S, Keller ERJ (2003). Investigating a new cryopreservation protocol for yams. (Dioscorea spp.) Plant Cell Reports 21:1159–1166.

Malaurie B, Trouslot MF, Engelmann F, Chabrillange N (1998). Effect of pretreatment conditions on the cryopreservation of in vitro-cultured yams (Dioscorea alata 'Brazo Fuerte' and D. bulbifera 'Noumea Imboro') shoot apices by encapsulation dehydration. Cryoletters 19:15–26.

Mandal BB (2000). Cryopreservation of yams apices: a comparative study with three different techniques. In: Cryopresevation of tropical plant germplasm, current progress and application. Eds. Engelmann F, Takagi H IPGRI, Rome, pp. 233–237.

Mandal BB, Chandel KPS, Dwivedi S (1996). Cryopreservation of yam (Dioscorea spp.) shoot apices by encapsulation-dehydration. Cryoletters 17:165–74.

Mantell SH (1987). Development of microtuber production systems for yam to enable direct field planting of micropropagated clone sections. In: Summaries of the Final Reports of Research Projects of 1st Programme, 1983-86, of Science and Technology for Development: Tropical and Subtropical Agriculture, Wagenigen, The Netherlands, CTA, Analytical. pp. 187-192.

Mantell SH, Haque SQ, Whitehall AP (1980). Apical meristem tip culture for eradication of flexuous rod viruses in yams (D. alata). Trop. Pest Manage. 26: 170-179.

Mantell SH, Hugo SA (1986). International germplasm transfer using micropropagules. In: Proceedings of a training workshop and symposuim on micropropagation and meristem culture and vegetative propagation, CSC Technical Publication Series; 205: 88-98.

Mantell SH, Hugo SA (1989). Effects of Photoperiod, mineral medium strength, Inorganic ammonium, Sucrose and Cytokinin on root, shoot and Microtuber development in shoot Cultures of Discorea alata L. And Dioscroea bulbifera L. Yams. Plant Cell, Tissue And Organ Culture 16: 23-37.

Murashige T, Skoog F (1962). A revised medium for rapid growth and bioassays with tobbaco tissue cultures. Physiologia Plantarium 15: 473-497.

Ng NQ, Daniel IO (2000). Storage of pollens for long-term conservation of yam genetic resources. In: Cryopreservation of tropical plant germplasm. Eds. Engelmann, F. and H. Takagi. JIRCAS/IPGRI, Rome, Italy. pp. 136-139.

Ng NQ, Ng SYC (1994). Approaches for yam germplasm conservation. In: Root crops for Food security in Africa. Ed. M.O. Akoroda, ISTRC-AB; pp. 135-140.

Ng SYC (1984). Meristem culture and multiplication. In: International Institute of Tropical Agriculture Annual Report, 1983, Ibadan, Nigeria, pp. 133-134.

Ng SYC (1988). In vitro tuberization in white yam (Dioscorea rotundata Poir). Plant Cell, Tissue Organ Cul. 14: 121-128.

Ng SYC (1992). Micropropagation of white yam (D. rotundata. poir), In: Biotechnology in Agriculture Forestry, High-tech and micropropagation III. Eds. Y.P.S. Bajaj. Berlin Heidelberg, Springer - Ver lag, 19: 135-159.

Ng SYC, Mantell SH (1996). Final Report of ODA Project R4886 (H) on Technologies for Germplasm Conservation and Distribution of Pathogen-free Dioscorea yams to National Root crop Research Programs, Wye College/IITA/SARI, p. 73.

Ng SYC, Ng NQ (1997). Germplasm conservation in food yams (Dioscorea spp): Constraints, Application and Future prospects. In: conservation of plant Genetic resources in vitro. Volume 1: General

Aspects. Eds. Razdan MK, Cocking EC. Science publishers Inc. U.S.A. pp. 257-286.

Ng SYC, Ng NQ (2000). Cryopreservation of cassava and yam shoot-tips by fast freezing. In: Cryopreservation of tropical plant germplasm. Eds. Engelmann, F. and H. Takagi. JIRCAS/IPGRI, Rome, Italy. pp. 418-420

Nweke FI, Ugwu BO, Asadu CLA, Ay P (1991). Production costs in the yam-based cropping systems of Southwestern Nigeria. Resource and Crop Management Division Research Monograph No. 6. IITA, Ibadan, Nigeria. p. 29.

Nwosu NA (1975). Recent developments in vegetative propagation of edible yam (Dioscorea species). Proc. Agric. Soc. Nig. 12: 15.

Okoli OO (1991). Yam germplasm diversity: uses and prospects for crop improvement in Africa. In: crop Genetic Resoruces of Africa, Vol. II. Eds. N.Q. Ng, P. Perrino, F. Attere and H. Zedan IITA/IBPGR (UNEP/CNR. Ibadan. Nigeria pp. 109-117.

Okoli OO, Igbokwe MC, Ene LSO, Nwokoye JU (1982). Rapid multiplication of yam by the minisett technique. Research Bulletin 2. National Root crops research Institute (NRCRI), Unudike, Nigeria p.12.

Okonkwo SNC (1985). The botany of the yam plant and its exploitation in enhanced productivity of the crop. In: The Biochemistry and Technology Of The Yam Tuber. Eds. Osuji. Biochemical Society Of Nigeria And Anambra State University Of Technology, pp. 3-29.

Onjo M, Okamoto S, Hayashi M (1999). Studies on The Development and Thickening Growth of Tubers in Yams (Dioscorea spp.) 3. Effects of gibberellins on growth, enlargement and dormancy of tubers in D. alata L. Japanese J. Trop. Agric. 43(2): 65-77.

Onjo M, Park BJ, Hayashi M (2001). Effects of plant growth regulators on plantlet growth and enlargement of microtubers of water yam (Dioscorea alata) in vitro. Jpn. J. Trop. Agric. 45(2): 145-147.

Orkwor GC, Asadu CLA (1998). Agronomy. In: Food yams. Advances in research. Eds. G.C. Orkwor, R. Asiedu and I.J. Ekanayake. IITA / NRCRI. pp. 105-141.

Park BJ, Onjo M, Tominaga S, Shiwachi H, Hayashi M (2003). Relationship between the dormancy and its release and external factors in tubers of water yam (Dioscorea alata L) Jpn. J. Trop. Agric. 47(1): 42-50.

Pennycooke JC, Towil LE (2001). Medium alterations improve regrowth of sweet potato (Ipomea batatas (L) Lam) shoot tips cryopreserved by vitrification and encapsulation-dehydration. CryoLetters 22: 381–389.

Piccioni E (1997). Plantlets from encapsulated micropropagated buds of M.26 apple rootstock. Plant Cell, Tissue and Organ Culture 47: 255-260.

Quin FM (1998). An overview of Yam Research. In. Food Yams; Advances in Research. Eds. G.C. Orkwor, R. Asiedu and I.J. Ekanayake IITA/NRCRI. pp. 215-229.

Rao PS, Suprasanna P, Ganapathi TR, Bapat VA (1998). Synthetic seeds: concepts, methods and application. In: Plant tissue culture and molecular biology. Ed. Srivastava PV. Narosa, India, pp. 607–619.

Sengupta J, Mitra GC, Sharma AK (1984). Organogenesis and tuberization in cultures of Dioscorea floribunda. Plant Cell, Tissue Organ Cul. 3(4): 325-331.

Shiwachi H, Ayankanmi T, Asiedu R (2002). Effect of daylength on the development of tubers in yams (Dioscorea spp.) Trop. Sci. 42: 162-170.

Sreekantan L, George S, Nair KH (1995). Sprouting and yied of yam (Dioscorea species) planted through minisetts. Indian J. Agron. 40: 149-50.

Standardi A, Piccioni E (1998). Recent perspective on synthetic seed technology using nonembryogenic in vitro-derived explants. Int. J. Plant Sci. 159: 968–978.

Taylor M (1996a). Field conservation of root and Tubber crop in the south Pacific. Paper presented at the consultation meeting on the management of Field and in vitro genebank, 15-20 January, 1996. CIAT, Cali, Colombia.

Taylor M (1996b). In Vitro conservation of root and tuber crops in the south pacific. Paper presented at the consultation meeting on the management of field and in vitro Genebank, 15-20 January, 1996 GAT,Cali, Coplombia.

Tessereau H, Florin B, Meschine MC, Thierry C, Pétiard V (1995). Cryo-preservation of Somatic Embryos: A Tool for Germplasm Storage and Commercial Delivery of Selected Plants. Ann. Bot. (1994) 74: 547-555.

Tschannen ABO, Girardin C, Nindjin D, Daouda Z, Farah P, Stamp, Escher F (2003). Improving the application of gibberellic acid to prolong dormancy of yam tubers (*Discorea spp*) J. Sci. Food Agric. 83: 787-796.

Uragami A (1993). Cryopreservation of cultured cells and organs of vegetables. In: Cryopreservation of Plant Genetic Resources. Japan International Cooperation Agency, JICA Ref. No. 6. pp.111-135.

Vick BA, Zimmerman DC (1984). Biosynthesis of Jasmonic acid by Several Plant Species. Plant Physiol. 75: 458-461.

Wang Q, Batuman O, Li P, Bar-Joseph M, Gafny R (2002). A simple and efficient cryopreservation of *in vitro* shoot tips of "Troyer" citrange [*Poncirus trifoliate* (L.) Raf. X *Citrus sinensis* (L.) Osbeck] by encapsulation-vitrification. Euphytica 128:135–142

Wickham LD, Passam HC, Wilson LA (1981). Tuber sprouting and early growth in four edible *Dioscorea* species. Ann. Bot. 47: 87-95.

Wickham LD, Passam HC, Wilson LA (1982). The origin, development and sprouting of bulbils in two *Dioscorea* species. Ann. Bot. 50: 621-627.

Wickham LD, Passam HC, Wilson LA (1984). Dormancy responses to post-harvest application of growth regulators in *Dioscorea* species. 1. Responses of bulbils, tubers and tuber pieces of three *Dioscorea* species. J. Agric. Sci. 102: 427-432.

Zamura AB, Paet CN (1996). *In vitro* genebanking activities, Institute of plant Breeding, College of Agriculture, University of the Philippines at Los Banos. Paper presented at the consultation meeting on the management of Field *In vitro* Genebank, 15-20 January 1996. CIAT, Cali, Colombia.

22

Biodiversity and conservation of medicinal and aromatic plants in Africa

Okigbo, R N.[1*], Eme, U E. [2] and Ogbogu, S. [3]

[1,2]Department of Botany, Nnamdi Azikiwe University, Awka, Anambra State, Nigeria.
[3]Dpartment of Zoology,Obafemi AwolowoUniversity, Ile-Ife, Nigeria.

Medicinal and aromatic plants (MAPS) represent a consistent part of the natural biodiversity endowment of many countries in Africa. The role and contributions of medicinal plants to healthcare, local economies, cultural integrity and ultimately the well-being of people, particularly the rural poor, have been increasingly acknowledged over the last decade. The demands of the majority of the populace for medicinal plants have been met by indiscriminate harvesting of spontaneous flora, including those in forests. This has resulted in severe loss of habitat and genetic diversity. The utilization of medicinal and aromatic plants (MAPs) as a source of fuel, building material, food, fodder, and fibre, in African countries has, however, led to a resurgence of natural product- based industries and pharmaceutical products. This had been spurred by the interests of the developed countries for traditional medicine and natural products. Furthermore, many African medicinal plants are well-known in the international markets, e.g. *Ancistrocladus abbreivatus*, a Cameroun plant with anti-HIV potential. Therefore, sustainable management and conservation of these endangered medicinal plant species are important not only because of their value as potential therapeutics, but also due to worldwide reliance on traditional medicinal plants for health. Effective conservation strategies for medicinal plant should take place within four main areas: *in-situ* and *ex-situ* conservation, education and research. Saving Africa's medicinal plant resources from extinction calls for intensive management and conservation, more research and increased level of public awareness about our vanishing heritage.

Key words: African, health care delivery, medicine, harvesting.

INTRODUCTION

Medicinal and aromatic plants represent a consistent part of the natural biodiversity endowment of many countries in Africa, as well as the world at large. Medicinal plants are plants containing inherent active ingredients tending or used to cure disease or relieve pain. Aromatic plants on the other hand, have strong characteristic smell or fragrance (King, 1992). Plants represent a huge store-house of drugs: they produce more than 10,000 different compounds to protect themselves from predators. These compounds could be potential drugs (King, 1992; Izuakor, 2005).

Biodiversity is the variety and variability of living organisms and biological communities in which they live, plus the ecological and evolutionary processes that keep

them functioning. It is often a varietals measure of the health of biological systems indicating the degree to which the aggregate of historic species are viable versus extinct (UNESCO, 1994a). Conservation, on the other hand, involves a careful preservation and protection of something, especially planned management of a natural resource to prevent neglect, over-exploitation or even destruction.

Historically, plant medicines were discovered by trial and error. Just as people learnt to exploit plants for food, so they learnt to use plants as medicine (UNESCO, 1994a). For example, our ancestors noticed that aches and pains went away when they drank tea made from the bark of a willow tree, *Salix* sp. Later, scientists discovered that willow bark contains salicylic acid, the active ingredient in aspirin. Other plant medicines such as *Cinchona, Opium, Belladona*, and *Aloe* were selected for use based on empirical evidence as gathered by traditional

*Corresponding author. E-mail: okigborn17@yahoo.com.

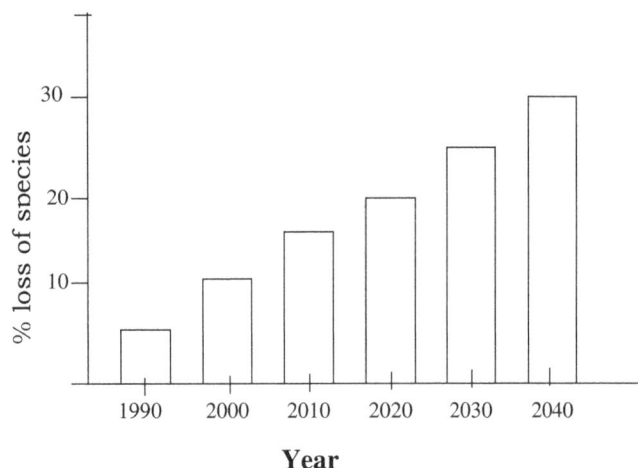

Figure 1. Projected gradual loss in tropical plant species (After Roche, 1992).

practitioners (Okigbo and Mmeka, 2006).

Traditional medicine, as a major African socio-cultural heritage, has been in existence for hundreds of years (Elujoba et al., 2005). It was once believed to be primitive and wrongly challenged by foreign religions dating back during the colonial rule in Africa, and subsequently by the conventional or orthodox medical practitioners (Elujoba et al., 2005). The populations of developing countries world-wide continue to rely heavily on the use of traditional medicines as their primary source of healthcare (Cunningham, 1993). Ethnobotanical studies carried out throughout Africa confirm that native plants are the main constituent of traditional African medicines (Cunningham, 1997). Furthermore, an increasing reliance on the use of medicinal plants in the industrialized societies has been traced to the extraction and development of several drugs and chemotherapeutics from these plants, as well as from traditionally-used rural herbal remedies (UNESCO, 1994a).

The demand of the majority of the populace of Africa and beyond for medicinal plants has led to indiscriminate harvesting of spontaneous flora, including those in forests (Cunningham, 1997). Moreover, the continent of Africa is estimated to have about 215, 634, 000 hectares of closed forest areas, and with a calculated annual loss of about 1%. Due to deforestation, many medicinal plants and other genetic materials become extinct before they are ever documented (Tuley de Silva, 1997). Habitat conversion threatens not only the loss of plant resources, but also traditional community life, cultural diversity, and the accompanying knowledge of the medicinal value of several endemic species (UNESCO, 1994b). As medicinal plant supplies diminish, constructive resource management and conservation strategies based on clear knowledge of the surrounding medicinal plant use must be designed. This study seeks, primarily, to respond to three central questions:

Of what importance are medicinal plants to developing countries?
What are the causes of the depletion of wild populations of medicinal plants species in Africa?
What can be done to ensure the effective conservation of all medicinal plant species?

CONSERVATION OF MEDICINAL AND AROMATIC PLANTS IN AFRICA

Reasons for conservation

Well thought-out arguments have been made many years ago to raise public awareness on the destruction of tropical rain forest and seasonally dry monsoon forest. The arguments were largely ignored earlier but today, dramatic efforts are being made to conserve biodiversity (Azimahtol et al., 1998).

The current cry of ecological genocide, genetic erosion, environmental degeneration, fragmen-tation and destruct-tion as well as extinction of our biological heritage, is consequence of inaction (Krikorian, 1998). As long as the destruction of forest continues, medicinal plants and their natural habitats will remain under the threat of over-exploitation than ever before (Walter and Gillett, 1998).

The ultimate goal of conservation is to preserve the natural habitats of vulnerable medicinal plant species and to achieve their sustainable exploitation in less vulnerable areas (Cunningham, 1993). A projected gradual loss in tropical plants shows that tropical forest species likely to go into extinction in coming decades (Figure 1).

The graph (Figure 1) showed the progressive rise in loss of species and if care is not taken in terms of conversion of the bio-resources (including medicinal and aromatic plants), their complete extinction in the coming decades will be inevitable.

Conservation strategies

Effective conservation strategy for medicinal plants should take place within four main areas:- *in-situ* conservation, *ex-situ* conservation, education and research.

In-situ conservation

In-situ conversation involves protection and establishment of plants and other biological resources in the location of their natural occurrence. In order to ensure that representative of wild populations of vulnerable medicinal plant species are maintained, core conserva-tion areas or other protected habitats that will allow natural processes to continue undisturbed should be designated (Cunningham, 1997). Since, it is only in nature that plant diversity at genetic, species and ecosystem levels can be

Table 1. Selected medicinal plant families of threatened species (Walter and Gillett, 1998).

Family	No of genera	No of species	% of total species threatened	Main uses	Examples of over-harvested species.
Rosaceae	100	3000	14.4	Stone fruit crops and medicinals	*Prunus africana*
Lauraceae	35-50	2000	13	Timber, medicines, cinnamon.	*Ocotea bullata*
Menispermaceae	70	400	9.5	Medicines, dyes.	*Stephania spp*
Apocynaceae	168-200	2000	7.5	Medicines	*Holarrhena floribunda*
Guttiferae	50	1200	13.3	Dyes, medicines, fruits, chewing sticks.	The West African *Garcinia* spp.
Legunminosae	590	12,000-14,200	18	Multiple uses: timber, . medicinal, forage, food.	*Dalbergia odorifera, Afzelia* spp.
Stangeriacea	1	1	100	Traditional medicine, symbolic	*Stangeria eriopus*
Canellaceae	6	20	35	Traditional medicine, moluscides.	*Warburgia elongate, W. salutaris*

conserved on a long term basis, identification of eco-systems with diverse medicinal plant species is very essential.

Ex-situ conservation

This involves establishment of plantations, maintenance of living collections in farm fields, home gardens, botanical gardens, and arboreta in location outside the zone of their natural occurrence (Roche, 1992). The essence of *ex-situ* conservation is the rapid development of alternative supply sources of medicinal plants through cultivation in large enough quantities and at low enough price in order to compete with prices obtained by gatherers of wild medicinal plant stocks (Cunningham, 1997). This will satisfy market demands, result in more secure jobs and provide fewer incentives to gather from the wild. If this does not occur, naturally occurring species will disappear from the wild, thereby undermining the local medicinal resource base.

Traditional Medicinal Practitioners (TMPs)

Traditionally, rural African communities have relied upon spiritual and practical skills of traditional medical practitioners, whose knowledge of the ecology of plant species are invaluable (Cunningham, 1997). Since very little goes unnoticed in communally owned areas where traditional medicinal practitioners or community leaders are likely to be, and since they are very aware of the conservation status of local medicinal plant resources, they can be influential in changing local opinion so as to limit over-exploitation (Cunningham, 1993; Marshall, 1998).

Botanical gardens and field gene banks

Seed and gene banks of vulnerable medicinal plant

species should be maintained as precaution and backup against extinction. Medicinal plants most likely to be collected are the slow-growing species where commercial cultivation is unlikely and wild populations are jeopardized (Cunningham, 1993). These gene banks offer:

• A source of variants in case a major crop or plant is felled by disease or environmental disaster;
• The return of endangered or extinct varieties to their native lands.
• The supply of genetic material from which researchers can fashion useful plants in the years to come, even after the species represented in the bank have become extinct.

Education and training

The conversation of medicinal plants is by necessity a long term project requiring the development of trained staff, supported by organizations and a general public that is aware of issues at stake. Improvement in national education standards is a key factor in the medicinal plants conservation issue which will come about only as a result of economic development in African nations (Cunningham, 1993).

To increase the awareness of the public on the value of medicinal plant resources, the following are suggested:
Campaigns that promote the importance of habitat and medicinal plant conservation and which encourage the cultivation of medicinal plants should be instituted. Target groups would include rural communities, government decision-makers, pharmaceutical companies such as Plantecam Medicam of France which works in Cameroun and Inverni della Beffa of Italy which works in Madagascar.

A media campaign through national radio networks to publicize information on the scarcity of popular medicinal plants should be implemented.

Information programme for decision makers in African government to link public health with medicinal plant conservation issues should be developed.

Studies and research information which identify threatened medicinal plants should be circulated through the International Board for Plant Genetic Resources (IBPGR) to regional gene banks.

Information relating to adverse toxic properties in medicinal plants should be circulated particularly to traditional medicinal practitioners and in primary healthcare training (Akerele, 1987; Good, 1987).

Research and monitoring

Research into the identification of areas of high biological diversity at the macro scale and research into the properties and usage of specific medicinal plants at the microscale should use the complementary skill of the TMPs and conservation biologists (Cunningham, 1993). Series of interactive discussions involving the TMPs, commercial gatherers and market-based traders to discover the perceived scarcity of species, sites of diversity, status of popular medicinal plant species, the perceived problems and solutions, should also be initiated (Cunningham, 1997). Moreover, studies should be initiated through the co-operative effort between African and European scientific institutions to study the genetic diversity of popular medicinal plant species like *Warburgia salutaris* and *Okoubaka aubervillei* in West Africa. This, carried out through isozyme electrophoresis, would help to identify the degree of genetic erosion taking place in areas of over-exploitation or habitat destruction (Cunningham, 1993).

Qverview of WHO'S guidelines for integrating African phytomedicine into the health scheme Guidelines for the institutionalization of traditional medicine into the health scheme as provided by WHO (1978) include the following.

Political Recognition

The government and Heads of States should be aware and help in the development of traditional medicine. This has already been achieved when the African summit of Heads of state declared 2001-2010 as 'Decade of African Traditional Medicine'.

Development of policy, legal and regulatory framework

Government should formulate national policies, legal framework and registration. WHO (1978) has provided guidelines for the assessment of herbal medicine.

Promoting scientific research on traditional medicine and collaboration work

Scientific research should be conducted on safety,

efficacy and quality of traditional medicine as proposed by WHO (Akerele, 1993).

Ensuring that intellectual property rights are protected

Intellectual property rights are a priority item on the agenda of member states to protect indigenous knowledge about traditional medicine (Elujoba et al, 2005; WHO, 1978) and legislation should be made on this (Calixto, 2000).

Disseminating appropriate information to the general public on the use of traditional medicine

Appropriate information should be given to the general public to empower them with knowledge and skills for the proper use of traditional medicine (WHO, 1978). This is achieved through organization of seminars to raise awareness as recorded by Makhubu (2006).

Providing a good economic environment

The government should ensure that a good economic, political and regulatory environment is established for local production by traditional herbal practitioners as well as develop industries that can produce standardized remedies to increase access (WHO, 1978).

Utilization of medicinal and aromatic plants in Africa

The use of medicinal and aromatic plants among Africans is widespread and has been in existence for many generations (Kokwaro, 1993). About 70% of the wild plants in North Africa are known to be of potential value in fields such as medicine, biotechnology and crop improvements (UNESCO, 1994a).

Therapeutic uses

Studies in the use of plant extracts for control of diseases have shown the importance of natural chemicals as possible sources of non-phytotoxic and easily biodegradable alternative fungicides and antibiotics (Akueshi et al., 2002; Okigbo and Nmeka, 2005). Virtually all native plant species are used for the treatment of one ailment or another. These involves the traditional medicinal use for despoil, preventive, curative and magical purposes (Osemeobo and Ujor, 1999). Some chemical substances in the plant tissues brought about the medicinal value of such drug plants. Phytomedicine have a wide range of therapeutic uses as shown in Table 2.

Economic benefits and industrial uses

African continent is mad up of many developing nations

Table 2. Some African medicinal and aromatic plants with their therapeutic uses (Sofowora, 1993).

Medicinal plant	Family	Active Ingredients	Therapeutic uses
Atropa Belladonna L., Datura strammonium L.	Solanaceae	Atropine, Hyoscine, Hyoscyamine.	Antispasmodic, mydriatic.
Digitalis purpurea L.	Scrophula-riaceae	Purpurea glycoside A and B, digoxin, digitoxin	Myocardia / stimulant.
Ephedra sinice Sprag.	Ephedraceae	Ephedrine	For relief of asthma and hay fever
Zingiber officinale Roscoe	Zingiberaceae	Volatile oil; gingerol.	As condiment and medicinally as carminative and aromatic
Papaver somniferum L.	Papaveraceae	Morphine, codeine, thebaine, narceine, papaverine.	Narcotic; Analgestic.
Rauwolfia serpentine Benth or R. vomiforia Afz.	Apiaceae	Volatile oil	Used in psychiatric cases and antihyper tensive.
Carum carvi L.	Apiaceae	Volatile oil	Flavouring agent and carminative.
Ricinus communis L.	Euphorbiaceae	Fixed oil	Purgative; vehicle for eye drops
Cinnamonum zeylanicum Blume.	Lauraceae	Volatile oil, tannin.	Stimulant, astringent, antiseptic, carminative, stops vomiting.
Cinchona succiruba PAV and other species.	Rubiaceae	Quinine, quinidine.	Bitter tonic, quinidine for atrial fibrilation.

Table 3. Phytotherapeutic Sales in World Market (Blumenthal 1999; Calixto 2000; Grunwald 1995; Robbers et al, 1996).

Year	Europe						America	Asia
	Germany	France	Italy	UK	Spain	Netherland	U.S.A	India
1995								$ 400 million
1996							$ 3.2 billion	
1997	$ 3.5 billion	$ 1.8 million	$ 700 million	$ 400 million	& 300 million	& 100 million		
1998								
1999							$ 5 billion	

with very limited resources like minerals or petroleum oil to sustain their economic development. Therefore, majority of the inhabitants of these nations depend on the natural vegetation as a source of necessities such as fuel, building material, food, fodder and fibre (Kokwaro, 1993). Presently, there is a resurgence of natural product-based industries and pharmaceutical products because of the increasing interest in traditional medicine and natural products in developing countries (Cunningham, 1997).

Phytomedicine has demonstrated its contributions to the reduction of excessive mortality, morbidity and disability due to diseases such as HIV/AIDS, malaria, tuberculosis, sickle-cell anemia, diabetes and mental disorders (Elujoba et al., 2005). It has also reduced poverty by increasing economic well-being of communities and developed health system by increasing the people's access to healthcare (Elujoba et al., 2005). The production, processing and sale of phytomedicine products create employment for the producing countries (Gunasena and Hughes, 2000). In the UK alone, herbal remedy trade is worth more than £200 (293 Euros) million per year (IUCN, 2005). The trend of sales of phytotherapeutic in the world market is shown below.

African medicinal and aromatic plant in world market

Africa is one of the main world producers of medicinal and aromatic plants (Table 4) and many of them are well known in international markets (Elujoba et al., 2005;

Table 4. Some African Medicinal plants in World Market (Okigbo and Mmeka, 2006).

Plant species	Action	Constituent	Indigenous countries	Source
Ancistrodadus abbreviatus	Anti-HIV	Michellamine B	Cameroun and Ghana	Sofowara, 1993; Boyd et al, 1994
Zingiber officinale	Spice, carminative and medicinal products.	Giingerol	Nigeria	Sofowora, 1993
Catharanthus roseus	Anti-Lenkemia and Hodgkin's disease	Triterpenoids, tannins and alkaloids.	Madagascar	Elujoba et al, 2005; Nayak and Pereira, 2006.
Cindona succirubra	Anti-malaria	Quinine	West African countries	Reiz and Lipp, 1982.
Agava sisalana	Corticosteroids and oral contraceptives.	Hecogenin	Tanzania	Elujoba *et al*, 2005.
Rauwolfia vomitoria	Tranquilizer and antihypertensive.	Reserpine, yohimbine	Nigeria, Zaria, Rwanda, Mozambique	Sofowara, 1993
Syzigium aromaticum	Dental remedy	Eugenols, terpendoids.	East africa countries. Madagascar.	Elujoba et al, 2005.
Chrysanthemum cinerariifollium	Insecticides	Fyrethrins	Ghana, Kenya, Rwanda, Tanzania South Africa	Wallis, 1967.

Sofowora, 1993). An Example is *Ancistrocladus abbreivatus,* a plant with anti-HIV potential and endemic to Cameroon (Sofowora, 1993).

Interest in phytomedicines in Africa

For years, public interest has increased for natural therapies (mainly phytomedicine) all over the world including Africa (Blumenthal, 1999). The pharmaceutical industry has come to consider traditional medicine as a source for identification of bio-active agents that can be used in the preparation of synthetic medicine (Cunningham, 1997). Many of the more pharmacologically (commercially) interesting medicinal plant species in use around the world are employed in more than one community, and often in more than one country, for multiple uses. The natural products industry in Europe and the United States is equally interested in traditional medicine (LeBeau, 1998). In Europe and America where phytomedicine industry is thriving, extracts from medicinal plants are sold in a purified form for the treatment and prevention of all kinds of diseases (WIPO, 1998).

According to Calixto (2000) and Grunwald (1995), there are several factors that lead to the preference and growth of phytotherapeutic market worldwide. They include:

- Preference of consumers for natural therapies.
- Great interest in alternative medicine
- High cost of synthetic drugs

- The belief that phytomedicine is used for the treatment of certain diseases where conventional medicine fails.
- The belief that phytomedicine is devoid of side effects since million of people all over the world have been using phytomedicine for thousands of years.
- Improvement in the quality, proof of efficacy and safety of phytomedicine.

Benefits of phytomedicine over synthetic drugs

Although synthetic or chemical drugs can have greater or quicker effects than do equivalent phytomedicines, they present a higher degree of side effects and risks (Okigbo and Mmeka, 2006). For instance, psycho-pharmacological products are associated with undesirable side effects such as uncoordinated motor skills and drowsiness, but phytomedicine acts on the body by regulating and balancing its vital processes rather than stopping or combating certain symptoms (Okigbo and Mmeka, 2006).

Phytomedicine have a wide range of therapeutic uses and are suitable for chronic treatments (Calixto, 2000). They are said to be gentle, effective and often specific in function in organs or systems of the body (Iwu et al., 1999). Plants like *Cimicifuga racemosa, Angelica sinensis* and *Agnus castus* are specifically useful for premenstrual syndrome, PMS (excessive estrogen) as recorded by Schellenburg (2001) and Wuttke (2000).

Phytomedicines are good dietary supplements, which are nutritive and replenish the body. For example, sunflower (*Helianthus annuus*) provides vitamin B_6 (Pyridoxine) as reported by MacDougall (2000).

Phytomedicines are effective in treating infection diseases as well as limit side effects associated with synthetic antimicrobial drugs (Iwu et al., 1999). Plants like *A. abbreviatus* from Cameroon has been reported to show a strong anti-HIV activity due to michellamine B and has been developed for treating people living with HIV/AIDS (Sofowora, 1993).

The symptoms of phytomedicine often extend beyond symptoms and treatment of diseases. For example, *Hydrastis canadensis* not only has antimicrobial properties but also promotes optimal activity of the spleen in releasing compounds as reported by Murray (1995). Finally, they are usually less expensive than the synthetic drugs (Calixto, 2000).

CONCLUSION AND RECOMMENDATION

Owing to the unsustainable exploitation of medicinal plant species in Africa for multiple uses such as grazing, fuels, food, timber and medicine, and the decline of natural vegetation due to unmonitored trade of these plant species, the survival of African medicinal and aromatic plants is in jeopardy. The Saving Africa's medicinal plants resource calls for more protection, management, research and an increasing level of public awareness about the vanishing heritage. Following the uncertainties in demographic and urbanization trends, the demand for traditional medicines is set to rise, and would mount increased pressure on the remaining areas of natural vegetation (Cunningham, 1997). There should be a shift in focus from conserving primarily conspicuous plants to the need to conserve all kinds of plants as well as their ecosystems. African countries' governments should give priority to species inventory aimed at documenting the various medicinal plants in use as therapeutics; establish local botanical gardens to ensure sustainable supply of safe, effective and affordable phytomedicines in each country; and make policies at both the international and national levels. These would ensure the success of an overall conservation strategy through elimination of wealth inequalities between nations. Furthermore, international conservation agencies, in conjunction with governments and other NGOs, need to determine a mechanism whereby those benefiting from the conservation of biotic diversity also contribute towards the costs of conserving it. The future of medicinal and aromatic plants in African rests on the ability to resolve the current conflict between conservation and resource use, as well as a shift towards more resource based agriculture that is already being challenged by the globalization of economics.

REFERENCES

Akerele. O (1993). Summary of WHO guidelines for the assessment of herbal medicines. *Herbalgram*, 28:13-17.

Azimahtol HLP, Soliman W (1998). A gift of Biodiversity: an anti cancer compound from traditional herbal plant. In: Nair MNB, Nathan G (Eds.). Medicinal Plants Cure for the 21 century. University Putra, Serdang, Malaysia. pp. 152-153.

Blumenthal M (1999). Harvard study estimates consumers spend $5.1 billion on herbal products. Herbalgram 45: 68.

Calixto JB (2000). Efficacy, safety, quality control, marketing and regulatory guidelines for herbal medicines (Phytotherapeutic agents). Brazillian J. Med. Biol. Res. 33 (2): 179-189.

Cunningham AB (1993). African Medicinal Plants: setting priorities at the Interface between conservation and primary healthcare. People and Plants Working paper I. UNESCO, Paris 92p.

Cunningham AB (1997). An Africa-wide overview of Medicinal Plant Harvesting, Conservation and Healthcare, Non-Wood Forest Products. In: Medicinal plants for Forest Conservation and Healthcare. FAO, Italy.

Elujoba AA, Odeleye OM, Ogunyemi CM (2005). Traditional Medical Development for medical and dental primary healthcare delivery system in Africa. Afr. J. Traditional, Complementary and Alternate Med. 2(1): 46-61.

Grunwald J (1995). The European Phytomedicine Market: figures, trends, analysis. *Herbalgram*, 38: 60-65.

Gunasena HPM, Hughes A (2000). Food for the future I. Tamarind (*Tamarindus Indica* L.) International Center for Under-utilized Crops (IUCN), Southampton, U.K. www.iucn.org/places/medoffice/nabp/med_arom.html -19k

IUCN (2005). Medicinal Plants in North Africa : Linking Conservation and Livelihoods. Malaga, 18 April, 2005 Press Release. IUCN Center for Mediterranean Corporation.

Iwu MW, Duncan AR, Okunji CO (1999). New antimicrobials of Plant Origin. In: J, Janick (Ed). Perspectives in New crops and New Uses. ASHS Press, Alexandra V.A. pp. 457-462.

Izuakor TM (2005). Bioresources conservation: The Role of Agroforestry. Heritage Printers, Nigeria. pp. 1-10.

King SR (1992). Conservation and Tropical Medicinal Research. Shaman Pharmaceutical Incorporated p. 650.

Kokwaro JO (1993). Current status of Utilization and Conservation of Medicinal Plants in Africa, South of the Sahara. Acta Hort. (ISHS) 332: 121-130.

Krikorian AD (1998). Medicinal plants and Tropical forest: Some orthodox and some not so orthodox musing. In: Nair MNB, Nathan G (Eds). Medicinal Plant Cure for the 21st century University Putra, Serdang, Malaysia. pp. 32-110.

Lebeau D (1998). Urban Patients' Utilization of Traditional medicine Upholding Culture and Tradition. University of Namibia, Windhock, Namibia.

Makhubu L (2006). Traditional medicine: Swaziland. Afr. J. Traditional, Complementary and Alternative Med. 5(2): 63-71.

Murray M (1995). The Healing Power of Herbs. Prima Publishing. Rocklin, C.A.

Okigbo RN, Mmeka EC (2006). An Appraisal of Phytomedicines in Africa. KMITL Sci. J. (Thailand) 4(1): 1-7

Okigbo RN, Nmeka IA (2005). Control of yam tuber rot with leaf extracts of *Xylopia aethiopica* and *Zingiber officinale*. Afri. J. Biotechnol. 4(8): 804-807.

Osemeobo GJ, Ujor G(1999). Non Wood forest products in Nigeria. Forest statistic and data collection AFDCA/TN/06

Robbers J, Speedie M, Tyler V (1996). Pharmacogosy and Pharmabiotechnology. Willams and Wilkins, Baltimore.

Roche L (1992). Guidelines for the Methodology of Conservation of Forest Genetic Resources. In: The Methodology of Conservation of Forest Genetic Resources- Report on a Pilot Project. FAO. Rome. pp. 201-203.

Schellenburg R (2001). Treatment for Premenstrual Syndrome wit *Agnus castus* fruit extract: prospective, randomized, placebo controlled study. Biochem.Med. J. 322: 134 - 137.

Sofowara A (1993). Medicinal Plants and Traditional Medicinal in Africa. 2nd Edition. Spectrum Books, Ibadan, Nigeria. pp. 26-100.

Tuley de S. (1997). Industrial utilization of medicinal plant in developing Countries, Non-wood First Products 11= Medicinal Plant for forest conservation and health care. FAO, Rome, Italy.

UNESCO (1994a). Traditional knowledge in Tropical Environment

Nature and Resource, 39(1) UNESCO, Paris.

UNESCO (1994b). Traditional knowledge into the Twenty-first century , Nature and Resources, 30 (2). UNESCO, Paris.

Walter KS, Gillett HJ (1998). 1997 IUCN Red List of Threatened Plants. IUCN, Gland, Switzerland.

World Health Organization (WHO) (1978). The Promotion and Development of Traditional Medicine. Technical Report series. p. 622.

World Intellectual Property Rights Organization (WIPO) (1998). Asian Regional Seminar on Intellectual Property Issues in the Field of Traditional Medicines. WIPO, New Delhi.

Wuttke W (2000). Phytotheraphy in the treatment of Mastodynia, Premenstrual symptoms and mental cycle disorders. Gynakoloque, 33(1): 36-39.

www.peopleandplants.org/storage/working-papers/pdf/wp1e.pdf (Retrived on 29th October, 2008).

23

External, extrinsic and intrinsic noise in cellular systems: analogies and implications for protein synthesis

Pratap R. Patnaik

Institute of Microbial Technology Sector 39-A, Chandigarh-160036, India. IMTECH Communication no.034/2006
E-mail: pratap@imtech.res.in

Multicellular systems, typically in bioreactors with one or more feed streams, are under the influences of intrinsic (intra-cellular), extrinsic (inter-cellular) and external (environmental) noise. Of these, intrinsic noise is relatively less important in determining protein synthesis and reactor behavior. Although extrinsic noise and external noise have different origins and controls, they have similarities and interactions. The interactions make it important to control both kinds of noise optimally to enhance the gene expression of a desired protein, and the similarities enable this to be done. These aspects are discussed to evolve a comprehensive noise filtering and control strategy for large bioreactors operated in realistic (noisy) environments.

Key words: cellular noise sources, analogies, interactions, protein synthesis

Table of contents

1.0 INTRODUCTION

Both molecular level and macroscopic (bioreactor level) studies have shown that noise is a ubiquitous feature of microbial processes. On a macroscopic level, noise enters a cultivation vessel (or bioreactor) mainly through one or more inlet streams. Its presence is seen as fluctuations in the flow rates (Chen and Rollins, 2000; Liden, 2001; Rohner and Meyer, 1995). These fluctuations usually increase with the size of the reactor. Scale-up rules lead to the corollary inference that the influx of noise increases with the flow rate of an inlet stream.

Noise at the molecular level is linked more intimately, but not exclusively, with gene expression and coding for useful proteins. Stochasticity in gene expression is a prime cause of phenotypic variations in a population of cells (Blake et al., 2003; Elowitz et al., 2002; Rao et al.,

2002; Thattai and van Oudenaarden, 2004), which in turn confers robustness to environmental disturbances and thus enhances the survival of viable cells for protein synthesis (Kaern et al., 2005; Stelling et al., 2004).

Although noise at the cellular level and at a process level arise from different sources and have different causes, one significant similarity between them is that uncontrolled noise is detrimental whereas judiciously controlled noise can be beneficial. Apart from inducing phenotypic diversity, controlled genetic noise may also favor resistance to certain diseases (Kaern et al., 2005; Seldman and Seldman, 2002) and induce cooperative inter-cellular dynamics that enhances the dominance of favorable phenotypes in adverse conditions (Chen et al., 2005). Likewise, controlled inflow of noise in the feed streams entering a bioreactor containing a microbial cult-

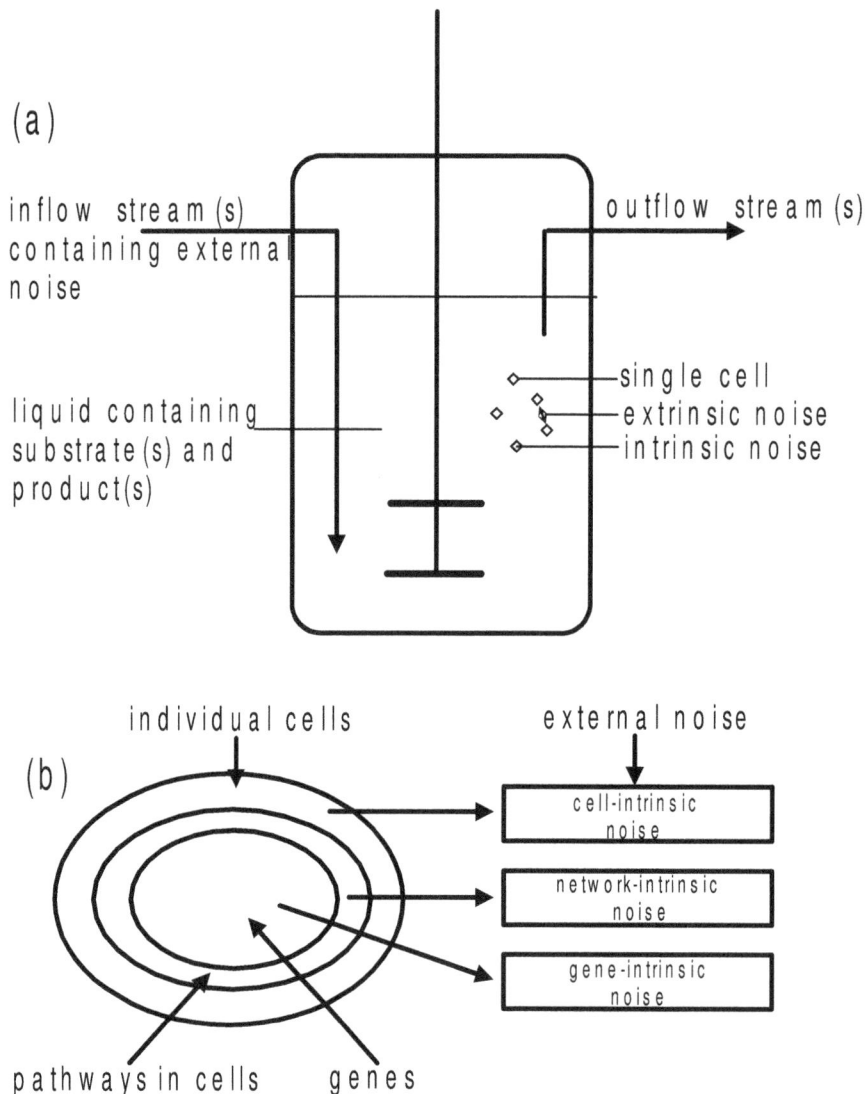

Figure 1. Schematic representations of (a) the noise sources associated with a microbial culture in a continuous flow bioreactor and (b) the locations and interactions among different sources, both within and outside the cells. Part (b) is adapted from Kaern et al. (2005) with permission from Macmillan Publishers Ltd, © 2005.

ure has been shown to increase product formation in a variety of fermentations (Patnaik, 1999, 2003a, 2004, 2006a). Interestingly, stochastic resonance seems to be at the heart of both forms of induced improvements (Chen et al., 2005; Patnaik, 2003b, 2006b), underlining a possible coherence between intra-cellular and extra-cultural fluctuations.

Recent studies (Kiss et al., 2003; Patnaik, 2006b) have demonstrated that optimal filtering of feed stream noise in continuous cultures of *Saccharomyces cerevisiae* helps to restore stable oscillations from chaotic behavior. Even if the noise is not strong enough to create chaos, a change in the variance can drive a culture from a nonoscillating (monotonic) state to an oscillating state or vice versa. These phenomena have remarkable similari-

ties with observations for gene-intrinsic noise. For instance, the galactose-utilization network in *S. cerevi-siae* has two positive and one negative feedback loops. The positive-feedback loops contribute to the establish-ment of two stable expression states, and negative feed-back controls the rate at which the cells switch between these states in a randomly changing environment (Acar et al. 2005; Thattai and van Oudenaarden, 2004).

An active population of cells in a bioreactor experie-nces noise from within the cells and from the outer environment. Because substrates enter the cells for fur-ther processing, any noise present in the inflow streams also penetrates and interacts with intra-cellular noise sources (Figure 1a, b). This possibility, the similarities between the two major sources of noise, and their effects on the dynamics of a multi-cellular system deserve careful analysis to devise methods to harness the noise in a manner that best promotes the desired protein synthesis functions of a target genetic network. Presently, the effects of feed stream noise on bioreactor performance have been studied by biochemical engineers (Chen and Rollins, 2000; Patnaik, 2004; Schmidt, 2005) while noise in genetic and metabolic networks has engaged the attention of biologists and biochemists (Blake et al., 2003; Kaern et al., 2005; Thattai and van Oudenaarden, 2004). The present overview seeks to unify these two streams of research by deriving similarities and compatibilities between external, extrinsic and intrinsic noises so as to evolve a comprehensive method to harness them to maximize the expression of desired proteins by a genetic cascade or network. The three kinds of noise are illustrated in Figure 1 and are described briefly in the next section.

2.0. Noise sources in microbial cultures

Microbial cells cultivated in a bioreactor are subject to noise within the cells as well as that from the environment. Here we designate noise from the environment as external noise, and this enters a culture medium mainly as fluctuations in the flow rates of inlet (or feed) streams. Carbon and nitrogen substrates as solutions are common feed streams; the noise they carry usually increases with the flow rates, largely because economic, practical and technological constraints place limits on the extent of control and filtering that may be employed. Data from both experimental (Montague and Morris, 1994; Rohner and Meyer, 1995; Schmidt, 2005) and simulated (Patnaik, 2003a, 2004; Riascos and Pinto, 2004; Zhang et al., 2004) fermentations indicate that feed stream noise may be characterized by a set of Gaussian distributions with time-dependent mean values and different variances. This noise has auto-correlation times from several minutes to about an hour.

If allowed to enter without any modulation, noise in the inlet streams can cause serious changes in the performance of a cellular system. Noise may displace a culture from a monotonic stable state to an unstable state or to an oscillating state or vice versa (Liden, 2001; Patnaik, 2005; Zamamiri et al., 2001; Zhang et al., 2004). Even if the displacement is to a second stable state, the latter may not retain all the relevant functional features of the original state, referred to as a loss of robustness (Kitano, 2004). Moreover, during an inter-state transition the cell culture may digress far away from both states, with consequent damage to the cells. These risks underline the importance of proper filtering of the inflow of environmental noise.

Noise entering through feed streams permeates the broth and impinges on the cells and the organelles inside,

where it encounters intra-cellular noise (Figure 1a). The latter may be of either of two types, and their difference may be explained with reference to their experimental measurement. This is done by using two green fluorescence protein (GFP) reporter genes under the control of promoters regulated by the Lac repressor (Elowitz et al., 2002). The genes encode the cyan and yellow forms of GFP, which are quantified by the fluorescence intensity of their respective emission peaks. Differences between the expressions of the two genes are indicative of *intrinsic noise*, i.e. noise inside a cell. The other kind of noise, *extrinsic noise*, affects both reporter genes equally within a given cell but generates differences between cells. These differences are attributed to variations in other proteins that affect GFP gene expression.

Intrinsic noise itself may have one or more of three locations (Kaern et al., 2005). Gene-intrinsic noise pertains to molecular level fluctuations in the reaction steps associated with gene expression. Network-intrinsic noise is generated by fluctuations in signal transduction. Both these contribute to cell-intrinsic noise; other factors include metabolite concentrations, cell size and cell age. As Figure 1b shows, all these sources of noise, including noise from the environment, may interact, thereby complicating cellular behavior.

Apart from their nature and sources, one significant difference between external noise, on the one hand, and extrinsic or intrinsic noise is that the former increases with system size where as the latter two decrease. This means external noise is greater for large bioreactors than for small ones but extrinsic (or intrinsic) noise is more pronounced for systems with small numbers of (large) molecules (Kaern et al., 2005; Raser and O'Shea, 2004). Experimental data with *Bacillus subtilis* (Ozbudak et al., 2002), *Escherichia coli* (Rosenfeld et al., 2005) and *S. cerevisiae* (Blake et al., 2003) show that extrinsic noise is the dominant cause of variability in gene expression. Cultures of these organisms are also subject to external noise (Liden, 2001; Rohner and Meyer, 1995; Schmidt 2005), thereby emphasizing the importance of understandding cellular behavior under the simultaneous effects of both kinds of noise.

3.0. Effects of noise on protein synthesis

Some key observations emerge from a comparison of external noise and intra-cellular noise. Of the two main kinds of intra-cellular noise, extrinsic noise is the major contributor to stochasticity in gene expression (Blake et al., 2003; Kaern et al., 2005; Ozbudak, et al. 2002; Stelling et al., 2004). For *E. coli*, the auto-correlation time for extrinsic noise is ~40 min (Rosenfeld et al., 2005), which is four times that for intrinsic noise and approximately equal to that of external noise (Montague and Morris, 1994; Patnaik, 2003a). Preliminary analysis of the time-domain profiles of cell, product and substrate concentra-

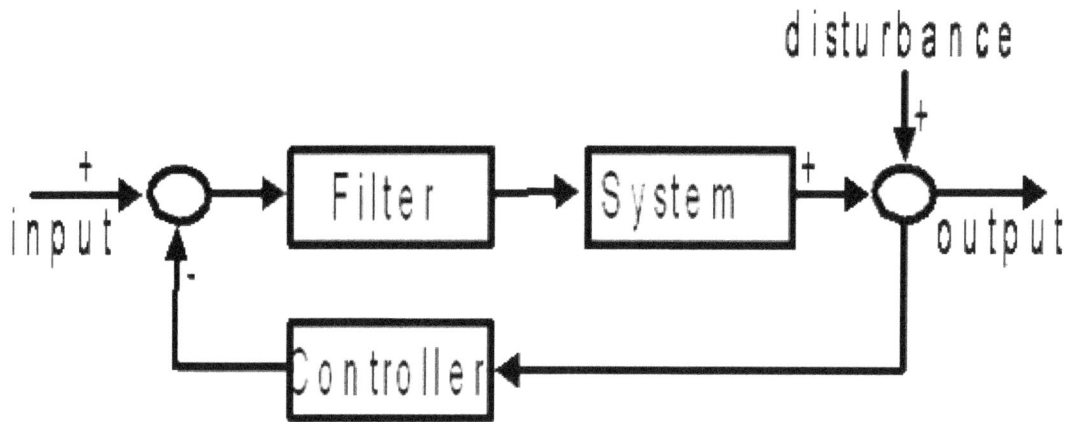

Figure 2. Information flow diagram for noise filtering and control, applicable to both external and extrinsic noise. The noisy inputs pass through a noise filter, typically a software device, and the filtered signals enter the biological system. External disturbances may impinge on this system. The outputs are fed back to a controller, which in turn regulates the operation of the filter.

trations suggest that the comparability of auto-correlation times for *E. coli* are also sustained for *B. subtilis* and *S. cerevisiae*. Variations in cellular output are symptomatic of different kinds of noise (Raser and O'Shea, 2005).

Feedback is a critical aspect of both extrinsic and external noise. A feedback stream may be either an inherent component of a reaction network or introduced as an external arrangement. Figure 2 depicts an information flow diagram common to both. The system box may be a genetic or metabolic network or a pair of cells or a multicellular fermentation broth. Complex systems may have more than one feedback loop and, correspondingly, many controllers and noise filters. Such systems tend to be robust to the impact of noise but may also be fragile and difficult to design (Kaern et al., 2005; Kitano, 2004; Stelling et al., 2004). For genetic networks, negative feedback generally provides a mechanism to reduce noise and increase stability (Becskei and Serrano, 2000; Rao et al., 2002). Similarly, negative feedback of output signals through a noise filter located upstream of a fermentation vessel improves filtering efficiency and reactor stability (Patnaik, 2004, 2006a; Dochain and Perrier, 1997).

Negative auto-regulation also minimizes fluctuations in downstream processes. By functioning effectively as a low pass filter, negative feedback allows the slower downstream processes to perceive only a time-averaged, less fluctuating signal (Kaern et al., 2005; Simpson et al., 2003). However, negative feedback can also destabilize and generate oscillations if it involves a time delay. By anology, positive feedback creates phenotypically distinct populations of cells, bistability and stochastic transitions between these states (Becskei et al., 2001; Ozbudak et al., 2004).

These effects have interesting similarities with control policies for bioreactors. Control theory teaches that negative feedback helps to return a perturbed system to its original state in a decaying oscillatory manner (Doch-

ain and Perrier, 1997), whereas positive feedback has the opposite effect. Kitano (2004) invokes this concept in his explanations of robustness of cellular systems, thereby strengthening the correspondence between microscopic feedback in genetic or metabolic networks and macroscopic feedback in bioreactor operations. A robust (cellular) system, according to Kitano, either returns to its current attractor or moves to a new attractor that presserves the system's functions. An attractor may be either static or periodic.

Continuous cultures of *S. cerevisiae* provide a lucid example of both kinds of attractors at the genetic as well as reactor levels. Isaacs et al.(2003) and Becskei et al. (2001) studies with a single-gene autocatalytic network illustrate bistability arising through positive feedback regulation. One of these stables may be oscillatory, and gene-level fluctuations can drive transitions between the states at a rate governed by a negative feedback loop (Acar et al., 2005). Analogously, a continuous flow bioreactor may exhibit either oscillating or non-oscillating (monotonic) outputs according to the dilution rate, i.e. the flow rate per unit volume of the fermentation broth (Beuse et al., 1998; Jones and Kompala, 1999).

Like the bistability in a genetic network with positive feedback, two oscillatory and one monotonic state are possible in a bioreactor with recycle (Figure 3) (Zamamiri et al., 2001) with the intermediate state being unstable. However, just as genetic networks may be strongly sensitive to certain small perturbations that persist long enough but less sensitive to frequently occurring large fluctuations (Kitano, 2004; Rosenfeld et al., 2005), the three states in a bioreactor may have different sensitivities under different conditions. Therefore, in a highly sensitive region, noise in a feed stream may propel a culture from one state to another, and even cause chaotic behavior (Patnaik, 2005). Such undesirable transitions are avoided by using noise filters. Whereas negative auto

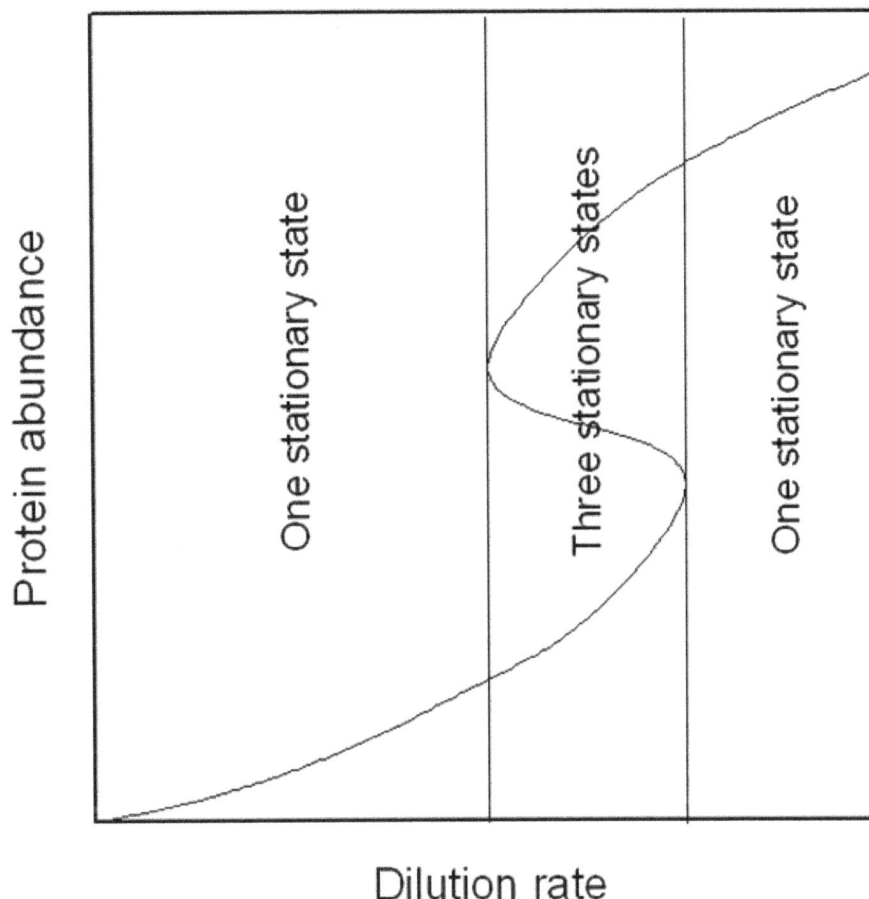

Figure. 3. Multiplicity pattern for a continuous flow bioreactor for different dilution rates. Depending on the starting conditions and the dilution rate, a continuous flow microbial bioreactor may have one or three stationary states. In a regime of three states, sufficiently strong noise may displace a culture from an existing state to another state, the latter usually less desirable. Noise filtering and control are employed to avoid this.

-regulation has an inherent filtering effect at a genetic level (Kaern et al., 2005; Simpson et al., 2003), specific filtering devices are needed for bioreactors (Patnaik, 1999, 2003a, 2003b). Interestingly, a low pass filter is one common device that functionally resembles the filtering by a negative genetic feedback.

In microbial cultures it often becomes necessary to seek a trade-off between sensitivity and productivity. This means a stationary state at which the cells synthesize a desired protein very efficiently may be vulnerable even to short duration disturbances of low intensity whereas another state that is somewhat less productive may offer a better combination of fragility and robustness. Then in a realistic (noisy) environment it may be prudent to operate at a less productive state with mild noise filtering and control.

Finally, we note that genetic buffering, through either chaperones or networks (Kitano, 2004), is a fundamental mechanism to provide robustness at a cellular level.

Buffering by the network topology is particularly effective when it is robust against external perturbations.

This feature and its origin in the modularity of gene regulation are attractively similar to the buffering effect explained by certain compartment models of microbial cells. For instance, Nielsen et al. (1991) proposed the four-compartment model shown in Figure 4 for recombinant *E. coli*. Compartment A contains mRNA, tRNA and ribosomes, P contains the plasmid DNA, the recombinant protein is in compartment E, and the rest of the cell mass, comprising mainly genomic DNA and structural material, is lumped into another compartment (G). Since A accounts for about 60% of the cell mass, it buffers any noise that enters through the substrate stream.

4.0. Concluding remarks

Noise in cellular systems is generally perceived as undesirable but unavoidable. However, if the mechanisms and processes underlying the noise are properly understood, it may be possible to harness noise intelligently with beneficial results.

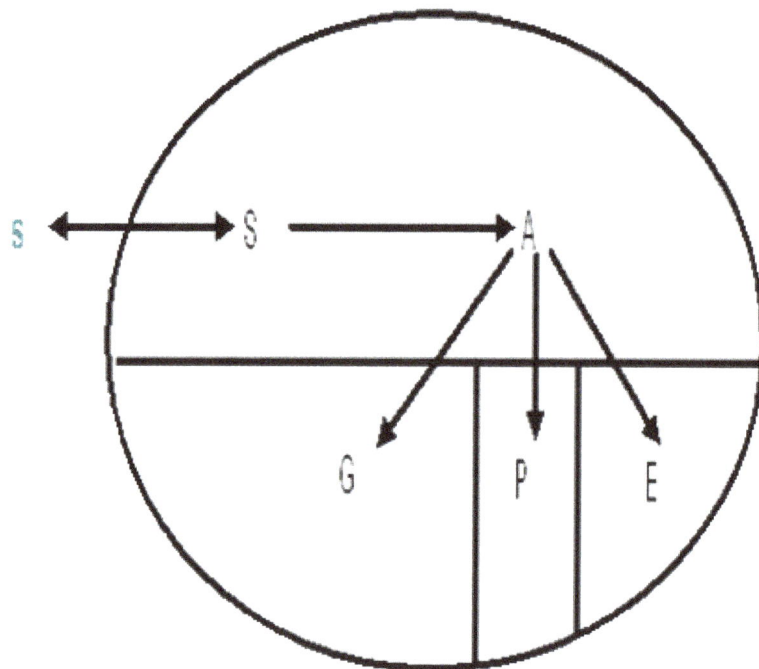

Figure. 4. Four compartment of a cell, according to Nielsen et al. (1999) s=substrate in the fluid around the cells; S=substrate inside the cells. A, E, G and P are explained in the text. Note the interactions between the substrate outside and inside the cells, and between intra-cellular substrate and the key components of a cell. The model is thus conceptually consistent with the noise sources in Figure 1 and their description in section 2.

At the cellular level, extrinsic (intra-cellular) noise is more significant than intrinsic (intra-cellular) noise. Like the external noise that enters a cell culture through a feed stream, the effects of extrinsic noise are governed by feedback loops, which may create two or three expression states. In both cases, negative feedback has a stabilizing and noise-reducing effect. Both kinds of noise have similar auto-correlation times and both may displace a culture from one state to another with different features.

These similarities and the likelihood of interactions between extrinsic and external noise, since the latter permeates the culture broth, emphasize the importance and the feasibility of noise filtering and process control strategies that, unlike most current methods, account for both kinds of noise. However, such a strategy is yet to be evolved. This is understandable since a detailed understanding of the mechanisms of biological noise generation and their effects has itself come recently. The few models proposed so far have focussed on either single cells or groups of cells (Chichigina et al., 2005; Kiehl et al., 2004), or the effects of external noise coupled with the hydrodynamics in the bioreactor (Tian et al., 2002; Patnaik, 2003a, 2004), ignoring fluctuations inside the cells. However, the similarities of techniques among these separate studies and the analogies outlined above lead to the expectation that a model encompassing both aspects is possible.

5.0 References

Acar MB, Becskei A, van Oudenaarden A (2005). Enhancement of cellular memory by reducing stochastic transitions. Nature 435: 228-231.

Becskei A, Seraphin B, Serrano L (2001). Positive feedback in eukaryotic gene networks: cell differentiation by graded to binary response conversion. EMBO J. 20: 2528-2535.

Becskei A, Serrano L (2000). Engineering stability in gene networks by autoregulation. Nature 405: 590-593.

Beuse M, Bartling R, Kopmann A, Diekmann J, Thoma M (1998). Effect of the dilution rate on the mode of oscillation in continuous cultures of Saccharomyces cerevisiae. J. Biotechnol. 61: 15-31.

Blake WJ, Kaern M, Cantor CR, Collins JJ (2003). Noise in eukaryotic gene expression. Nature 422: 633-637.

Carson JM, Doyle J (2002). Complexity and robustness. Proc. Natl. Acad. Sci. USA 99(Suppl.1): 2538-2545.

Chen VCP, Rollins DK (2000). Issues regarding artificial neural network modeling of reactors and fermenters. Bioproc. Biosyst. Eng. 22: 85-93.

Chen L, Wang R, Zhou T, Aihara K (2005). Noise-induced cooperative behavior in a multicell system. Bioinformatics 21: 2722-2729.

Chichigina O, Valenti D, Spagnolo B (2005). A simple noise model with memory for biological systems. Fluct. Noise Lett. 5: L243-L250.

Dochain D, Perrier M (1997). Dynamic modeling, analysis, monitoring and control of nonlinear bioprocesses. Adv. Biochem. Eng. Biotechnol. 56, 147-197.

Elowitz M, Levine A, Siggle E, Swain P (2002). Stochastic gene expression in a single cell. Science 297: 1183-1186.

Isaacs FJ, Hasty J, Cantor CR, Collins JJ (2003). Prediction and measurement of an autoregulatory genetic module. Proc. Natl. Acad. Sci. USA 100: 7714-7719.

Jones KD, Kompala DS (1999). Cybernetic model of growth dynamics of Saccharomyces cerevisiae in batch and continuous cultures. J. Biotechnol. 71: 105-131.

Kaern M, Elston TC, Blake WJ, Collins JJ (2005). Stochasticity in gene expression: from theories to phenotypes. Nature Rev. Genet. 6: 451-464.

Kiehl TR, Mattheyses RM, Simmonds MK (2004). Hybrid simulation of cellular behavior. Bioinformatics 20: 316-322.

Kiss IZ, Zhai Y, Hudson JL, Zhou C, Kurths J (2003). Noise enhanced phase synchronization and coherence resonance in sets of chaotic oscillators. Chaos 13: 267-278.

Kitano H (2004). Biological robustness. Nature Rev. Genet. 5: 826-837.

Liden G (2001). Understanding the bioreactor. Bioproc. Biosyst. Eng. 24: 273-279.

Montague GA, Morris AJ (1994). Neural network contributions in biotechnology. Trends Biotechnol. 12: 312-324.

Nielsen J, Pedersen AG, Strudshlom K, Villadsen J (1991). Modeling of fermentations with recombinant microorganisms: formulation of a structured model. Biotechnol. Bioeng. 37: 802-808.

Ozbudak EM, Thattai M, Kurtser I, Grossman AD, van Oudenaarden A (2002). Regulation of noise in the expression of a single gene. Nature Genet. 31: 69-73.

Ozbudak EM, Thattai M, Lim HN, Shraiman BI, van Oudenaarden A (2004). Multistability in the lactose utilization network of Escherichia coli. Nature 427: 737-740.

Patnaik PR (1999). Coupling of a neural filter and a neural controller for improvement of fermentation performance. Biotechnol. Tech. 13: 735-738.

Patnaik PR (2003a). An integrated hybrid neural system for noise filtering, simulation and control of a fed-batch recombinant fermentation. Biochem. Eng. J. 15: 165-175.

Patnaik PR (2003b). On the performances of noise filters in the restoration of oscillatory behavior in continuous yeast cultures. Biotechnol. Lett. 25: 681-685.

Patnaik PR (2004). Neural and hybrid neural modeling and control of fed-batch fermentation for streptokinase: comparative evaluation under nonideal conditions. Can. J. Chem. Eng. 82: 599-607.

Patnaik PR (2005). Application of the Lyapunov exponent to detect noise-induced chaos in oscillating microbial cultures. Chaos Solitons Fractals 26: 759-765.

Patnaik PR (2006a). Enhancement of PHB biosynthesis by Ralstonia eutropha in fed-batch cultures by neural filtering and control. Food Bioprod. Proc. 84(C2): 150-156.

Patnaik PR (2006b). Hybrid filtering to rescue stable oscillations from noise-induced chaos in continuous cultures of budding yeast. FEMS Yeast Res. 6: 129-138.

Rao CV, Wolf DM, Arkin AP (2002). Control, exploitation and tolerance of intracellular noise. Nature 420: 231-237.

Raser JM, O'Shea EK (2004). Control of stochasticity in eukaryotic gene expression. Science 304: 1811-1814.

Raser JM, O'Shea EK (2005). Noise in eukaryotic gene regulation: origins, consequences, and control. Science 309: 2010-2013.

Riascos CAM, Pinto JM (2004). Optimal control of bioreactors: a simultaneous approach for complex systems. Chem. Eng. J. 99: 23-34.

Rohner M, Meyer H.-P. (1995). Application of modeling for bioprocess design and control in industrial production. Bioproc. Eng. 13: 69-78.

Rosenfeld N, Young JW, Alon U, Swain PS, Elowitz MB (2005). Gene expression at the single cell level. Science 307: 1962-1965.

Schmidt FR (2005). Optimization and scale up of industrial fermentation processes. Appl. Microbiol. Biotechnol. 68: 425-435.

Seldman JG, Seldman C (2002). Transcription factor haploinsufficiency: when half a loaf is not enough. J. Clin. Invest. 109: 451-455.

Simpson ML, Cox CD, Sayler GS (2003). Frequency domain analysis of noise in autoregulated gene circuits. Proc. Natl. Acad. Sci. USA 100, 4551-4556.

Stelling J, Sauer U, Szallasi Z, Doyle III FJ, Doyle J (2004). Robustness of cellular functions. Cell 118: 675-685.

Thattai M, van Oudenaarden A (2004). Stochastic gene expression in fluctuating environments. Genetics 167: 523-530.

Tian Y, Zhang J, Morris AJ (2002). Optimal control of a fed-batch bioreactor based upon an augmented recurrent neural network model. Neurocomputing 48: 919-927.

Zamamiri AM, Birol G, Hjortso MA (2001). Multiple steady states and hysteresis in continuous, oscillating cultures of budding yeast. Biotechnol. Bioeng. 75: 305-312.

Zhang S, Chu J, Zhuang Y (2004). A multiscale study of industrial fermentation processes and their optimization. Adv. Biochem. Eng. Biotechnol. 87:97-150.

24

Molecular Chaperones involved in Heterologous Protein Folding in *Escherichia coli*

E. BETIKU

Department of Chemical Engineering, Obafemi Awolowo University,Ile-Ife, Osun State, Nigeria

The Gram-negative bacterium *Escherichia coli* is one of the most attractive host employed in the heterologous production of proteins. However, these target proteins are deposited as insoluble aggregates known as inclusion bodies (IBs) and hence are biologically inactive. The ubiquitous molecular chaperones, a group of unrelated classes of polypeptides help in the mediation of proper folding of the target protein. However, the choice of chaperone(s) is still based on a trial-and-error procedure. Wrong choice of chaperone(s) will affect both the host micro-organism and product stability, negatively. Recent advances in the mechanisms and substrate specificities of the major chaperones and their roles in the chaperone-network now gives some ideas for more rational choice of the chaperone(s) for co-expression. Here, the functions and mechanisms of interactions between the major molecular chaperones are presented.

Key words: molecular chaperones, inclusion bodies, heterologous, aggregates, protein folding

Table of Content

1.0   INTRODUCTION

The heterologous production of proteins in the bacterium host *Escherichia coli* is a widely used techniques both in research and for commercial purposes. However, a fraction of these proteins are deposited in insoluble form. These proteins form aggregates that accumulate into inclusion bodies (IBs). IBs are refractile protein aggregates with porous structures (Taylor et al. 1986; Rinas et al. 1992; Carrió and Villaverde 2001). They have high density (Hwang 1996) and are known to be in non-native form and hence biologically inactive (Goloubinoff et al. 1999; Hoffmann and Rinas 2004). The need to

---

*Correspondence author. E-mail: ebetiku@oauife.edu.ng

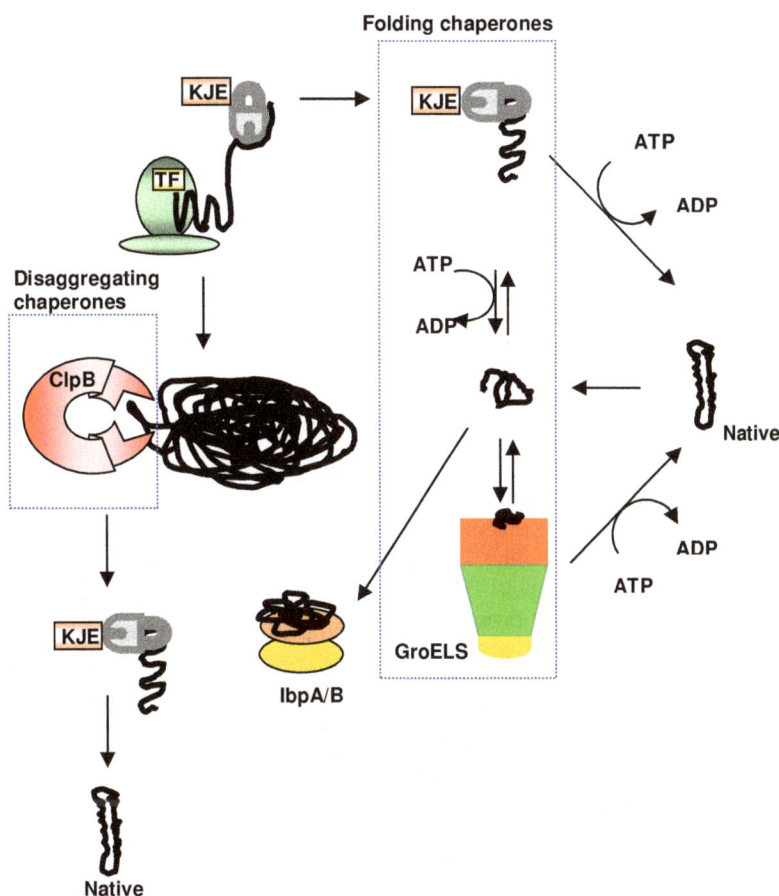

Figure 1. A model of molecular chaperone-mediated protein folding in the cytoplasm of the bacterium *Escherichia coli*. Newly synthesized polypeptides first interact with Trigger Factor or DnaKJE. The intermediate formed may reach native protein or interacts with GroELS before reaching native form. The intermediate may also form aggregates known as inclusion bodies (IBs), which may need to interact with ClpB for disaggregation before reaching native form after interaction with DnaKJE. IbpAB binds partially folded proteins until disaggregating chaperone ClpB becomes available.

avoid formation of aggregates during heterologous production of proteins in *E. coli* is not only informed by the increase demand for cellular "quality control" machinery (Hoffmann and Rinas 2004) which, may lead to low productivity but also has to do with the involvement of aggregates in some unrelated diseases such as Alzheimer's disease, bovine spongiform encephalopathy and type II diabetes (Haper and Lansdury 1997; Azriel and Gazit 2001).

Molecular chaperones are ubiquitous and highly conserved proteins that shepherd other polypeptides to fold properly and are not themselves components of the final functional structures (Hartl 1996; Baneyx and Palumbo, 2003). There are ~ 20 families of this class of proteins which have different molecular weights, structures, cellular locations and functions (Radford, 2000). They were originally identified by their increased abundance as a result of heat shock (Bukau and Horwich, 1998). Molecular chaperones work as networks in protein folding in the cytoplasm of *E. coli* (Figure 1).

2.0 Methods for preventing or decreasing protein aggregation.

Several methods have been suggested or shown to prevent or decrease aggregation during overproduction of recombinant protein in the host cell. Some of these methods include: rate of synthesis, fusion proteins, mutations in the target protein, cultivation conditions and coexpression of molecular chaperones.

2.1 Control of the rate of synthesis of expressed proteins

Typically, the more rapid the intracellular product accumulation, the greater the probability of product aggregation. The expression rate and the correct folding of the product are among other parameters determined by the level of gene induction, promoter strength, the efficiency of translation initiation and mRNA stability (Swartz 2001). Best results are usually obtained by low cultivation temperature (18-25°C) and application of low gene dosage (Kopetzki et al. 1989; Swartz 2001). Hence, high soluble protein yield depends on low specific protein synthesis rate and sustained production period (Kopetzki et al. 1989; LaVallie et al. 1993).

2.2   Use of fusion proteins

Unrelated proteins originally were constructed together (at genetic level) to facilitate protein detection/purification and immobilization (Uhlen et al. 1983), and to couple the activity of enzymes acting in a single metabolic pathway. However, expression of a set of foreign genes e.g. protease domain of human urokinase UKP as a fusion to ubiquitin gene in yeast showed improved yield of recombinant protein (Butt et al. 1989). Some fusion partners often employed include prokaryotic *Staphylococcus* protein A (Abrahmsén et al. 1986); maltose-binding protein (Sachdev and Chirgwin 1998), thioredoxin (Lavillie et al. 1993) and DsbA (Winter et al., 2000) from *E. coli*. The order of fusion partners often determines the solubility level of the target protein (Sachdev and Chirgwin 1998). As a periplasmic protein, maltose-binding protein directs by its native signal peptide the whole fusion to the periplasmic space of the cells. This positive influence of maltose-binding protein can be attributed to both its molecular characteristics and its interaction with the target proteins. Although, most protein fusions are soluble but the target proteins are not always correctly folded (Sachdev and Chirgwin 1998) and IBs can still be formed (Strandberg and Enfors 1991). The disadvantages of fusion-protein technologies include: liberation of the passenger proteins requires expensive proteases (such as Factor Xa), cleavage is rarely complete leading to reduction of yields, additional steps may be required to obtain an active product e.g. formation and isomerisation of disulfide bonds (Banyex 1999).

2.3   Mutations in the target protein

Many reports have been presented to show the effects of mutations in target proteins overproduced in *E. coli*. King and co-workers employed genetic techniques to identify second-site suppressor mutations of temperature sensitive folding mutants of the P22 tailspike protein, which when placed in a wild-type background, give the phenotype of decreased IBs content compared to wild-type (Fane and King 1991). Mutations in the hFGF-2 gave different results; no soluble hFGF-2 was formed when cysteines 26 and 93 were replaced with serines, while a single substitution of cysteine 70 by serine decreases the fraction of soluble hFGF-2 significantly (Rinas et al. 1992). Recombinant production of interferon gamma protein (IFN-γ) in *E. coli* at 37°C results in over 90% of the total accumulated gene product into IBs and

in addition, mutations in the protein show mutants that retain high biological activity and are localized almost entirely in the soluble fraction (Wetzel et al. 1991). Observations in a series of mutations in the human interleukin-1 beta (IL-1β) show no strong correlations between extent of IB formation and either thermodynamic or thermal stability (Chrunyk et al. 1993). Replacement of Lys[97] by Val produces substantially more IL-1β in IBs than in wild type despite generating a protein more thermodynamically stable than wild type (Chrunyk et al. 1993).

2.4   Optimisation of cultivation conditions

IB formation during high-level recombinant production may be reduced or avoided by optimising culture conditions. Growth temperatures have been shown to affect formation of IBs e.g. β-lactamase (Valax and Georgiou 1993) and hFGF-2 (Squires et al. 1988). High cultivation temperature leads to recombinant protein aggregation (Schein and Noteborn 1988; Schein 1989). Low temperatures can greatly reduce the formation of IBs (Chalmers et al. 1990). This is corroborated by the work of Piatak et al. (1988), soluble and fully functional Ricin A chain was produced at 37°C, but the one produced at 42°C was aggregated. pH of the culture medium also affects inclusion body formation. Formation of IBs also depends on the level of induction. By using 0.01 mM IPTG for induction of alkaline phosphate when produced in *E. coli*, more than 90% of the product could be recovered from the periplasm in soluble form, whereas when induction was made at 1 mM IPTG, most of the secreted alkaline phosphate formed IBs (Choi et al. 2000). The expression of α-glucosidase depends upon the inducer concentration as well as on the period of induction (Kopetzki et al. 1989). The type of medium employed for recombinant production has influence on the level of IBs formed. It has been shown that growth on glycerol (Kopetzki et al. 1989) or on complex medium (Winter et al. 2000) can be advantageous for solubility and folding of the recombinant product. The ratio of soluble to aggregated β-lactamase can be increased by growing cells in the presence of certain non-metabolizable sugars (Bowden and Georgiou 1988) and it was also shown that the inhibition of aggregation depends on the concentration of the sugar in the growth medium (Bowden and Georgiou 1990). Wunderlich and Glockshuber (1993) reported a five-fold increase in correctly folded target protein after adding reduced and oxidized glutathione to the growth medium. Glycine also influences the folding of aggregation-prone proteins (Kaderbhai et al. 1997). However, optimization of these various parameters (such as pH, host strains, media, temperature) is required for prevention of aggregation and for the production of soluble and active products (Kopetzki et al. 1989; Winter et al. 2000).

Table 1. Enhancing soluble production by co expression of molecular chaperones

Chaperone	Protein product	Results	References
GroEL/GroES	Rubisco	Increased production of assembled and active Rubisco proteins from various species is observed.	Goloubinoff et al., 1989
	Protein-tyrosine kinase P^{50csk}	>50% of P^{50csk} is soluble following GroELS overexpression.	Amrein et al., 1995
Trigger Factor (TF)	Endostatin	>80% of Endostatin is soluble following TF overexpression.	Nishihara et al., 2000
DnaK	Human growth factor	Co expression of DnaK inhibits human growth factor IB formation and increases the amount of soluble product from 5% to >85%.	Blum et al., 1992
GroEL/GroES and TF	ORP150	86% of ORP150 is soluble following GroELS/TF overexpression.	Nishihara et al., 2000
GroEL/GroES and DnaK	Cryj2	Co expression of GroELS/DnaK resulted in marked stabilization and accumulation of Cryj2 without extensive aggregation	Nishihara et al., 1998

Source: Betiku, 2005

2.5   Co-expression of molecular chaperones

Anfinsen's observation that all information necessary for a protein to adopt the unique three-dimensional structure is contained in the amino acids sequence (Anfinsen 1973) remains unchallenged, in the last decade this view of cellular protein folding has changed considerably. Protein folding in the vicious and crowded environment of the cell is very different from *in vitro* processes in which a single protein is allowed to refold at low concentration in an optimised buffer (Baneyx and Palumbo 2003). The initial folding of proteins and assembly of multiprotein complexes can be helped and sometimes required the participation of chaperones. By binding exposed hydrophobic patches on the protein, they prevent proteins from aggregating into insoluble, non-functioning IBs and help them reach their stable native state (Wickner et al. 1999). Chaperones do not provide specific steric information for the folding of the target protein, but rather inhibit unproductive interactions (Walter and Buchner 2002).

3.0   Molecular chaperones

The major chaperones implicated in *de novo* protein folding are the trigger factor (TF), and the DnaK and the GroEL chaperone systems (Horwich et al. 1993; Bukau et al. 2000). Other chaperones involved in folding of recombinant proteins include the AAA+ chaperone ClpB and IbpA/IbpB. These molecular chaperones have been reported to enhance soluble production of recombinant proteins in *E. coli* (Table 1).

3.1   The Trigger Factor (TF)

TF was originally identified by its activity to stimulate membrane translocation of the precursor of the outer-membrane protein A (preOmpA) *in vitro* (Crooke and Wickner 1987). In the *E. coli* cytosol, nascent polypeptides interact first with TF (Valent et al. 1995; Hesterkamp et al. 1996). TF has both peptidyl-prolyl cis-trans isomerase activity and chaperone-like function (Crooke and Wickner 1987; Hesterkamp et al. 1996). The enzymatic mechanism of TF follows the Michaelis-Menten kinetic (Scholz et al. 1997). TF binds to the ribosome at proteins L23/L29 near the polypeptide exit site (Kramer et al. 2002). TF's peptidyl-prolyl cis-trans isomerase activity is not essential for protein folding in *E. coli* (Kramer et al. 2004). It is composed of three domains: an N-terminal domain, which mediates association with the large ribosomal subunit; a central substrate binding and peptidyl-prolyl cis-trans isomerase (PPIase) domain with homology to FKBP [(FK506 Binding Protein)(FK506 is a macrolide lactone, Tacrolimus also called Fujimycin)]; and a C-terminal domain of unknown function (Hesterkamp and Bukau 1996; Hesterkamp et al. 1997). TF affinity for substrate is very low compared to most chaperones and is ATP independent, suggesting that rapid binding to and release from TF may be critical for elongating polypeptide chains (Maier et al., 2001). The binding motif of TF has been identified as a stretch of eight amino acids, enriched in aromatic residues and with a positive net charge (Patzelt et al. 2001). TF cooperates with the DnaK system in folding of nascent polypeptides. They share an overlapping substrate pool (Teter et al. 1999; Deuerling et al. 1999, 2003). Both chaperones help in multidomain protein folding but at the expense of folding speed (Agashe et al. 2004). They can compensate for one another; however, their combined deletion is lethal at temperatures above 30°C (Deuerling et al. 1999; Teter et al., 1999). Overproduction of GroEL chaperone system could efficiently suppress the growth defect as a result of *tigdnak* deletion (Genevaux et al. 2004). TF function together with GroEL-GroES in selective degradation of certain polypeptides (Kandror et al. 1995) and *in vivo*, TF associate with GroEL to promote its binding to certain unfolded proteins (Kandror et al. 1997). TF prevents the aggregation of recombinant proteins either in combination with the chaperonin GroEL-GroES or alone (for example, lysozyme, Nishihara et al. 2000).

3.2   The DnaK System

The DnaK is the most general molecular chaperone. It is also known as heat shock protein 70 (Hsp70). The structural and mechanistic features of the *E. coli* DnaK chaperone system have been reviewed (Bukau and Horwich 1998). DnaK works in cooperation with its cochaperones – DnaJ and GrpE (Liberek et al. 1991). Structural features of DnaK are required for interaction with DnaJ (Suh et al. 1999). The importance of these features for substrate binding has been shown by mutational analysis (Mayer et al. 2000). The rate of ATP hydrolysis is accelerated by DnaJ (Laufen et al. 1999). This stimulation is disrupted by mutation of conserved leucine residues of DnaK located in the linker between substrate binding and ATPase domains, resulting in considerable loss of chaperone activity (Han and Christen 2001). DnaJ also targets the substrates to DnaK (Liberek et al. 1995), and substrates with low affinity for DnaK are not able to stimulate the ATPase and chaperone activity of DnaK without DnaJ (Mayer et al. 2000). The cochaperone GrpE accelerates the exchange of ADP with ATP, resulting in the release of the unfolded substrate and completion of the chaperone cycle (Packschies et al. 1997). Besides the promiscuous binding of aggregation-prone substrate proteins, DnaK – targeted by DnaJ (Liberek et al. 1995) – specifically recognizes a "region C" of the heat-shock sigma factor σ^{32} (Nagai et al. 1994). Abundant free DnaK-DnaJ inhibits σ^{32}-dependent gene expression (Tatsuta et al., 1998). Furthermore, the C-terminal part of σ^{32} becomes accessible to the protease FtsH, resulting in rapid degradation of σ^{32} (Blaszczak et al. 1999), therefore, DnaK negatively regulates the heat-shock response. Under stress conditions, misfolded proteins withdraw DnaK from σ^{32}, which regains activity and stability, resulting in enhanced transcription of σ^{32}-dependent heat-shock genes, including *dnaK*, until sufficient amounts of DnaK accumulate to bind both the misfolded proteins and σ^{32} (Bukau 1993). The increase in the level of σ^{32} is accelerated, when, additional to the titration of DnaK by misfolded proteins, high temperatures stimulate

translation of the *rpoH* mRNA (Morita et al. 2000). The DnaK chaperone acts on different levels: *de novo* folding of protein (Bukau and Horwich 1998), rescue or degradation of denatured proteins and reversion of aggregation (Hoffmann and Rinas 2004). In addition to the role in ATP-dependent unfolding, DnaK can prevent aggregation by longterm binding to thermolabile substrates when higher temperatures reduce the affinity of DnaK for both DnaJ and GrpE (Diamant and Goloubinoff 1998), thereby preventing aggregation or stabilizing the substrates for refolding by the GroEL chaperone system (Buchberger et al. 1996). DnaK binds preferentially newly synthesized proteins in the size range of 16-167 kDa with an enrichment of proteins larger than 60 kDa (Deuerling et al. 2003).

3.3   Cooperation of DnaK system with other Chaperones

Beside TF, DnaK also cooperates with the *E. coli* Hsp31 ("holdase") in management of protein misfolding under severe stress conditions (Mujacic et al. 2004). *In vitro* and *in vivo* experiments show that cooperation between DnaK and the AAA+ chaperone ClpB is needed for prevention and reversion of aggregation in prokaryotes (Mogk et al. 1999; Zolkiewski 1999). Heat-inactivated proteins released by the DnaK-ClpB bichaperone system are recognized as non-native folding intermediates by the chaperonins system (Watanabe et al. 2000). The disaggregating activity of the ClpB-DnaK chaperone network exhibits broad substrate specificity; at least 75% of thermally aggregated *E. coli* proteins in cell extract are solubilised (Mogk et al. 1999). The mechanism of solubilisation and refolding of protein aggregates by this bichaperone network is sequential (Goloubinoff et al. 1999). It has been proposed that ClpB interacts directly with protein aggregates prior to the DnaK on protein substrates (Weibezahn et al. 2003). Schlee et al. (2004) have shown that a specific interaction between ClpB and DnaK exists, and the affinity of the complex formed is weak and their interaction is nucleotide-dependent (Schlee et al. 2004). Hsp104/ClpB was first described as a heat-inducible protein conferring thermo-tolerance to yeast (Sanchez and Lindquist 1990). ClpB is ATP-dependent (Woo et al. 1992) and belongs to the Hsp100/Clp family of AAA+ (ATPase associated with a variety of cellular activities) and is composed of an N-terminal domain and two AAA domains that are separated by a "linker" region (Schirmer et al. 1996). The AAA domains mediate ATP binding and hydrolysis and are essential for ClpB oligomerization (Mogk et al. 2003a). The function of the N domain and the "linker" segment are currently unknown. While the N domains are dispensable for the disaggregating activity of ClpB, the linker region has an essential function in this process (Mogk et al. 2003a).

3.4   The Small Heat-Shock Proteins (sHsps)

The small heat-shock proteins (sHsps) are ATP-independent proteins, grouped as a family of heat-shock proteins based on a low degree of homology in a core region of about 85 amino acids (the α-crystallin domain), their ability to be induced by cellular stress, and their low protomer molecular weight, which usually ranges between 15-30 kDa (Shearstone and Baneyx 1999). The *E. coli* homology is IbpA/IbpB with molecular weight of 14- and 16-kDa, respectively, co-transcribed during stress by the bacterial heat shock transcription factor σ^{32} (Allen et al. 1992). IbpB consists mainly of β-pleated secondary structure (Shearstone and Baneyx 1999). In *E. coli*, IbpA and IbpB are found associated with endogenous proteins that aggregate intracellularly during heat shock (Laskowska et al. 1996) and with non-native recombinant proteins in inclusion bodies (Allen et al. 1992). However, they are not found in inclusion bodies of partially soluble proteins (Valax and Georgiou 1993; Hoffmann and Rinas 2000). Over-production of IbpA/IbpB can increase stress tolerance in *E. coli* (Thomas and Baneyx, 1998). Despite the high sequence homology between IbpA and IbpB, the two proteins behave differently upon over-expression in *E. coli*; whereas IbpA is found in the insoluble S-fraction, IbpB is mainly soluble when produced in the absence of IbpA, but co-migrates to the aggregated fraction upon co-production with IbpA (Kuczyńska-Wiśnik et al. 2002). Generally, sHsps bind substrate proteins exposing hydrophobic surfaces and for refolding, a transfer to ATP-dependent chaperones is required (Hoffmann and Rinas 2004). IbpA/IbpB cooperate with the bichaperone (DnaK and ClpB) both *in vivo* and *in vitro*, in reversing aggregated proteins, and they become essential at 37°C if DnaK levels are reduced (Mogk et al. 2003b).

3.5   The GroEL System

The GroEL-GroES system (i.e. chaperonins) are currently the molecular chaperone system, for which there is the most structural and mechanistic information (Braig et al. 1994; Rye et al. 1997; Sigler et al. 1998). They are essential for cell viability at all temperatures (Fayet et al. 1989; Horwich et al. 1993). During cellular stress, 30% of newly translated polypeptides depend on the GroEL chaperone (Horwich et al. 1993). GroEL is also known as heat shock protein 60 (Hsp60) and is a homo-oligomer of 14 subunits, each of relative molecular mass of 57 kDa, arranged into two heptameric rings, forming a cylindrical structure with two large cavities (Braig et al. 1994). Substrate protein, with hydrophobic amino acid residues exposed, binds in the central cavity of the cylinder, engaging the hydrophobic surfaces exposed by the apical GroEL domain (Fenton et al. 1994). Folding usually occurs with the aid of GroES, a dome-shaped ring containing seven subunits of 10 kDa (Hunt et al. 1996).

Binding of GroES to the polypeptide-containing ring of GroEL in an ATP-dependent reaction results in the displacement of polypeptide into an enclosed cage, defined by the GroEL cavity and the dome of GroES, in which aggregation is prevented and folding to native state is possible (Weissman et al. 1994). After the GroEL-bound ATP has been hydrolysed to ADP, ATP binding to the opposite ring of GroEL results in the dissociation of GroES and folded protein from GroEL (Figure 2). Some proteins require multiple chaperonin cycles for folding (Hartl 1996; Sigler et al. 1998). GroEL preferentially interacts with newly synthesized polypeptides with the size range between 10-55 kDa (Ewalt et al. 1997), but most GroEL substrates are larger than 20 kDa (Houry et al. 1999). GroEL substrates consist of two or more domains with αβ-folds, which contain α-helices and buried β-sheets with extensive hydrophobic surfaces (Houry et al. 1999). The oligomeric structure of GroEL-GroES is required for biologically significant chaperonin function in protein folding (Weber et al. 1998) and the maximum size of substrate protein that can be encapsulated in the GroEL-GroES cavity is ~ 57 kDa (Sakikawa et al. 1999). GroEL-GroES can also mediate folding of substrate protein, which are too large to be enclosed within this cavity (Chaudhuri et al. 2001). Co-production of GroEL-GroES can increase solubility of some recombinant proteins (Goloubinoff et al. 1989; Amrein et al. 1995). GroEL-chaperone system cooperates with other molecular chaperones, for example DnaK (Buchberger et al. 1996; Nishihara et al. 1998, 2000), and TF (Nishihara et al. 2000) in increasing solubility of certain recombinant proteins. Betiku (2005) show that GroELS can prevent inclusion bodies formation during recombinant production of human basic Fibroblast Growth Factor (hFGF-2) in *E. coli*. GroELS of *E. coli* are the rate-limiting cellular determinant of growth at lower temperatures (Ferrer et al. 2003).

4.0   CONCLUSION

Several methods have been suggested or shown to prevent or decrease aggregation during overproduction of recombinant protein in the host cell. Some of these methods include: controlling the rate of synthesis, use of fusion proteins, mutations in the target protein, optimisation of cultivation conditions and co-expression of molecular chaperones. The use of chaperones is just evolving and more studies are still needed to understand how they function. Hitherto, the use of chaperones in improving recombinant production of protein is by trial-and-error procedure. Future research should be focused on the interactions/cooperation between these chaperones and the various target protein. When there is enough data, it will then be possible to choose the right combination of molecular chaperones to co-express with the individual target protein in order to avoid formation of aggregation and hence, increase the yield.

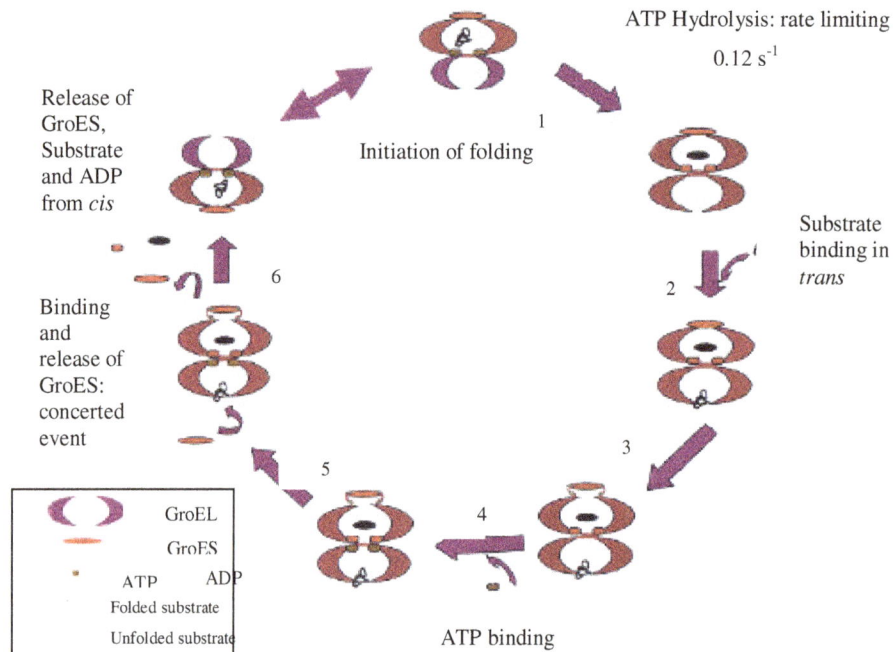

Figure 2. The GroEL-GroES reaction cycle. (1) Binding of substrate protein stimulates ATP and GroES binding in *cis*, which leads to the substrate protein being released in the cavity, and initiation of folding. (2) Substrate protein binds to the *trans* ring only after ATP hydrolysis takes place in the *cis* ring. (3) In the presence of substrate in the *trans* ring, there is a fast structural rearrangement in the ADP and GroES-bound *cis* ring that primes it for releasing GroES. (4) The binding of substrate protein in the *trans* ring stimulates ATP binding in *trans*. (5) The subsequent binding of GroES to the *trans* ring is simultaneous with the release of GroES from the *cis* ring. (6) The GroES- and ATP- bound *trans* ring causes structural rearrangements in the *cis* ring leading to release of ADP and substrate protein. Upon completion of one folding cycle, the next cycle is initiated in the alternate ring (adapted from Bhutani and Udgaonkar, 2002).

5.0   ACKNOWLEDGEMENT

The author wishes to acknowledge the financial support for this work by DAAD. This work was carried out at the National Research centre for Biotechnology, Braunschweig, Germany.

6.0   REFERENCES

Abrahmsén L, Moks T, Nilsson B, Uhlen M (1986). Secretion of heterologous gene products to the culture medium of *Escherichia coli*. Nucleic Acids Res. 14: 7487-7500.

Agashe VR, Guha S, Chang HC, Genevaux P, Hayer-Hartl M, Stemp M, Georgopoulos C, Hartl FU, Barral JM (2004). Function of trigger factor and DnaK in multidomain protein folding: increase in yield at the expense of folding speed. Cell. 117: 199-209.

Allen SP, Polazzi JO, Gierse JK, Easton AM (1992). Two novel heat shock genes encoding proteins produced in response to heterologous protein expression in *Escherichia coli*. J. Bacteriol. 174: 6938-6947.

Amrein KE, Takacs B, Stieger M, Molnos J, Flink NA, Burn P (1995). Purification and characterization of recombinant human p50csk protein-tyrosine kinase from an *Escherichia coli* expression system overproducing the bacterial chaperones GroES and GroEL. Proc. Natl. Acad. Sci. USA. 92: 1048-1052.

Anfinsen CB (1973). Principles that govern the folding of protein chains. Science. 181: 223-230.

Azriel R, Gazit E (2001) Analysis of the Minimal Amyloid-forming Fragment of the Islet Amyloid Polypeptide. J. Biol. Chem. 276: 34156

-34161.

Baneyx F (1999). Recombinant protein expression in *Escherichia coli*. Curr. Opin. Biotechnol. 10: 411-421.

Baneyx F, Palumbo JL (2003). Improving heterologous protein folding via molecular chaperone and foldase co-expression. Methods Mol. Biol. 205: 171-197.

Betiku E (2005) Heterologous Production of the Human Basic Fibroblast Growth Factor and the Glucosyltransferase-S in *Escherichia coli* with and without Coproduction of Molecular Chaperones (PhD thesis) ISBN 3-86537-367-4.

Bhutanib N, Udgaonkar JB (2002). Chaperonins as protein-folding machines. Curr. Sci. 83: 1337-1351 2002

Blaszczak A, Georgopoulos C, Liberek K (1999). On the mechanism of FtsH-dependent degradation of the sigma 32 transcriptional regulator of *Escherichia coli* and the role of the DnaK chaperone machine. Mol. Microbiol. 31: 157-166.

Blum P, Velligan M, Lin N, Matin A (1992). DnaK-mediated alterations in human growth hormone protein inclusion bodies. Bio/Technology. 10: 301-304.

Bowden G, Georgiou G (1988). Effects of Sugars on β-lactamase Aggregation in *Escherichia coli*. Biotechnol. Prog. 3: 97-101.

Braig K, Otwinowski Z, Hegde R, Boisvert DC, Joachimiak A, Horwich AL, Sigler PB (1994). The crystal structure of the bacterial chaperonin GroEL at 2.8 Å. Nature. 371: 578-586.

Buchberger A, Schröder H, Hesterkamp T, Schönfeld H-J, Bukau B (1996). substrate shuttling between the DnaK and GroEL system indicates a chaperone network promoting protein folding. J. Mol. Biol. 261: 328-333.

Bukau B (1993). Regulation of the *Escherichia coli* heat-shock response. Mol. Microbiol. 9: 671-680.

Bukau B, Deuerling E, Pfund C, Craig EA (2000). Getting newly synthesized proteins in to shape. Cell. 101: 119-122.

Bukau B, Horwich AL (1998). The Hsp70 and Hsp60 chaperone machines. Cell. 92: 351-366.

Butt TR, Jonnalagadda S, Monia BP, Sternberg EJ, Marsh JA, Stadel JM, Ecker DJ, Crooke ST (1989). Ubiquitin fusion augments the yield of cloned gene products in Escherichia coli. Proc. Natl. Acad. Sci. USA. 86: 2540-2544.

Carrió MM, Villaverde A (2001). Protein aggregation as bacterial inclusion bodies is reversible. FEBS Lett. 489: 29-33.

Chalmers JL, Kim E, Telford JN, Wong EY, Tacon WC, Shuler ML, Wilson DB (1990). Effects of temperature on Escherichia coli overproducing beta–lactamase or human epidermal growth factor. Appl. Environ. Microbiol. 56: 104-111.

Chaudhuri TK, Farr GW, Fenton WA, Rospert S, Horwich AL (2001). GroEL/GroES-mediated folding of a protein too large to be encapsulated. Cell. 107: 235-246.

Choi JH, Jeong KJ, Kim SC, Lee SY (2000). Efficient secretory production of alkaline phosphatase by high cell density culture of recombinant Escherichia coli using the Bacillus sp. endoxylanase signal sequence. Appl. Microbiol. Biotechnol. 53: 640-645.

Chrunyk BA, Evans J, Lillquist J, Young P, Wetzel R (1993). Inclusion body formation and protein stability in sequence variants of interleukin-1 beta. J. Biol. Chem. 268: 18053-18061.

Crooke E, Wickner W (1987). Trigger factor: a soluble protein that folds pro-OmpA into a membrane-assembly-competent form. Proc. Natl. Acad. Sci. USA. 84: 5216-5220.

Deuerling E, Patzelt H, Vorderwülbecke S, Rauch T, Kramer G, Schaffitzel E, Mogk A, Schulze-Specking A, Langen H, Bukau B (2003). Trigger factor and DnaK posses overlapping substrate pools and binding specificities. Mol. Microbiol. 47: 1317-1328.

Deuerling E, Schulze-Specking A, Tomoyasu T, Mogk A, Bukau B (1999). Trigger factor and DnaK cooperate in folding of newly synthesized proteins. Nature. 400: 693-696.

Diamant S, Goloubinoff P (1998). Temperature-controlled activity of DnaK-DnaJ-GrpE chaperones: protein-folding arrest and recovery during and after heat shock depends on the substrate protein and the GrpE concentration. Biochem. 37: 9688-9694.

Ewalt KL, Hendrick JP, Houry WA, Hartl FU (1997). In vivo observation of polypeptide flux through the bacterial chaperonin system. Cell. 99: 491-500.

Fane B, King J (1991). Intragenic suppressors of folding defects in the P22 tailspike protein. Genetics. 127: 263-277.

Fayet O, Ziegelhofer T, Georgopoulos C (1989). The groES and groEL heat shock gene products of Escherichia coli are essential for bacterial growth at all temperatures. J. Bacteriol. 171: 1379-1385.

Fenton WA, Kashi Y, Furtak K, Horwich AL (1994). Residues in chaperonin GroEL required for polypeptide binding and release. Nature. 371: 614-619.

Ferrer M, TN Chernikova, MM Yakimov, PN Golyshin and KN Timmis (2003). Chaperonins govern growth of Escherichia coli at low temperatures. Nature Biotechnol. 21: 1266 – 1267.

Genevaux P, Keppel F, Schwager F, Langendijk-Genevaux PS, Hartl FU, Georgopoulos C (2004). In vivo analysis of the overlapping functions of DnaK and trigger factor. EMBO Reports. 5: 195-200.

Goloubinoff P, Gatenby AA, Lorimer GH (1989). GroE heat-shock proteins promote assembly of foreign prokaryotic ribulose bisphosphate carboxylase oligomers in Escherichia coli. Nature. 337: 44-47.

Goloubinoff P, Mogk A, Ben-Zvi AP, Tomoyasu T, Bukau B (1999). Sequential mechanism of solubilization and refolding of stable protein aggregates by a bichaperone network. Proc. Natl. Acad. Sci. USA. 96: 13732-13737.

Harper JD and PT Lansbury (1997) Models of Amyloid Seeding in Alzheimer's Disease and Scrapie: Mechanistic Truths and Physiological Consequences of the Time-Dependent Solubility of Amyloid Proteins Ann. Rev. Biochem. 66: 385-407.

Hartl FU (1996). Molecular chaperones in cellular protein folding. Nature. 381: 571-579.

Hesterkamp T, Bukau B (1996). Identification of the prolyl isomerase domain of Escherichia coli trigger factor. FEBS Lett. 385: 67-71.

Hesterkamp T, Deuerling E, Bukau B (1997). The amino-terminal 118 amino acids of Escherichia coli trigger factor constitute a domain that is necessary and sufficient for binding to ribosomes. J. Biol. Chem. 272: 21865-21871.

Hesterkamp T, Hauser S, Lutcke H, Bukau B (1996). Escherichia coli trigger factor is a prolyl isomerase that associates with nascent polypeptide chains. Proc. Natl. Acad. Sci. USA. 93: 4437-4441.

Hoffmann F, Rinas U (2000). Kinetics of heat-shock response and inclusion body formation during temperature-induced production of basic fibroblast growth factor in high-cell density cultures of recombinant Escherichia coli. Biotechnol. Prog. 16: 1000-1007.

Hoffmann F, Rinas U (2004). Roles of heat-shock chaperones in the production of recombinant proteins in Escherichia coli. Adv. Biochem. Eng. Biotechnol. 89: 143-161.

Horwich AL, Low KB, Fenton WA, Hirshfield IN, Furtak K (1993). Folding in vivo of bacterial cytoplasmic proteins: role of GroEL. Cell. 74: 909-917.

Houry WA, Frishman D, Eckerskorn C, Lottspeich F, Hartl FU (1999). Identification of in vivo substrates of the chaperonin GroEL. Nature. 402: 147-154.

Hwang SO (1996) Effect of inclusion bodies on the buoyant density of recombinant Escherichia coli . Biotechnol. Techniques. 10: 157 – 160.

Han W, Christen P (2001). Mutations in the interdomain linker region of DnaK abolish the chaperone action of the DnaK/DnaJ/GrpE system. FEBS Lett. 497: 55-58.

Hunt JF, Weaver AJ, Landry SJ, Gierasch L, Deisenhofer J (1996). The crystal structure of the GroES co-chaperonin at 2.8 Å resolution. Nature. 379: 37-45.

Kaderbhai N, Karim A, Hankey W, Jenkins G, Venning J, Kaderbhai MA (1997). Glycine-induced extracellular secretion of a recombinant cytochrome expressed in Escherichia coli. Biotechnol. Appl. Biochem. 25: 53-61.

Kandror O, Sherman M, Moerschell R, Goldberg AL (1997). Trigger factor associates with GroEL in vivo and promotes its binding to certain polypeptides. J. Biol. Chem. 272: 1730-1734.

Kandror O, Sherman M, Rhode M, Goldberg AL (1995). Trigger factor is involved in GroEL-dependent protein degradation in Escherichia coli and promotes binding of GroEL to unfolded proteins. EMBO J. 14: 6021-6027.

Kopetzki E, Schumacher G, Buckel P (1989). Control of formation of active soluble or

inactive insoluble baker's yeast alpha-glucosidase PI in Escherichia coli by induction and growth conditions. Mol. Gen. Genet. 216: 149-155.

Kramer G, Patzelt H, Rauch T, Kurz TA, Vorderwulbecke S, Bukau B, Deuerling E (2004) Trigger factor peptidyl-prolyl cis/trans isomerase activity is not essential for the folding of cytosolic proteins in Escherichia coli. J. Biol. Chem. 279: 14165-14170.

Kramer G, Rauch T, Rist W, Vorderwulbecke S, Patzelt H, Schulze-Specking A, Ban N, Deuerling E, Bukau B (2002). L23 protein functions as a chaperone docking site on the ribosome. Nature. 419: 171-174.

Kuczyńska-Wisnik D, Kędzierska S, Matuszewska E, Lund P, Taylor A, Lipińska B, Laskowska E (2002). The Escherichia coli small heat-shock proteins IbpA and IbpB prevent the aggregation of endogenous proteins denatured in vivo during extreme heat shock. Microbiol. 148: 1757-1765.

Laskowska E, Wawrzynó A, Taylor A (1996). IbpA and IbpB, the new heat-shock proteins, bind to endogenous Escherichia coli proteins aggregated intracellularly by heat shock. Biochimie. 78: 117-122.

Laufen T, Mayer MP, Beisel C, Klostermeier D, Mogk A, Reinstein J, Bukau B (1999). Mechanism of regulation of hsp70 chaperones by DnaJ cochaperones. Proc. Natl. Acad. Sci. USA. 96: 5452-5457.

LaVallie ER, DiBlasio EA, Kovacic S, Grant KL, Schendel PF, McCoy JM (1993). A thioredoxin gene fusion expression system that circumvents inclusion body formation in the E. coli cytoplasm. Bio/Technol. 11: 187-193.

Liberek K, Wall D, Georgopoulos C (1995). The DnaJ chaperone catalytically activates the DnaK chaperone to preferentially bind the sigma 32 heat shock transcriptional regulator. Proc. Natl. Acad. Sci. USA. 92: 6224-6228.

Liberek K, Marszalek J, Ang D, Georgopoulos C (1991). Escherichia coli DnaJ and

GrpE heat shock proteins jointly stimulate ATPase activity of the DnaK. Proc. Natl. Acad. Sci. USA. 88: 2874-2878.

Maier R, Scholz C, Schmid FX (2001). Dynamic association of trigger factor with protein substrates. J. Mol. Biol. 314: 1181-1190.

Mayer MP, Schroder H, Rudiger S, Paal K, Laufen T, Bukau B (2000). Multistep mechanism of substrate binding determines chaperone activity of Hsp70. Nat. Struct. Biol. 7: 586-593.

Mogk A, Schlieker C, Strub C, Rist W, Weibezahn J, Bukau B (2003a). Roles of individual domains and conserved motifs of the AAA+ chaperone ClpB in oligomerization, ATP hydrolysis, and chaperone activity. J. Biol. Chem. 278: 17615-17624.

Mogk A, Deuerling E, Vorderwülbeck S, Vierling E, Bukau B (2003b). Small heat shock proteins, ClpB and the DnaK system form a functional triade in reversing protein aggregation. Mol. Microbiol. 50: 585-595.

Mogk A, Tomoyasu T, Goloubinoff P, Rüdiger S, Röder D, Langen H, Bukau B (1999). Identification of thermolabile *Escherichia coli* proteins: Prevention and reversion of aggregation by DnaK and ClpB. EMBO J. 18: 6934-6949.

Morita MT, Kanemori M, Yanagi H, Yura T (2000). Dynamic interplay between antagonistic pathways controlling the sigma 32 level in *Escherichia coli*. Proc. Natl. Acad. Sci. USA. 97: 5860-5865.

Mujacic M, Bader MW, Baneyx F (2004). *Escherichia coli* Hsp31 functions as a holding chaperone that cooperates with the DnaK-DnaJ-GrpE system in the management of protein misfolding under severe stress conditions. Mol. Microbiol. 51: 849-859.

Nagai H, Yuzawa H, Kanemori M, Yura T (1994). A distinct segment of the sigma 32 polypeptide is involved in DnaK-mediated negative control of the heat shock response in *Escherichia coli*. Proc. Natl. Acad. Sci. U S A. 91: 10280-10284.

Nishihara K, Kanemori M, Kitagawa M, Yanagi H, Yura T (1998). Chaperone co expression plasmids: differential and synergistic roles of DnaK-DnaJ-GrpE and GroEL-GroES in assisting folding of an allergen of japanese cedar pollen, Cryj2, in *Escherichia coli*. Appl. Environ. Microbiol. 64: 1694-1699.

Nishihara K, Kanemori M, Yanagi H, Yura T (2000). Overexpression of trigger factor prevents aggregation of recombinant proteins in *Escherichia coli*. Appl. Environ. Microbiol. 66: 884-889.

Packschies L, Theyssen H, Buchberger A, Bukau B, Goody RS, Reinstein J (1997). GrpE accelerates nucleotides exchange of the molecular chaperone DnaK with an associative displacement mechanism. Biochem. 36: 3417-3422.

Patzelt H, Rüdiger S, Brehmer D, Kramer G, Vorderwülbecke S, Schaffitzel E, Waitz A, Hesterkamp T, Dong L, Schneider-Mergener J, Bukau B, Deuerling E (2001). Binding specificity of *Escherichia coli* trigger factor. Proc. Natl. Acad. Sci. USA. 98: 14244-14249.

Piatak M, Lane JA, Laird W, Bjorn MJ, Wang A, Williams M (1988). Expression of soluble and fully functional ricin A chain in *Escherichia coli* is temperature-sensitive. J. Biol. Chem. 263: 4837-4843.

Radford SE (2000). Protein folding: progress made and promises ahead. Trends Biochem Sci. 25: 611-618.

Rinas U, Tsai LB, Lyons D, Fox GM, Stearns G, Fieschko J, Fenton D, Bailey JE (1992). Cysteine to serine substitutions in basic fibroblast growth factor: Effect on inclusion body formation and proteolytic susceptibility during in vitro refolding. Bio/Technol. 10: 435-440.

Rye HS, Burston SG, Fenton WA, Beechem JM, Xu Z, Sigler PB, Horwich AL (1997). Distinct actions of cis and trans ATP within the double ring of the chaperonin GroEL. Nature. 388: 792-798.

Sachdev D, Chirgwin JM (1998). Order of fusions between bacterial and mammalian proteins can determine solubility in *Escherichia coli*. Biochem. Biophys. Res. Commun. 244: 933-937.

Sakikawa C, Taguchi H, Makino Y, Yoshida M (1999). On the maximum size of proteins to stay and fold in the cavity of GroEL underneath GroES. J. Biol. Chem. 274: 21251-21256.

Sanchez Y, Lindquist SL (1990). HSP104 required for induced thermotolerance. Science. 248: 1112-1115.

Schein CH (1989). Production of soluble recombinant proteins in bacteria. Bio/Technology. 7: 1141-1149.

Schein CH, Noteborn MHM (1988). Formation of soluble recombinant proteins in *Escherichia coli* is favored by lower growth temperature.

Bio/Technol.. 6: 291-294.

Schirmer EC, Glover JR, Singer MA, Lindquist S (1996). HSP100/Clp proteins: a common mechanism explains diverse functions. Trends Biochem. Sci. 21: 289-296.

Schlee S, Beinker P, Akhrymuk A, Reinstein J (2004). A chaperone network for the resolubilization of protein aggregates: direct interaction of ClpB and DnaK. J. Mol. Biol. 336: 275-285.

Scholz C, Stoller G, Zarnt T, Fischer G, Schmid FX (1997). Cooperation of enzymatic and chaperone functions of trigger factor in the catalysis of protein folding. EMBO J. 16: 54-58.

Shearstone JR, Baneyx F (1999). Biochemical characterization of the small heat shock protein IbpB from *Escherichia coli*. J. Biol. Chem. 274: 9937-9945.

Sigler PB, Xu Z, Rye HS, Burston SG, Fenton WA, Horwich AL (1998). Structure and function in GroEL-mediated protein folding. Ann. Rev. Biochem. 67: 581-608.

Squires CH, Childs J, Eisenberg SP, Polverini PJ, Sommer A (1988). Production and characterization of human basic fibroblast growth factor from *Escherichia coli*. J. Biol. Chem. 263: 16297-16302.

Strandberg L, Enfors SO (1991). Factors influencing inclusion body formation in the production of a fused protein in *Escherichia coli*. Appl. Environ. Microbiol. 57: 1669-1674.

Suh WC, Lu CZ, Gross CA (1999). Structural features required for the interaction of the Hsp70 molecular chaperone DnaK with its cochaperone DnaJ. J. Biol. Chem. 274: 30534-30539.

Swartz JR (2001). Advances in *Escherichia coli* production of therapeutic proteins. Curr. Opin. Biotechnol. 12: 195-201.

Tatsuta T, Tomoyasu T, Bukau B, Kitagawa M, Mori H, Karata K, Ogura T (1998). Heat shock regulation in the ftsH null mutant of *Escherichia coli*: dissection of stability and activity control mechanisms of sigma32 in vivo. Mol. Microbiol. 30: 583-593.

Taylor G, Hoare M, Gray DR, Marston FAO (1986). Size and density of protein inclusion bodies. Bio/Technology. 4: 553-557.

Teter SA, Houry WA, Ang D, Tradler T, Rockabrand D, Fischer, G, Blum P, Georgopoulos C, Hartl FU (1999). Polypeptide flux through bacterial Hsp70: DnaK cooperates with trigger factor in chaperoning nascent chains. Cell. 97: 755-765.

Thomas JG, Baneyx F (1998). Roles of the *Escherichia coli* small heat shock proteins IbpA and IbpB in thermal stress management: comparison with ClpA, ClpB, and HtpG in vivo. J. Bacteriol. 180: 5165-5172.

Uhlen M, Nilsson B, Guss B, Lindberg M, Gatenbeck S, Philipson L (1983). Gene fusion vectors based on the gene for staphylococcal protein A. Gene. 23: 369-378.

Valax P, Georgiou G (1993). Molecular characterization of □-lactamase inclusion bodies produced in *Escherichia coli* 1. composition. Biotechnol. Prog. 9: 539-547.

Valent QA, Kendall DA, High S, Kusters R, Oudega B, Luirink J (1995). Early events in preprotein recognition in *E. coli*: interaction of SRP and trigger factor with nascent polypeptides. EMBO J. 14: 5494-5505.

Walter S, Buchner J (2002). Molecular chaperones – cellular machines for protein folding. Angew Chem. Int. Ed. Engl. 41: 1098-1113.

Watanabe YH, Motohashi K, Taguchi H, Yoshida M (2000). Heat-inactivated proteins managed by DnaKJ-GrpE-ClpB chaperones are released as a chaperonin-recognizable non-native form. J. Biol. Chem. 275: 12388-12392.

Weber F, Keppel F, Georgopoulos C, Hayer-Hartl MK, Hartl FU (1998). The oligomeric structure of GroEL/GroES is required for biologically significant chaperonin function in protein folding. Nat. Struct. Biol. 5: 977-985.

Weibezahn J, Schlieker C, Bukau B, Mogk, A (2003). Characterization of a Trap mutant of the AAA+ chaperone ClpB. J. Biol. Chem. 278: 32608-32617.

Weissman JS, Kashi Y, Fenton WA, Horwich AL (1994). GroEL-mediated protein folding proceeds by multiple rounds of binding and release of nonnative forms. Cell. 78: 693-702.

Wetzel R, Perry LJ, Veilleux C (1991). Mutations in human interferon gamma affecting inclusion body formation identified by a general immunochemical screen. Bio/Technology. 9: 731-737.

Wickner S, Maurizi MR, Gottesman S (1999). Posttranslational quality control: folding, refolding, and degrading proteins. Science. 286: 1888-1893.

Winter J, Neubauer P, Glockshuber R, Rudolph R (2000). Increased production of human proinsulin in the periplasmic space of *Escherichia coli* by fusion to DsbA. J. Biotechnol. 84: 175-185.

Woo KM, Kim KI, Goldberg AL, Ha DB, Chung CH (1992). The heat-shock protein ClpB in *Escherichia coli* is a protein-activated ATPase. J. Biol. Chem. 267: 20429-20434.

Wunderlich M, Glockshuber R (1993). *In vivo* control of redox potential during protein folding catalyzed by bacterial protein disulfide-isomerase (DsbA). J. Biol. Chem. 268: 24547-24550.

Xu Z, Horwich A, Sigler P (1997). The crystal structure of the asymmetric GroEL–GroES–(ADP)$_7$ chaperonin complex. Nature. 388: 741-750.

Zolkiewski M (1999). ClpB cooperates with DnaK, DnaJ, and GrpE in suppressing protein aggregation. A novel multichaperone system from *E. coli*. J. Biol. Chem. 274: 28083-28086.

Permissions

All chapters in this book were first published in BMBR, by Academic Journals; hereby published with permission under the Creative Commons Attribution License or equivalent. Every chapter published in this book has been scrutinized by our experts. Their significance has been extensively debated. The topics covered herein carry significant findings which will fuel the growth of the discipline. They may even be implemented as practical applications or may be referred to as a beginning point for another development.

The contributors of this book come from diverse backgrounds, making this book a truly international effort. This book will bring forth new frontiers with its revolutionizing research information and detailed analysis of the nascent developments around the world.

We would like to thank all the contributing authors for lending their expertise to make the book truly unique. They have played a crucial role in the development of this book. Without their invaluable contributions this book wouldn't have been possible. They have made vital efforts to compile up to date information on the varied aspects of this subject to make this book a valuable addition to the collection of many professionals and students.

This book was conceptualized with the vision of imparting up-to-date information and advanced data in this field. To ensure the same, a matchless editorial board was set up. Every individual on the board went through rigorous rounds of assessment to prove their worth. After which they invested a large part of their time researching and compiling the most relevant data for our readers.

The editorial board has been involved in producing this book since its inception. They have spent rigorous hours researching and exploring the diverse topics which have resulted in the successful publishing of this book. They have passed on their knowledge of decades through this book. To expedite this challenging task, the publisher supported the team at every step. A small team of assistant editors was also appointed to further simplify the editing procedure and attain best results for the readers.

Apart from the editorial board, the designing team has also invested a significant amount of their time in understanding the subject and creating the most relevant covers. They scrutinized every image to scout for the most suitable representation of the subject and create an appropriate cover for the book.

The publishing team has been an ardent support to the editorial, designing and production team. Their endless efforts to recruit the best for this project, has resulted in the accomplishment of this book. They are a veteran in the field of academics and their pool of knowledge is as vast as their experience in printing. Their expertise and guidance has proved useful at every step. Their uncompromising quality standards have made this book an exceptional effort. Their encouragement from time to time has been an inspiration for everyone.

The publisher and the editorial board hope that this book will prove to be a valuable piece of knowledge for researchers, students, practitioners and scholars across the globe.

List of Contributors

Kenichi Yoshida
Department of Life Sciences, Meiji University School of Agriculture, 1-1-1 Higashimita, Tama-ku, Kawasaki, Kanagawa 214-8571, Japan

Bruce A. Rosa
Biorefining Research Initiative, Lakehead University, 955 Oliver Road, Thunder Bay ON, Canada, P7B 5E1
Department of Biology, Lakehead University, 955 Oliver Road, Thunder Bay ON, Canada, P7B 5E1

Lada Malek
Department of Biology, Lakehead University, 955 Oliver Road, Thunder Bay ON, Canada, P7B 5E1

Wensheng Qin
Biorefining Research Initiative, Lakehead University, 955 Oliver Road, Thunder Bay ON, Canada, P7B 5E1
Department of Biology, Lakehead University, 955 Oliver Road, Thunder Bay ON, Canada, P7B 5E1

F. O. Atanu
Department of Biochemistry, Kogi State University, Anyigba, Nigeria

U. G. Ebiloma
Department of Biochemistry, Kogi State University, Anyigba, Nigeria

E. I. Ajayi
Department of Biochemistry, Osun State University, Osogbo, Nigeria

A. Salihu
Department of Biochemistry, Ahmadu Bello University, Zaria, Nigeria

I. Abdulkadir
2Department of Chemistry, Ahmadu Bello University, Zaria, Nigeria

M. N. Almustapha
Department of Pure and Applied Chemistry, Usman Danfodio University, Sokoto, Nigeria

Monde Ntwasa
School of Molecular and Cell Biology, University of the Witwatersrand, Wits, 2050. South Africa

Sam Zwenger
University of Northern Colorado, School of Biological Sciences, Greeley, Colorado, 80639, USA

Chhandak Basu
University of Northern Colorado, School of Biological Sciences, Greeley, Colorado, 80639, USA

Biswadeep Chaudhuri
Department of Biotechnology and Medical Engineering, National Institute of Technology – Rourkela - 769008, India

Krishna Pramanik
Department of Biotechnology and Medical Engineering, National Institute of Technology – Rourkela - 769008, India

K. O. Soetan
Department of Veterinary Physiology, Biochemistry and Pharmacology, University of Ibadan, Nigeria

M. O. Abatan
Department of Veterinary Physiology, Biochemistry and Pharmacology, University of Ibadan, Ibadan, Nigeria

Town Mohammad Hussain
Department of Horticulture, Hamelmalo Agricultural College, Keren, P.O. Box 397, Eritrea, North East Africa

Thummala Chandrasekhar
Department of Botany, Sri Venkateswara University, Tirupati-517 502, Andhra Pradesh, India

Mahamed Hazara
School of Life Sciences, Department of Molecular Biology, University of Skövde, Skövde, Sweden

Zafar Sultan
Department of Plant Protection, Hamelmalo Agricultural College, Keren, P.O. Box 397, Eritrea, North East Africa

Brhan Khiar Sale
Department of Horticulture, Hamelmalo Agricultural College, Keren, P.O. Box 397, Eritrea, North East Africa

Ghanta Rama Gopal
Department of Botany, Sri Venkateswara University, Tirupati-517 502, Andhra Pradesh, India

Kay Parker
Department of Biology, College of Science and Technology, North Carolina Central University, Durham, NC 27707, USA

Michelle Salas
Department of Biology, College of Science and Technology, North Carolina Central University, Durham, NC 27707, USA

Veronica C. Nwosu
Department of Biology, College of Science and Technology, North Carolina Central University, Durham, NC 27707, USA

Irfan-ur-Rauf Tak
Centre of Research for Development, University of Kashmir, Srinagar-190 006, India

Jehangir Shafi Dar
Centre of Research for Development, University of Kashmir, Srinagar-190 006, India

B. A. Ganai
1Centre of Research for Development, University of Kashmir, Srinagar-190 006, India

M. Z. Chishti
Centre of Research for Development, University of Kashmir, Srinagar-190 006, India

R. A. Shahardar
Department of Veterinary Parasitology, SKUAST, Kashmir, India

TowsiefAhmad Tantry
Centre of Research for Development, University of Kashmir, Srinagar-190 006, India

Masarat Nizam
Centre of Research for Development, University of Kashmir, Srinagar-190 006, India

Shoaib Ali Dar
Department of Zoology, Punjabi University Patiala

J. Amudha
Central Institute for Cotton Research, Indian Council of Agricultural Research, Post Box 2, Shankar Nagar, Nagpur, Maharashtra-440 010, India

G. Balasubramani
Central Institute for Cotton Research, Indian Council of Agricultural Research, Post Box 2, Shankar Nagar, Nagpur, Maharashtra-440 010, India

Benard Chirende
Key Laboratory for Terrain-Machine Bionic Engineering (Ministry of Education), Jilin University, Changchun 130022, P.R. China

Jianqiao Li
Key Laboratory for Terrain-Machine Bionic Engineering (Ministry of Education), Jilin University, Changchun 130022, P.R. China

Janani Kumar
Vellore Institute of Technology, VIT Vellore, India

Sowmiya Jayaraman
Vellore Institute of Technology, VIT Vellore, India

Nandhitha Muralidharan
Vellore Institute of Technology, VIT Vellore, India

Adinarayana Kunamneni
Departamento de Biocatálisis, Instituto de Catálisis y Petroleoquímica, CSIC, C/ Marie Curie 2, Cantoblanco, 28049– Madrid, Spain
Pharmaceutical Biotechnology Division, University College of Pharmaceutical Sciences, Andhra University, Visakhapatnam - 530 003, India

Bhavani Devi Ravuri
Departamento de Biocatálisis, Instituto de Catálisis y Petroleoquímica, CSIC, C/ Marie Curie 2, Cantoblanco, 28049– Madrid, Spain

Poluri Ellaiah
Departamento de Biocatálisis, Instituto de Catálisis y Petroleoquímica, CSIC, C/ Marie Curie 2, Cantoblanco, 28049– Madrid, Spain

Taadimalla Prabhakhar
Departamento de Biocatálisis, Instituto de Catálisis y Petroleoquímica, CSIC, C/ Marie Curie 2, Cantoblanco, 28049– Madrid, Spain

Vinjamuri Saisha
Departamento de Biocatálisis, Instituto de Catálisis y Petroleoquímica, CSIC, C/ Marie Curie 2, Cantoblanco, 28049– Madrid, Spain

U. M. Okeh
Department of Industrial Mathematics and Applied Statistics, Ebonyi State University, Abakaliki Nigeria

Richard A. Manfready
Department of Biology, Massachusetts Institute of Technology, Cambridge, MA, USA
Whitehead Institute for Biomedical Research, Cambridge, MA, USA

Sonia Plaza-Wüthrich
Institute of Plant Sciences, University of Bern, Altenbergrain 21, 3013 Bern, Switzerland

Zerihun Tadele
Institute of Plant Sciences, University of Bern, Altenbergrain 21, 3013 Bern, Switzerland

K. Sobha
Department of Biotechnology, RVR and JC College of Engineering, Chowdavaram, Guntur – 522 019, Andhra Pradesh, India

K. Surendranath
Department of Physics, RVR and JC College of Engineering, Chowdavaram, Guntur – 522 019, Andhra Pradesh, India

V. Meena
Department of Chemical Engineering, Andhra University, Visakhapatnam – 530 003, Andhra Pradesh, India

T. Keerthi Jwala
Department of Biotechnology, RVR and JC College of Engineering, Chowdavaram, Guntur – 522 019, Andhra Pradesh, India

N. Swetha
Department of Biotechnology, RVR and JC College of Engineering, Chowdavaram, Guntur – 522 019, Andhra Pradesh, India

K. S. M. Latha
Department of Biotechnology, RVR and JC College of Engineering, Chowdavaram, Guntur – 522 019, Andhra Pradesh, India

Morufat Oloruntoyin Balogun
Institute of Agricultural Research and Training, Obafemi Awolowo University, P. M. B. 5029, Ibadan, Nigeria

R N. Okigbo
Department of Botany, Nnamdi Azikiwe University, Awka, Anambra State, Nigeria

U E. Eme
Department of Botany, Nnamdi Azikiwe University, Awka, Anambra State, Nigeria

S. Ogbogu
Dpartment of Zoology,Obafemi AwolowoUniversity, Ile-Ife, Nigeria

Pratap R. Patnaik
Institute of Microbial Technology Sector 39-A, Chandigarh-160036, India

E. BETIKU
Department of Chemical Engineering, Obafemi Awolowo University,Ile-Ife, Osun State, Nigeria